The Ogooué Delta

The Ogooué Delta

Jean P. Vande weghe & Tariq Stévart (editors)

Eric Lowry (translator)

Missouri Botanical Garden

Parcs Gabon

This work was produced by the Missouri Botanical Garden (MBG) as part of the Support Project for the Sustainable Management of Critical Wetland Ecosystems in Gabon (PAZH) implemented by the General Directorate of Environment and Nature Protection (DGEPN) and the National Agency for National Parks (ANPN) with support from the Global Environment Fund (GEF) through the World Bank.

ISBN: 978-1-935641-22-3
Library of Congress Control Number: 2020940198

Publisher: Liz Fathman
Managing Editor: Allison M. Brock
Editor: Lisa J. Pepper
Press Coordinator: Amanda Koehler

Typesetting: Jean P. Vande weghe, Emily Rogers

4344 Shaw Blvd.
St. Louis, MO 63110, U.S.A.
www.mbgpress.info

Printed in Canada.

Contents

Cover:
The Ogooué River and its upper deltaic plain (Photo: Jean P. Vande weghe).

Page 1:
1.1. The Ogooué River northeast of Lambaréné with Lake Zilé at the front and Lambaréné at the far left (Photo: Jean P. Vande weghe).

Pages 2 and 3:
1.2. The heart of the upper deltaic plain with Lake Nyayé, often visited by manatees (Photo: Jean P. Vande weghe).

Page 4:
1.3. The upper deltaic plain with its trees and shrubs growing along an old channel of the river (Photo: Jean P. Vande weghe).

Page 6:
1.4. Eleven buffalo in the swamps at the center of the upper deltaic plain (Photo: Jean P. Vande weghe).

1.5. The southern delta, with the Djembagombé and Adeloué rivers crossing old mangroves and Raphia groves. Although they are of marine origin, these environments are now flooded with freshwater (Photo: Jean P. Vande weghe).

Preface

The Lower Ogooué wetlands, including Africa's third-largest and best-preserved river delta, is a complex of diverse, permanently and seasonally flooded habitats, which contain a treasure trove of biodiversity. While parts of the region have long been explored and studied, sections are effectively impenetrable and represent some of the last unexplored ecosystems in Gabon.

Early explorers described huge pods of hippos and scores of massive Nile crocodiles basking on the extensive sand banks that emerge when the waters are down. Sadly, this is a thing of the past—there has been intense hunting in parts of the delta, and large mammal populations have suffered; the era of the monofilament fishing net, now outlawed, may have exterminated the sawfish, which were once abundant, but otherwise the biodiversity of the delta is thriving.

This book is the first to comprehensively document the rich history of the human populations and the biodiversity of the Lower Ogooué system. It completes my 15-year collaboration with Jean-Pierre Vande weghe to document Gabon's rich biodiversity. This time he has assembled a multidisciplinary team of researchers who have undertaken original research as well as synthesizing existing knowledge. The end result is an attractive and informative volume that will long remain the reference for this fascinating region and which completes Vande weghe's legacy as one of Gabon's and Central Africa's preeminent natural historians.

This volume along with the database that was created during its conception lays the foundation for us to develop a comprehensive management plan for the region, which was gazetted as a Ramsar site, a wetland of international significance, in 2008 and is Gabon's largest terrestrial protected area.

Professor Lee White CBE

Introduction

Tariq Stévart, Olivier S. G. Pauwels & Jean P. Vande weghe

The wetlands of the Lower Ogooué represent the third-largest deltaic complex in Africa after those of the Nile and Niger rivers. Freshwater covers 80–90% of their area. Nevertheless, they were not mentioned in the compilation of African wetlands published by ORSTOM (Burgiss & Symoens, 1987). An early, typewritten version of their work states that there are no internationally significant water bodies in Gabon, and the version currently available online does not even mention Gabon. Even in a recent report by the IUCN on the conservation of Central African freshwater bodies, the wetlands of Gabon's coastal sedimentary basin do not appear (Brooks *et al.*, 2008).

These wetlands are clearly smaller than those of the central basin of the Congo. Nevertheless, the coastal area of Gabon probably has nearly a million hectares of wetlands of significant conservation value. Some habitats are unique to Central Africa, and Lower Ogooué wetlands represent the last remaining large delta in Africa with nearly intact landscapes. The Nile Delta (26,000 km²) has been cultivated for over 2,000 years and is currently inhabited by 40 million people, while the Niger Delta (70,000 km²) has been heavily degraded since the start of oil drilling and now has 31 million inhabitants. Although the Ogooué Delta is much smaller (7,170 km²), it is nearly intact–at least in terms of its landscapes–with a population barely reaching 200,000 inhabitants. Paradoxically, while several hundreds of thousands of aquatic birds still inhabit the Nile Delta, the fauna of the Lower Ogooué, especially mammals, is in a precarious state, and the impact of human activity is slowly but steadily growing.

From time to time, more disturbing threats from development are appearing and could severely affect the entire complex in the near future, especially since large-scale activities are rarely compatible with conservation objectives, which generally end up absent or nearly so from the planning process. This lack of consideration of biodiversity in development is often attributed to a lack of information regarding aquatic species, but this is often just a pretext. Impact studies are in fact financed by companies seeking to develop an area, and are sometimes biased from the start. Very often they are actually subdivided into a multitude of substudies in an effort to conceal global effects.

It is thus important to implement protective measures for this unique site, starting with raising awareness in order to thwart arguments regarding a "lack of knowledge," which are too often used by developers.

The idea of protecting this unique set of landscapes is not new. As early as 1990, proposals were made for the creation of two protected areas in the Lower Ogooué region, one along the Gulf of Guinea and the other centered around the Great Lakes to the southwest of Lambaréné (Wilks, 1990). When Gabon's national parks were created in 2002, it was unfortunately not possible to include even part of the wetlands due to the human population which, though sparse, is omnipresent and depends heavily on the environment's natural resources for fishing and hunting. Furthermore, some sectors of the complex were part of oil and forest exploitation concessions. Moreover, the Ogooué was already a major communications and transport route. In parallel with conservation projects, studies were carried out in order to valorize this delta. In the 1990s, a very discreet plan was suggested to allocate the vast alluvial and upper deltaic plains for extensive agricultural use.

Nevertheless, the eastern part of the complex was recognized as a Ramsar site in 2009 (site #1851). Unfortunately, this 850,000 ha site–one of the largest in the world–only covers the eastern part of the deltaic complex, and 70% of it is comprised of dry land, which bears no relationship with the wetlands and is actually part of

1.6. The Inguéssi Plain, with its patchwork of dry and humid savannas, permanent marshes, Pandanus *groves, and swamp forests. Large mammal tracks are omnipresent, especially at the boundary between wet savannas and water (Photo: Jean P. Vande weghe).*

a very different ecosystem. Furthermore, the recognition as a Ramsar site has no real impact on management and resource allocation for the site. Rather, it is just a declaration of intent with absolutely no commitment—as shown by recent plans to channel part of the river, allowing for year-round navigation in association with planned mining operations.

Freshwater wetlands are the most threatened environments on the planet, more so than marine and terrestrial environments (Revenga & Kura, 2003; Leveque *et al.*, 2005; Dudgeon *et al.*, 2006; Thieme *et al.*, 2010a). The seriousness of this situation is difficult to grasp because over 80% of the world's wetlands have been transformed for agriculture or into arid lands. According to Davidson (2016), 87% of wetlands have been destroyed since 1700, and according to Vörösmarty *et al.* (2010), 65% of the floodplains remaining today are under severe threat. According to Postel & Richter (2003), approximately half of remaining wetlands are degraded.

This destruction happened mainly in order to expand agricultural land, although some governments drained wetlands for political reasons. For example, in the heart of Hungary, 30,000 km² of floodplains were drained by the Austrians during the second half of the 19th century at a time when Hungarians were fighting for independence. In the south of Italy, the Pontine Marshes were destroyed by Mussolini's regime. In the Soviet Union, the communist regime drained the vast Pinsk Marshes in the west and the Aral Sea in the south. In Iraq, the enormous marshes of the Tigris and Euphrates were drained by Sadam Hussein at the end of the 20th century. Officially, these drainages were considered to be humankind's victory over nature—just as the extermination of wolves and bears or the fight against cholera, smallpox, or plague. But in fact, it was also a good way to silence "recalcitrant" human populations. Many other wetlands were also destroyed to provide work for large civil engineering companies. In South America, some dream of this kind of "development" for the Pantanal–the largest marsh in the world, covering 220 km²–and the United Nations is financing studies to divert the waters of the Congo River to Lake Chad and thus drain the center of the Congo Basin, which already suffers from a lack of water.

Even without considering biodiversity, wetlands offer considerable environmental benefits. A study estimated the global value of services rendered by wetlands at $7 trillion (Costanza *et al.*, 1997) or between $12,000 and $28,000 per hectare. Despite this, the environmental benefits of wetlands receive little to no consideration in regional development plans. Consequently, humans continue to degrade and destroy wetlands, believing their actions are productive (Postel & Carpenter, 1997; Postel, 2003; Wallace *et al.*, 2003). Many examples show, however, that it is more economical to maintain healthy wetlands than to have to invest in restoration operations in order to recover just a portion of the environmental services they offered before degradation (Jackson *et al.*, 2001; Blamford *et al.*, 2002; Postel, 2003; Emerton & Bos, 2004).

Conservation efforts for the marshes that remain today, which are the most productive environments on the planet, are up against both a profound ignorance of how wetland ecosystems function and serious political and financial obstacles. From a strictly technical perspective, they face major difficulties due to their aquatic nature. The waters that run through these environments do not simply disperse nutrients, but also any harmful elements present. For example, if highly toxic materials were dumped into the Ogooué at Ndjolé, they could kill the fish and other lifeforms all the way to the Great Lakes and the area near Port-Gentil. In addition to issues of water quality, major changes to the rivers that feed into the wetlands would profoundly affect them. Building a major dam at Lopé could completely destroy the Lower Ogooué complex 200 km downstream. Many wetlands are in fact heavily dependent on fluctuations in water level, whether seasonal or tidal. These fluctuations can both determine the size of a wetland and also distribute nutrients throughout the ecosystem, as mentioned above. Yet today, such variations in the water level are considered by most people to be a disaster that absolutely must be fixed. Rivers must be channeled and regulated at all costs! Often these projects do not really improve the fate of farmers, but they do serve to create jobs for major civil engineering companies. Furthermore, wetlands are sensitive to the cumulative effects of threats such as climate change, changes to water flow, pollution, hunting, and fishing, which can add up to have highly unexpected effects on wetlands. For example, species of fish weakened by overfishing will have a harder time defending themselves from introduced species.

One major problem in the conservation of wetlands is the fact that some species are present only during part of the year, or part of their lifespan, spending the rest of the year or their life elsewhere. This is the case for some species of fish, such as the giant African threadfin in the Ogooué Delta (page 227). It also holds for most aquatic birds, which move throughout the year according to changes in the environment and available resources (pages 319-320). For the most part, wading species, which feed on the mudflats of the Nazaré Bay, also nest in Eurasian tundras. Laridae, pratincoles, plovers, and skimmers nest on the sandbars of the Ogooué or the delta's sandbanks during the dry season and then move to the coast of the Gulf of Guinea when the water levels of the Ogooué are high. Herons, cormorants, anhinga, and pelicans, which also nest in the Lower Ogooué Delta during the dry season, disperse throughout the coastal basin or elsewhere in intertropical Africa during the rest of the year.

Many aquatic birds also enjoy different environments at different times of day, since they feed, sleep, and nest in different areas. For this reason, migratory or highly mobile species can only be protected by establishing a network of protected areas or "sustainably" managed sites. This is why the Ramsar Convention was established in Iran in 1971.

Another problem regarding wetland conservation, especially in Gabon, is that existing conservation efforts have mostly focused on forested areas. For this reason, few people, whether Gabonese or foreigners, know about the wetlands and their specific issues or have any experience regarding their conservation. Mudflats and especially mangroves are too often considered as unhealthy environments that should be sanitized.

Finally, one last point should be made regarding the Lower Ogooué wetlands. These environments are largely comprised of inundated or floodable forests surrounded by dryland forests that partly depend on the nearby presence of open wetlands. This may be less so today but was probably true in the past during periods with cooler and drier climates. Several observations even suggest that these forests constituted microrefuges (pages 175-177).

Regarding the knowledge acquired about the Lower Ogooué, to say that we know nothing would be false. In fact, studies were carried out in the 1960s. Initially they concerned pedology (Chatelin, 1962 and 1964), sedimentology (Giresse, 1969) and hydrology (Carré, 1978; Mahé *et al.*, 1990). A brief description of the Lower Ogooué complex has been published (Lebigre, 1983), and several studies have been carried out with a view to optimizing fisheries (Loubens, 1964a, 1964b, 1966). These first works were clearly oriented toward a "development" or at least a "sustainable exploitation" of the complex.

1.7. The northern delta, just above the Animba Plain, with its tidal rivers, vast low mangroves, narrow strips of high mangrove, tannes, and islets of savanna (Photo: Jean P. Vande weghe).

Starting in the 2000s, studies were rather oriented towards conservation. In 2003, a catalog of fish from Lower Ogooué was published (Mbega & Teugels, 2003). Some studies have concerned botanical aspects (Ondo Assoumou, 2006; Boupoya & Stévart, 2013); others fauna (Mbina, 2001; Latour, 2005; Motsaba & Aba'a, 2011; Louembet, 2005, 2006, 2012; OELO 2013-2016; Shirley, 2012; Fermon, 2015; Vande weghe & Christy, 2015). Still others were interested in socioeconomic aspects (Motsaba & Aba'a, 2011; Loundou *et al.*, 2013). Most were carried out by WWF and WCS, mainly under the Parks 2 program, funded by the Global Environmental Fund (GEF). Around the same time, however, a study of the sands of the Ogooué (Seranne *et al.*, 2008) and a sedimentological study of Mandji Island and the submarine canyons of the Ogooué Delta (Biscara, 2011) were also carried out. Finally, as part of impact studies on possible mining developments in the far east of the Ramsar site of Lower Ogooué, botanical, faunal, and hydrological studies were implemented.

The results of the vast majority of this work are confined to reports which will inevitably be locked in cupboards and will be forgotten in a few years. It was therefore both important and urgent to publish a summary showing the values of the last more or less intact great river delta of the African continent.

Bibliography

BALMFORD, A., BRUNER, A., COOPER, P., COSTANZA, R., FARBER, S., GREEN, R.E., JENKINS, M., JEFFERISS, P., JESSAMY, V., MADDEN, J., MUNRO, K., MYERS, N., NAEEM, S., PAAVOLA, J., RAYMENT, M., ROSENDO, S., ROUGHGARDEN, J., TRUMPER, K. & TURNER, R.K., 2002. Economic reasons for conserving wild nature. *Science*, 297: 950-953.

BISCARA, L., 2011. *Le Système Turbiditique de l'Ogooué (Gabon, Marge Ouest Africaine): Evolution Fini-Holocène de la Morphologie et de la Dynamique Sédimentaire*. Thèse de l'université Bordeaux I. Bordeaux.

BOUYPOYA, A. & T. STÉVART, 2013. *Première étude*

sur la flore et la végétation de la région du Lac Évaro (Moyen-Ogooué-Gabon). WCS, Libreville.
BROOKS, E.G.E., D.J. ALLE & W.R.T. DARWALL (Compilers), 2008. *The status and distribution of freshwater biodiversity in Central Africa*. IUCN, Gland and Cambridge.
BURGISS, M.J. & J.J. SYMOENS, 1987. *Zones humides et lacs peu profonds d'Afrique. Répertoire*. ORSTOM, Paris.
CARRÉ, P., 1978. *Fleuves et rivières du Gabon. Diversité des régimes hydrologiques*. CENAREST, Libreville.
CHATELIN, Y., 1962. *Note sur les sols de la presqu'île de Port-Gentil*. ORSTOM, Libreville.
CHATELIN, Y., 1964. *Les sols de la vallée de l'Ogooué en aval de Lambaréné*. ORSTOM, Libreville.
COSTANZA, R., D'ARGE, R., DE GROOT, R., FARBER, S., GRASSO, M., HANNON, B., LIMBURG, K., NAEEM, S., O'NEILL, R.V., PARUELO, J., *et al*., 1997. The value of the world's ecosystem services and natural capital. *Nature*, 387: 253–260.
DAVIDSON, N.C., 2016. Wetland Losses and the Status of Wetland-Dependent Species. Pp. 1–14 in : C.M. FINLAYSON, G.R. MILTON, R.C. PRENTICE & N.C. DAVIDSON (éds.). *The Wetland Book*. Springer Netherlands, Dordrecht.
DUDGEON, D., ARTHINGTON, A.H., GESSNER, M.O., KAWABATA, Z-I., KNOWLER, D.J., LÉVÈQUE, C., NAIMAN, R.J., PRIEUR-RICHARD, A-H., SOTO, D., STIASSNY, M.L.J. & SULLIVAN, C.A, 2006. Freshwater biodiversity: importance, threats, status and conservation challenges. *Biological Review*, 81: 163–182.
EMERTON L. & E. BOS, 2004. *Value. Counting Ecosystems as an Economic Part of Water Infrastructure*. IUCN, Gland, Switzerland and Cambridge, UK.
FERMON, Y., 2015. *Caractérisation Ichtyologique du Massif Forestier d'Evaro*. Rapport final de l'inventaire multi-ressources relatif aux résultats ichtyologique. Association Aimara-WWF.
GIRESSE, P. & KOUYOUMONTZAKIS, G., 1974. Observations sur le quaternaire côtier et sous-marin du Congo et des régions limitrophes: aspects eustatiques et climatiques, *Ass. Sénég. Et. Quater. Bull. Liaison*, 42–43: 45–61.
GIRESSE, P., 1969. Carte sédimentologique des fonds sous-marins du delta de l'Ogooué. *Cah. Océanogr., XXI (10)*: 965–994.
GIRESSE, P., D. MALOUNGUILA-NGANGA & G. DELIBRIAS, 1984. Rythmes de la transgression et de la sédimentation holocène sur les plate-formes sous-marines du Gabon et du Congo. *C.R. Acad. Sci. Fr*., Série 2, 7: 327–330.
JACKSON, R.B., CARPENTER, S.R., DAHM, C.N., MCKNIGHT, D.M., NAIMAN, R.J., POSTEL, S.L. & RUNNING, S.W. 2001. Water in a changing world. *Ecological Applications*, 11 (4): 1027–1045.
LATOUR, S., 2005. *Inventaire de faune autour des lacs du delta de l'Ogooué*. WCS, Libreville.
LEBIGRE, J.M., 1983. *Le littoral du Gabon : Aspects géomorphologiques et biogéographiques*. Institut Pédagogique National, Libreville.
LEGOUX, P., 1952. Un type nouveau de côte alluviale basse : la côte à formations parallèles ou côte de type gabonais. *C.R. Acad. Sc. Fr*., 234: 119–121.
LOUBENS, G., 1966. *Biologie de* Polydactylus quadrifilis *dans le bas Ogooué (Gabon). Peuplement en poisson des eaux saumâtres et son origine*. Thèses de la faculté des Sciences de l'université de Paris.
LOUBENS, G., 1964a. *Rapport technique sur les travaux effectués en 1963*. Eaux et Forêts, Lambaréné.
LOUBENS, G., 1964b. *Travaux et vue du développement de la pêche dans le bassin inférieur de l'Ogooué*. Publication 27, CTFC, Nogent-sur-Marne, France.
LOUEMBET, MBOUMA S., 2005. *Approche d'étude pour une connaissance de l'exploitation d'un mammifère aquatique : le lamantin* (Trichechus senegalensis) *dans la région des lacs du sud de l'Ogooue*. Mémoire ENEF, Libreville.
LOUEMBET, MOUMBA S., 2006. *Enquëte d'évaluation de la chasse aux lamantins et de la grande chasse, pour une stratégie de conservation au massif forestier d'Ebel Abanga*. WWF, Libreville.
LOUEMBET, S., 2012. *Quelques informations phares issues des études antérieures dans la zone d'Ebel Abanga et Lambaréné*. WWF, Lambaréné.
LOUNDOU, P., M. STARKEY, W. MBOMBÉ TSONGUÉ & B. BIMBISSA MINANGA, 2013. *Evaluation socioéconomique du complexe Evaro-Wonga-Wongué*. WCS, Libreville.
MAHÉ, G., J. LÉRIQUE & J.-C. OLIVRY, 1990. Le fleuve Ogooué au Gabon. *Hydrol. continent*., 5 (2): 105–124.
MBEGA, J.-D. & G.G. TEUGELS, 2003. *Guide de détermination des poissons du bassin inférieur de l'Ogooué*. Presses universitaires de Namur, Namur.
MBINA, C., 2001. *Evaluation and statut of manatees* (Trichechus senegalensis) *of Ogooue bassin in Gabon*. WCS, Libreville.
MOTSABA, P. & R. ABA'A, 2011. *Réserve présidentielle dxe Wonga-Wongué: étude pilote des grands mammifères et des activités anthropiques*. WCS, Libreville.
OELO, 2013–2016. *Rapports de vente illégale de la viande de brousse. Marché Isaac*. Lambaréné.
ONDO ASSOUMOU, E., 2006. *Dynamique des paysages végétaux du littoral centre-ouest du Gabon autour de Port-Gentil : Approche spatiale et analyse des données de terrain*. Thèse de l'université Paul Valéry, Montpellier 3. Montpellier.
POSTEL, S.L., 2003. Securing water for people, crops, and ecosystems: new mindset and new priorities. *Natural Resources Forum*, 27: 89–98.
POSTEL, S. & S.R. CARPENTER, 1997. Freshwater ecosystem services. In: G. Daily (ed.). *Nature's services*. IslandPress, Washington DC, USA.
POSTEL, S. & B. RICHTER, 2003. *Rivers for Life: Managing Water for People and Nature*. Island Press, Washington DC, USA.
REVENGA, C. & Y. KURA, 2003. Status and Trends of Biodiversity of Inland Water Ecosystems. *Secretariat of the Convention on Biological Diversity, Montreal, Technical Series*, 11.
SERANNE, M., O. BRUGUIER & M. MOUSSAVOU, 2008. U-Pb single zircon grain dating of Present fluvial and Cenozoic Aeolian sediments from Gabon: consequences on sediment provenance, reworking, and erosion processes on the equatorial West African margin. *Bull. Soc. Géol. Fr*., 179: 29–40.
SHIRLEY, M.H., 2012. *Inventaire des crocodiles dans les lacs et rivières du complexe Wonga-Wongué-Evaro*. WCS, Libreville.
THIEME, M.L., TURAK, E., MCINTYRE, P., DARWALL, W., TOCKNER, K., CORDEIRO, J. & BUTCHART, S.H.M., 2010a. Freshwater ecosystems under threat: the ultimate hotspot. Pp. 123–151 in: Mittermeier, R.A., Farrell, T.A., Harrison, I.J., Upgren, A.J. and Brooks, T.M. (eds.) *Fresh water: the essence of life*. CEMEX and ILCP, Arlington, Virginia, USA.
VANDE WEGHE, J.P. & P. CHRISTY, 2015. *Les oiseaux du complexe de milieu humides du bas Ogooué*. Rapport WWF, non publié, Libreville.
VÖRÖSMARTY, C.J., P.B. McINTYRE, M.O. GESSNER, D.D. DUDGEON, A. PRUSEVICH *et al*., 2010. Global threats to human water security and river biodiversity. *Nature*, 467: 555–561.
WALLACE, J.S., ACREMAN, M.C. & SULLIVAN, C.A. 2003. The sharing of water between society and ecosystems: from conflict to catchment-based co-management. *Philosophical Transactions of the Royal Society of London B*, 358: 2011–2026.
WILKS, C., 1990. *La Conservation des Ecosystèmes forestiers du Gabon*. UICN, Gland.

1.8. The northern delta, with the Atadali River in the foreground, surrounded by its high and low mangroves; the Kondjo River is visible in the background, through the haze (Photo: Jean P. Vande weghe).

1.9. *The beach of Mandji Island, with the northern end of the Ogooué sandbar. In the foreground, dying mangroves can be seen being replaced by swamp forest (Photo: Jean P. Vande weghe).*

Summary

The *Ogooué Delta* details the biodiversity of this sparsely populated 250 × 50 km wetland area, the third largest and best preserved of Africa's deltas. While exploration of the area's biodiversity began in the 19th century, systematic inventories started only in 1960. Numerous biodiversity studies have been conducted during the past 20 years, including fieldwork in remote areas never before explored, which revealed the unique value of the Lower Ogooué River for the conservation of Gabon's humid habitats. In 2009, part of the delta was designated as a Ramsar site, but despite this international recognition, it remained unprotected at the national level.

This book presents all available information on the biodiversity of this remarkable area, making it available to the general public as well as to conservation practitioners and decision makers, and it includes recommendations for strengthening conservation of the site. The current delimitation of the Ramsar site is also discussed and proposals are made to include highly diverse areas that currently fall outside the Ramsar site. The data presented in this book are also intended to support the recognition of the Lower Ogooué River as a new Biosphere Reserve and to contribute to the establishment of a management plan.

The book, edited by J.P. Vande weghe and T. Stévart, is co-authored by 21 contributors, and is available in both French and English. Its 11 richly illustrated chapters, written by specialists, cover the history, geography, vegetation, zoology, and flora of this unique area.

The Ogooué Delta is uniquely valuable for bird conservation because it harbors more than 1% of the global population of four species that nest on sandbanks: the "Ogooué" tern (95%), the African skimmer (8–13%), the gray pratincole (3–9%), and the *mechowi* subspecies of the white-fronted plover (7–10%). Furthermore, the delta plays a critical role for nesting populations of many bird species in Gabon. The area is also home to significant populations of manatees and hippopotamuses, the largest in Gabon, and probably of international importance. The waters of the Lower Ogooué are home to a very diverse fish fauna, with 27 endemic species, in addition to the taxa that remain to be described and no doubt also to be discovered. The delta is an important region for the growth and maturation of the giant African threadfin, *Polydactylus quadrifilis*, a species of great commercial value. Moreover, the population of long-snouted crocodiles, *Mecistops cataphractus*, which probably includes at least a thousand individuals, likely represents more than 1% of the total global population of this endangered Central African species.

The delta shelters a complex gradient of vegetation types, from the coastal meadows and pioneering mangroves to the swamp forests, flooded forests, and ultimately dryland forests. It thus harbors an unaltered succession that is unique within Gabon. Overall, although the Lower Ogooué contains only a small number of threatened habitats, the largest portion of Gabon's wetland habitats can be found there. The rarest and most endangered include *Ctenolophon* swamp forests, which play a unique role in maintaining the forested vegetation of the coastal part of Gabon; the vegetation of the coastal dune ridges, which includes a set of dryland forests that have a limited distribution and are highly threatened; humid forests and dry savannas, including savannas with *Brachystelma letestui*; and flooded savannas. Finally, the delta includes populations of at least 167 potentially threatened plant species, most of which have been documented to occur in dryland forests, but around 30 are found in wetland habitats.

The Ogooué Delta also is home to an ever-changing set of habitats and a sparse human population that is nevertheless highly dependent on natural resource exploitation, such as fishing. The delta includes two of the country's largest economic centers, and important economic activities are conducted, such as oil extraction and forestry. The conservation of this area's unique diversity of species and habitats is thus inextricably linked to the preservation of the fragile ecosystemic equilibrium that has shaped its landscape over thousands of years, from the mangroves of Port-Gentil to the lakes of the Lambaréné area. The Lower Ogooué wetlands represent a single, indivisible entity and offer a unique opportunity to promote the development and implementation of sustainable practices through its designation as a World Heritage site.

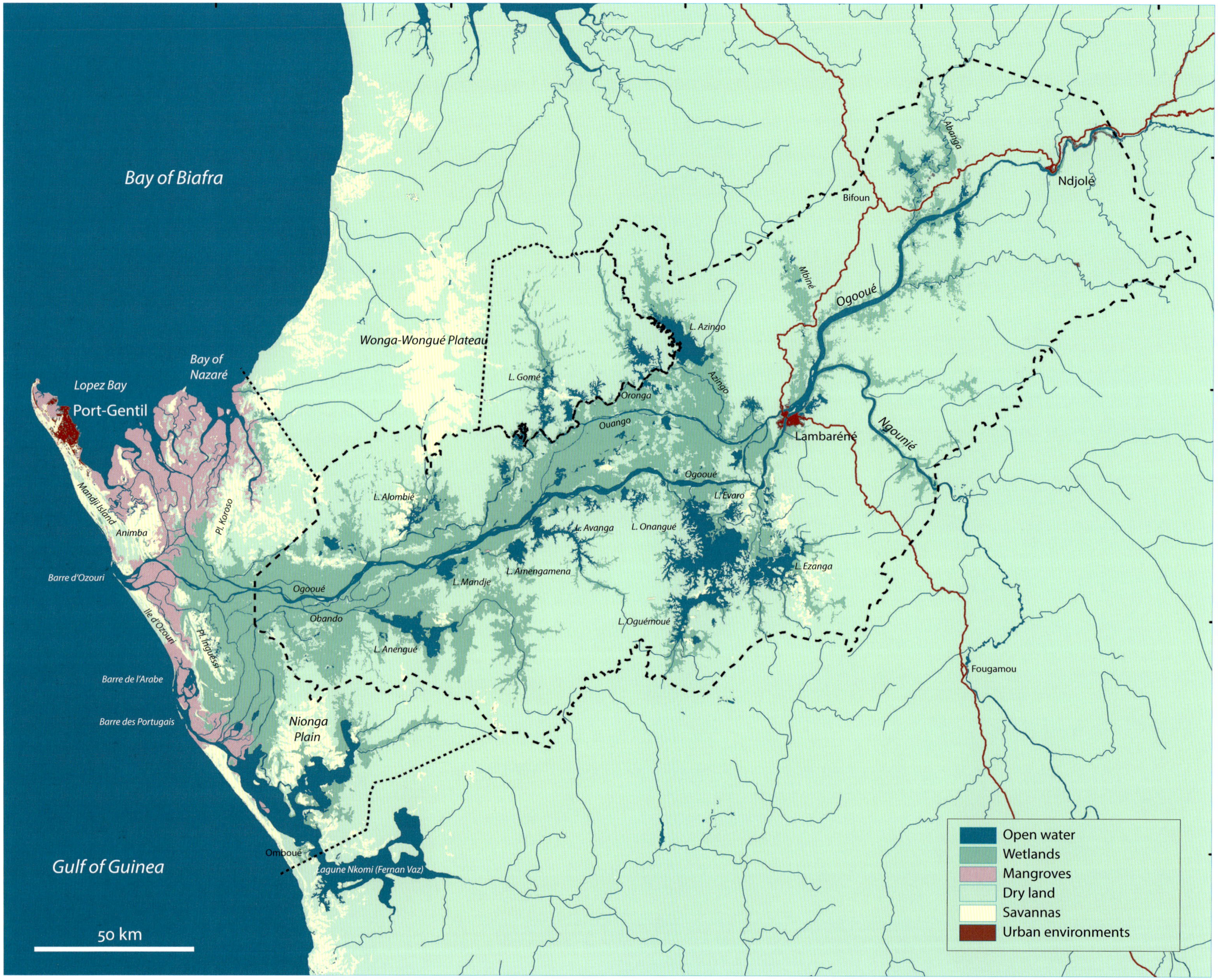
Bay of Biafra
Lopez Bay
Port-Gentil
Bay of Nazaré
Wonga-Wongué Plateau
Mandji Island
Animba
Pl. Koroso
Barre d'Ozouri
Ile d'Ozouri
Pl. Inguéssi
Barre de l'Arabe
Barre des Portugais
Nionga Plain
Omboué
Lagune Nkomi (Fernan Vaz)
Gulf of Guinea
50 km
L. Gomé
L. Azingo
Azingo
Oronga
Ouango
Ogooué
L. Alombié
L. Avanga
L. Onangué
L. Evaro
L. Amengamena
L. Mandje
L. Ezanga
Obando
L. Anengué
L. Oguémoué
Lambaréné
Ngounié
Mbiné
Bifoun
Abanga
Ndjolé
Fougamou
Open water
Wetlands
Mangroves
Dry land
Savannas
Urban environments

1. Structure and history of the Ogooué Delta

Jean P. Vande weghe

The Ogooué River

The Ogooué River is the main coastal river between the Congo and Niger rivers. Its headwaters are located in Congo, at about 850 m altitude in the region of Zanga on the Bateké Plateau, and it is approximately 900 km long (Lérique, 1983). Its two most important tributaries are the Ivindo on the right bank and the Ngouié on the left bank (Figure 1.2). The first drains a large part of the northeast of Gabon as well as a part of southern Cameroon and northwestern Congo. The second drains south-central Gabon, primarily the western flanks of the Chaillu Mountains.

The Ogooué watershed, upstream from Lambaréné, covers 203,500 km², which represents approximately 95% of the estimated 215,200 km² of the entire basin (Mahé *et al.*, 1900; Lérique, 1983). This watershed covers most of Gabon and, as mentioned above, some neighboring parts of Cameroon and Congo, as well as Equatorial Guinea (Figure 1.2). It is bisected by the geographic equator, but is mostly south of the climatic equator, which is situated near the Gabon-Cameroon border. Over 80% of the watershed is covered with dense forest, while savanna only covers 5–10% and anthropized environments

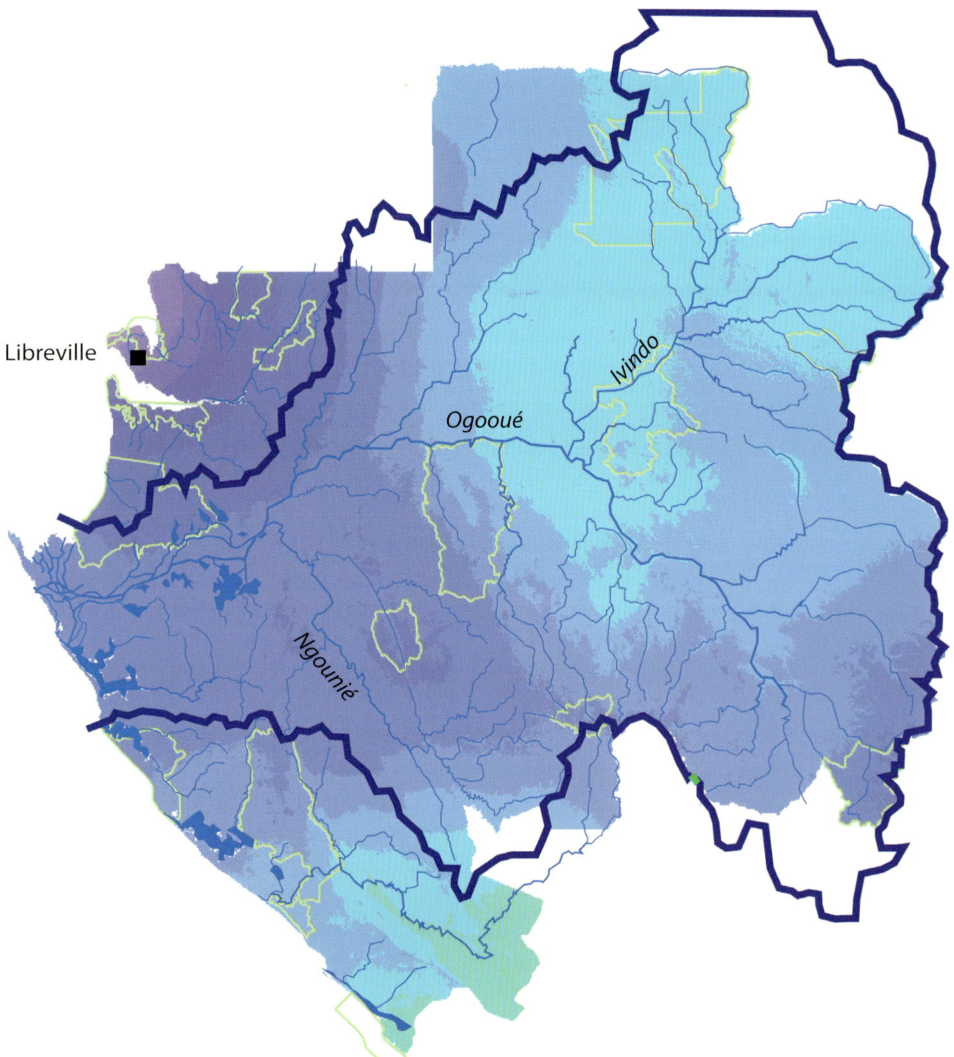

Page 18:
1.1. The entire Lower Ogooué deltaic complex, with both the current and proposed limits of the Ramsar site.

Page 19:
1.2. The Ogooué basin superimposed on average annual precipitation data, ranging from more than 3,000 mm in the Libreville region to less than 1,400 mm in the far south of the country.
> 3,000 mm
< 1,400 mm

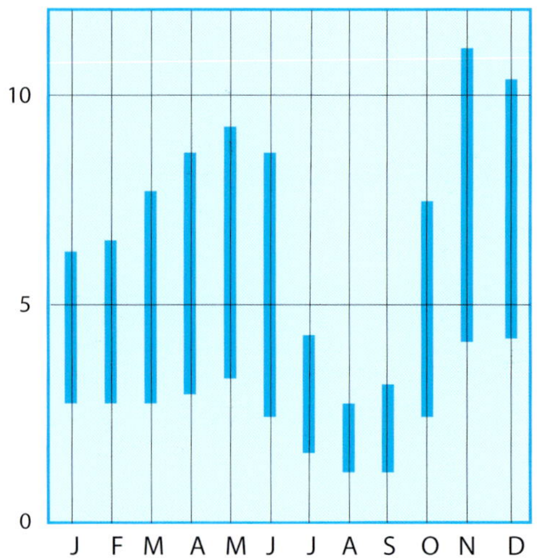

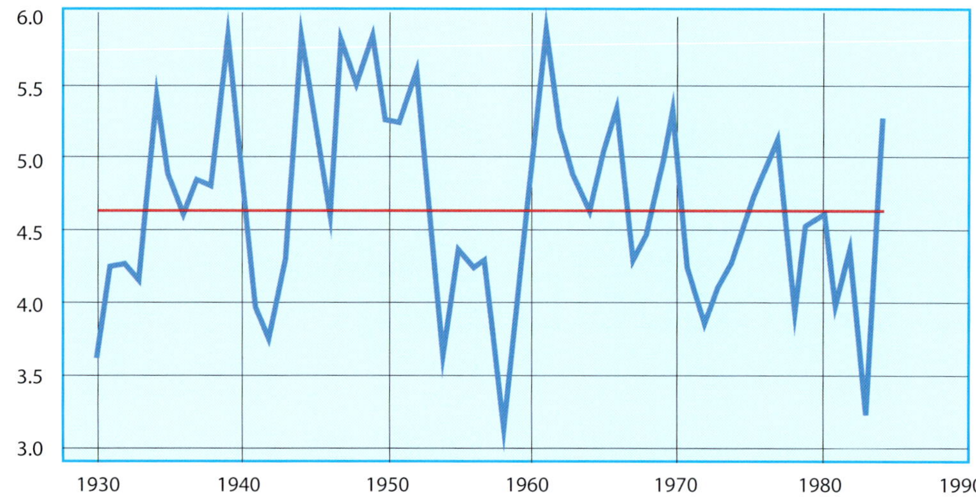

1.3. Monthly variation in the flow of the Ogooué River, expressed in m³/sec, according to Mahé et al. *(1990).*

1.4. Change in the annual flow of the Ogooué River from 1934 to 1984, expressed in m³/sec, according to Mahé et al. *(1990). The red line indicates the average annual flow.*

occupy another 5-10%. Average annual rainfall across the whole basin is 1,840 mm.

According to Mahé *et al.* (1990), the average annual flow rate, measured at Lambaréné, is 4,680 m³/s, which was quite high considering the size of the basin. The river's average flow rate is 23.1 l/s/km², which represents one of the highest values for an intertropical watershed of more than 200,000 km² (Fermon, 2012). This could be the result of the heavy rainfall over western and south-central Gabon (Figure 1.2). However, the flow is irregular and seems heavily influenced by the Ivindo River, which runs through an area with a bimodal (or equatorial) rainfall pattern that varies substantially from year to year due to the unpredictability of rainfall near the climatic equator. The climatic pattern is bimodal throughout the entire basin, with predominant austral influence: heavy flooding in May is followed by a decrease in June and July, with the lowest flow in August and September, and the greatest flooding in November and the start of December—mostly due to the Ngounié River—and reduced flow again in February and March (Figure 1.3). Thus, the longest low-water period corresponds to the austral summer. The average flow at low levels is 1,510 m³/sec, which is relatively high, mostly a result of the short dry season (three months) and because most of the watershed is covered in forest, which attenuates all variations. The average flow during flooding is 9,150 m³/sec and 13,000 m³/sec during 10-year floods.

Analysis of the flow rates observed monthly at Lambaréné since 1970 (Mahé *et al,* 1990) shows significant variation from year to year (Figure 1.4). Moreover, the austral rainfall pattern has undergone significant changes over time: flooding is now less pronounced and starts earlier in the year. From 1961 to 1984, the annual flow has followed a continuous downward trend. The causes of these changes are probably complex, but there is a correlation between the changes observed along the Ogooué and the reduction in seasonal resurgences of cold water along the Congolese and Gabonese coasts (Mahé *et al.*, 1990).

The Lower Ogooué wetlands: An overview

From the point where it enters into the littoral sedimentary basin, 200 km from the coast, the Ogooué River crosses a vast complex of wetlands covering approximately 715,000 ha. Its size is difficult to determine with precision due to annual variation of the extent of floodplains and because the border between forests situated on solid ground and wetlands can be very difficult to determine using satellite imagery.

This complex is the largest delta in Africa after those of the Nile and Niger, which have both been heavily degraded—for many millennia in the case of the Nile Delta, and since the start of fossil fuel exploitation in the case of the Niger Delta.*

The Lower Ogooué wetland complex comprises two structural units whose composition is of vastly different origin (Figure 1.5). The majority of the complex is comprised of a delta formed by river alluvium. Its apex is situated around Lambaréné,† and its discontinuous boundary stretches from Cape Lopez Bay to the north to Nkomi in the south. Areas upstream from Lambaréné are flooded by the river following alluvial deposition. These extend to where the river joins the littoral sedimentary basin towards the east, and include the lower parts of the Abanga, Ngounié, and Mbiné rivers (Figure 1.1).

All along the coast, this deltaic complex is bordered by large sand bars, formed of sand deposited by the littoral currents that follow the southern coast of Gabon towards the northwest and bend towards Annobon Island beyond Cape Lopez. They are also shaped by the dominant southwestern Atlantic swell. This current is often mistaken for the Benguela current (Lebigre, 1983; Ondo Assoumou, 2006), which is actually pushed away toward the middle of the Atlantic Ocean by the waters of the Congo River and runs more than 1,000 km off the coast of Gabon. The currents that follow the southern coast of Gabon actually carry relatively warm waters up from the estuary of the Congo River (Wauthy, 1977). These waters are only cooled during the dry season, from June to September, when cold water from the center of the Atlantic is brought to the surface by the Lomonosov undercurrent (Koleshnikov *et al.*, 1971; Hisard & Molière,

* The Niger Delta covers approximately 7 million hectares but is inhabited by 31 million humans and has suffered serious environmental damage following the beginning of oil exploitation in 1950. According to the government of Niger, 7,000 oil leaks occurred between 1970 and 2000. The Nile Delta covers approximately 2.6 million hectares and has been cultivated for at least 5,000 years. It suffered heavy environmental degradation following the construction of the Aswan Dam, which retains the river's alluvium. It is currently inhabited by 39 million people (for a density of 1,000 inhabitants/km²) but still houses large populations of aquatic birds in wintertime (Bahaa-El-Din, 1999).

† The Ogooué Delta is often reduced to its mostly marine section, characterized by the presence of mangroves. In fact, it starts in Lambaréné, just as the Nile starts in Cairo.

1973; Guilerm, 1975; Wauthy, 1977).

Regardless of their vastly different origins, both components influence each other, and as indicated below, the configuration of the lower part of the delta has been heavily influenced by littoral sand deposits for the last 4,000 to 6,000 years (pages 46 and 47).

Geology

The Lower Ogooué traverses Gabon's littoral sedimentary basin, which is formed by a slightly undulating plateau that once formed the bottom of the Atlantic Ocean. It is formed by sediments that become older from west to east (Figure 1.6), extending from the Permian to the Miocene. The basin is bisected by the Lambaréné horst, and the sediments are covered by Pliocene sand, forming the Wonga-Wongué Plateau and the Nionga Plain, as well as Quaternary and Holocene sands deposited by the Ogooué. Oceanic sand is found all along the coast. Further information on the geologic history of this region is provided below (pages 30 and 31).

The deltaic complex

General structure

The Ogooué traverses a littoral sedimentary basin for 200 air-km, starting approximately 30 km downstream from Ndjolé. With an initial width of about 80–200 m, the river quickly widens to approximately 1 km and reaches 2.3 km in width slightly upstream from Lambaréné. Approximately 41 km downstream from Ndjolé, it is joined by a small tributary on the right bank, the Abanga River, whose headwaters are in Equatorial Guinea and which drains a large part of the Crystal Mountains. Eleven kilometers upstream from Lambaréné, the Ogooué is joined by its main left bank tributary, the Ngounié River.

The river splits in two, the Ogooué proper and the Ouango, which cuts through the horst (Figure 1.6). Beyond, these rivers spread across the upper deltaic floodplains, formed mostly of a vast alluvial plain with small islands of dry land that escape flooding. Its maximum length is 107 km, calculated starting from the horst, and its peak width is 25 to 30 km. In its central section, this plain contains residual lakes, and along its periphery there are numerous flood lakes that extend into secondary plains, both to the north and south. Between the two branches of the horst, the river's main course diverts into the Agouma River (or Rivière des Pélicans) and the Ezanga River, both flowing into the Onangué and Ezanga lakes, which form part of the Great Lakes situated 30 km to the southwest of Lambaréné. These two rivers become two small inland deltas.

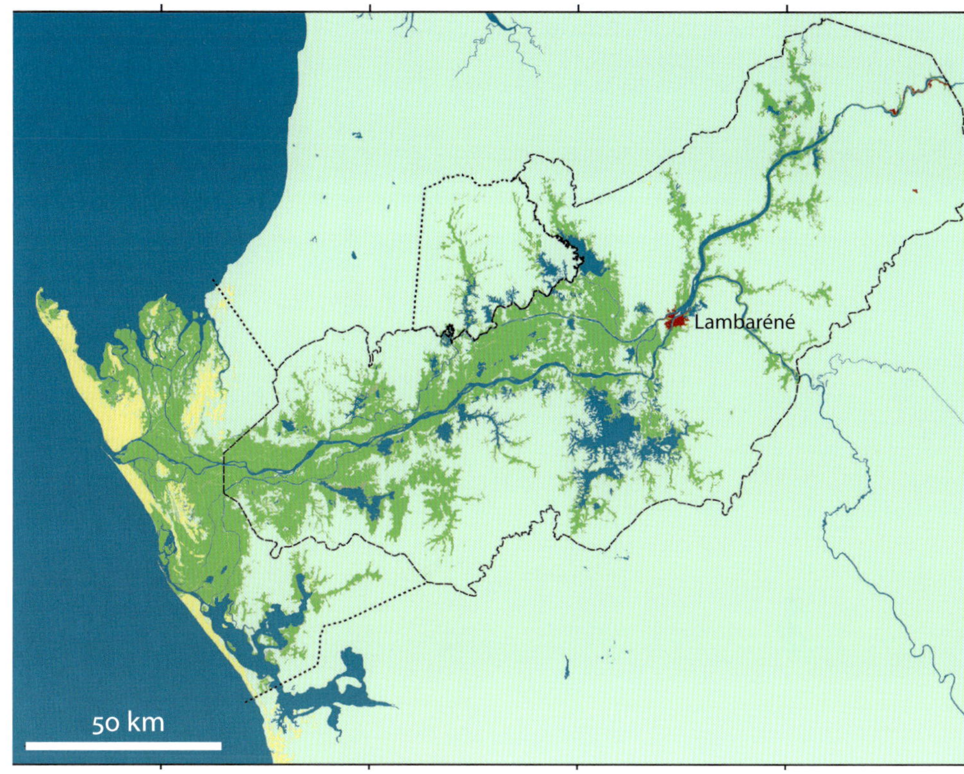

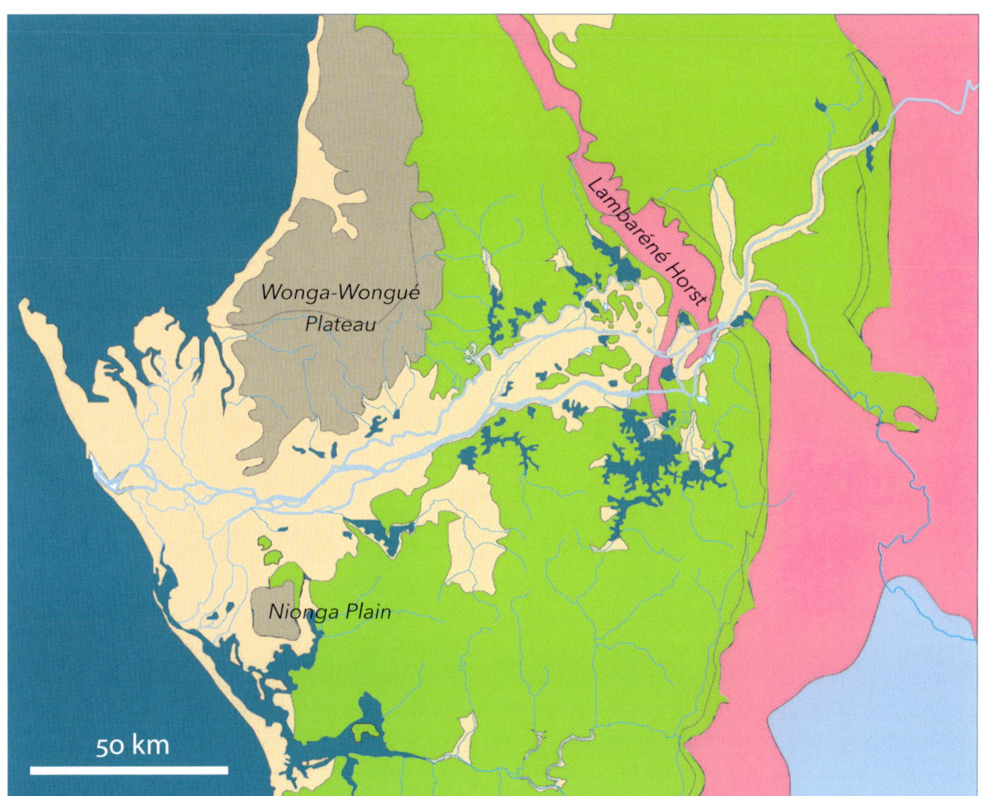

1.5. The two main structural groups of the deltaic complex: alluvial structures (in green) and coastal sand structures (in pale yellow).

1.6. The four main groups of geological structures:

- *Metamorphosed crystalline formations of the Archean and Paleoproterozoic*
- *Sedimentary formations resulting from the opening of the Atlantic rift*
- *Sand deposits from the Pliocene (Wonga-Wongué Plateau and Niongo Plain)*
- *Sedimentary deposits from the Pleistocene and Holocene*

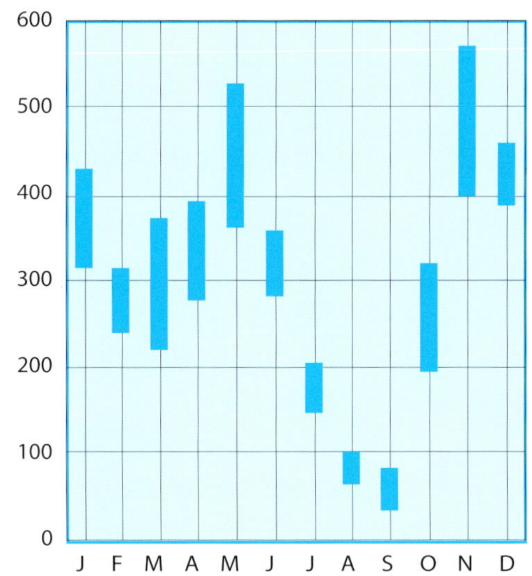

1.7. Monthly variation in the water level of Lake Ezanga, recorded over 3 years, according to Loubens (1966).

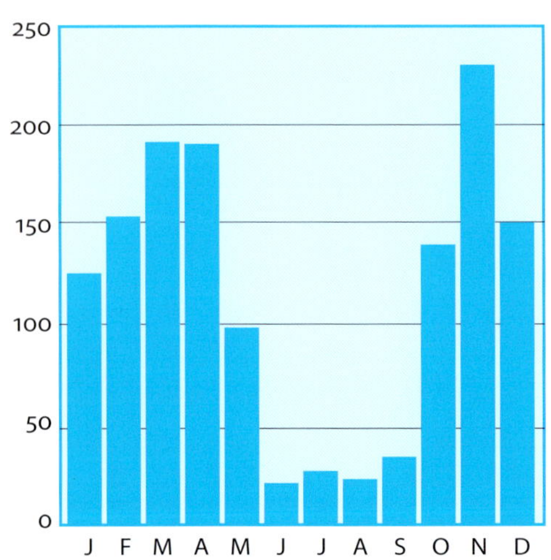

1.8. The rainfall in Port-Gentil, expressed in mm/month.

Roughly 140 km downstream from Lambaréné, both watercourses of the Ogooué join up and the upper deltaic plain progressively turns into the lower deltaic plain or tidal plain, which today constitutes the oceanic delta proper. This transition is quite gradual and not particularly visible. However, the lower deltaic plain comprises three very distinct sections. The main river continues its nearly linear course through the center of the delta towards the Gulf of Guinea. It forms an estuary delta, protected by an intermittent narrow sand bar 12 km in length. Two watercourses flow to the north, reaching the Cape Lopez Bay and the Bay of Nazaré east of Port-Gentil. They form a typical estuary delta. Finally, three watercourses flow south, cross the southern delta, and flow into the Nkomi Lagoon (or Fernan Vaz Lagoon) forming an arcuate delta. From there on, the waters of the Ogooué River and the lagoon flow together into the Gulf of Guinea via a single outlet, protected by a 15 to 18 km sand bar.

Profile of the Lower Ogooué

Between Ndjolé and Lambaréné, the slope is around 0.095 m/km, whereas in the upper delta, it is approximately 0.08 m/km. Such a gentle slope would normally create a sinuous course. This is not the case, however. The Ogooué flows in a very straight manner toward the sea, which is probably because this section of the river follows an old fault line related to the birth of the Atlantic Ocean (Lebigre, 1983). In the lower deltaic plain, the slope is 0.04 m/km.

Local climate

The Lower Ogooué region has an equatorial humid tropical climate. Average annual rainfall varies from 1,950 to 2,100 mm. The three-month dry season is from June to September. The rainy season is from September to May, with two peaks in rainfall around April and November, and a period of lower precipitation around January and February (Figure 1.8). The average monthly temperature varies little, but is approximately 2–3°C cooler during the dry season, which reduces evaporation and evapotranspiration.

Variations in water level

At Lambaréné, where the river passes through two bottlenecks, seasonal variation in water level reaches 6 m. At the southern Great Lakes (Onangué, Oguemoué, Ezanga, and Evaro), the variation is 3.4 m on average, but reached 5.5 m in 1961 (Loubens, 1966). In the upper deltaic plain, variation in level is largely buffered by the fact that the water spreads out.

In the lower deltaic plain, water level variation ranges from 1 to 1.5 m and is largely the result of tides. According to Loubens (1966), the tidal amplitude is 1.1 m on average downstream from the Nkomi Lagoon, ranging from 0.5 to 1.7 m. Variation seems to be slightly more pronounced at Cape Lopez bay. Thus, from east to west, seasonal variations in water level can be seen all along the river caused by rainfall in the watershed, which gives way downstream to daily low-amplitude variation caused by tides. These changes are not simply anecdotal. Not only do they strongly influence the vegetation, they could even be compared to a living organism's heartbeat: these changes modulate the flow of nutrients throughout the ecosystem, whose productivity depends directly on which nutrients are provided.

These nutrients come in part from the river's watershed, including the numerous small rivers that flow directly into the complex, though mainly from regular flooding of dry land directly within the deltaic complex. Thus, when heavy flooding occurs, alluvium is carried by the river from the upper deltaic plain and deposited in these flooded areas, contributing to the filling of the marshes. At the same time, organic matter built up in these marshes by fauna and vegetation is dissolved by high waters and added or returned to the ecosystem. During periods with low water levels, living organisms are concentrated in the remaining waters and the nutrients, previously dispersed during the rainy season, become more readily accessible for aquatic birds and for fishermen and women—which explains why they are most active during the dry season.

Water quality

There are few observations regarding the various biological and physical parameters of the water in the complex and only cover the main channels of the river between Ndjolé and Ngomo, a few lakes (Azingo and Onangué), and the section downstream from the Nkomi Lagoon (Loubens, 1966; Fermon *et al.*, 2017).

No measurements seem to have been made in the black waters of the flooded forests, the great swamps, the numerous small lakes of the lower deltaic plains, or the numerous marshes and small lakes of Mandji Island. However, the waters of the complex belong to two very different and usually quite separate types: (1) black waters and (2) brown waters, heavily loaded with alluvium. Curiously, no publications make reference to this fact. Yet in other tropical regions, notably the Amazon Basin, these two types of waters are usually

1.9. The black waters of the alluvial plain and the brown waters of the Ogooué in the center of the upper deltaic plain. On the left, the brown waters of the main river channel, opposite the village of Achouka; on the right, the black waters from Lake Nyayé. The Ntondji River, which connects the Ouango to the Ogooué proper, is reduced to a narrow stream, in the gallery that crosses the image in the background; it can only be entered by canoe when the water level is very high. It would be surprising if these waters all had the same physicochemical characteristics (Photo: Jean P. Vande weghe).

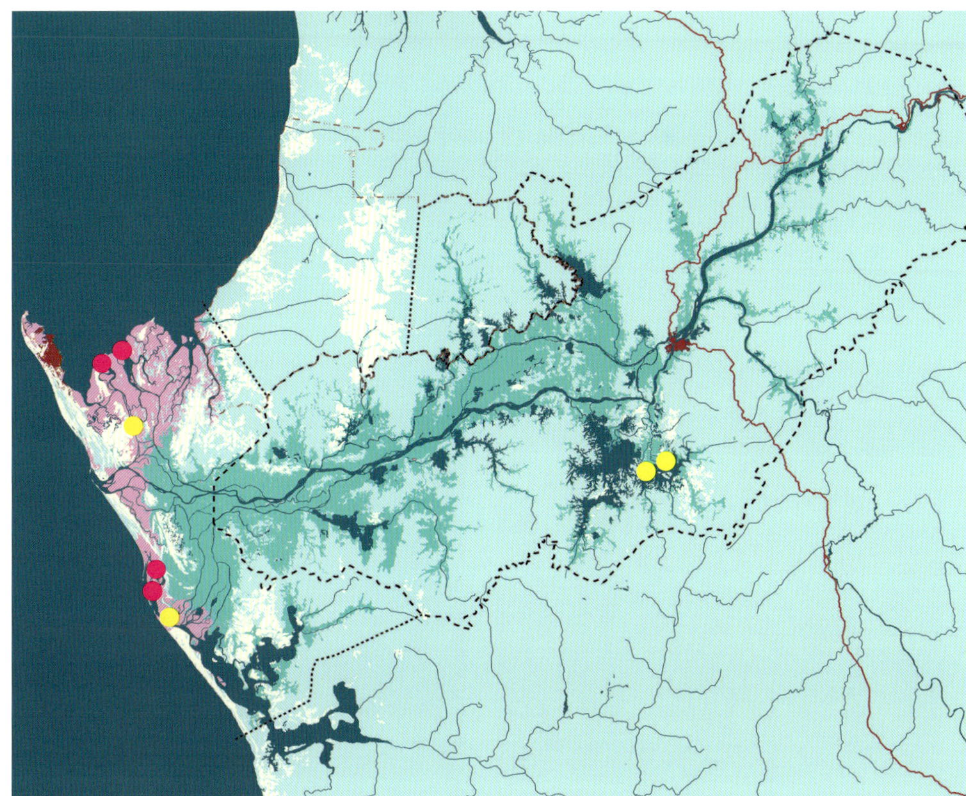

1.10. Observations regarding the salinity of the delta's waters (Loubens, 1963, 1966; OLAM, pers. comm.)

- *Freshwater*
- *Salt water*

Most of the lower deltaic plain has freshwater.

occupied by different types of fauna and flora (Klinge & Furch, 1991; Worbes, 1997). Is Gabon an exception? Based on our initial observations, we do not think so because the marshy forests that are bathed in black waters appear to have a clearly different flora from the forests that are flooded by waters rich with alluvium: *Anthostema aubryanum* (Euphorbiaceae) is dominant in areas with black waters, whereas *Uapaca heudelotii* (Phyllanthaceae) is dominant in areas with brown waters. This deserves to be studied in depth.

Alluvial load

Alluvial load appears to be very high in the river's main channels, but is absent in the black waters of lakes and channels that drain marshes. As seen from above, through satellite imagery or aerial photography, it is easy to discriminate between waters loaded with alluvium and black waters, and the limit between the two is usually very sharp (Figure 1.9). In some rivers, such as the Ambila, which drains Lake Onangué, black waters flow in parallel with waters loaded with alluvium and only mix very slowly (Figure 1.10).

Temperature

The surface temperature of the Ogooué upstream from Lambaréné is 26.6°C on average, with extremes of 21°C and 28.9°C (Mbega, 2004). A slightly higher average temperature of 28.6°C was recorded more recently at Ngomo (Fermon, 2015).

The temperature is stable throughout the year in the Great Lakes. The surface temperature is 25°C in July and reaches 30°C in March or April. Temperatures below the surface of these lakes are 3°C-4°C lower than at the surface. This difference is not as pronounced during the dry season. Thus, deeper waters are always at least 23°C (Loubens, 1966). According to Mbega (2004), the surface temperature is 26.8°C on average, with extremes of 23°C and 29.5°C. In the lower deltaic plains, the temperature is approximately the same: 25.9°C on average, with extremes of 23°C to 29.5°C (Mbega, 2004). Temperature variation is greater at the outlet of the Nkomi Lagoon because of a marked temperature decrease during the dry season caused by an intake of cool water from the ocean (Loubens, 1966).

Salinity

The presence of multiple species of marine fish near Lambaréné or even in the Abanga River suggests the possible presence of saltwater upwellings, although no measurements have confirmed this hypothesis. On the contrary, surface waters exhibit a salinity of 0.03-0.12‰ and 0.04‰ on average throughout most of the delta. The water thus remains entirely fresh, including in the lower deltaic plain. In the Nkomi Lagoon, salinity is only present between Ozoumboua and Olendé (Loubens, 1966), although it varies based on the season, tides, and depth (Loubens,

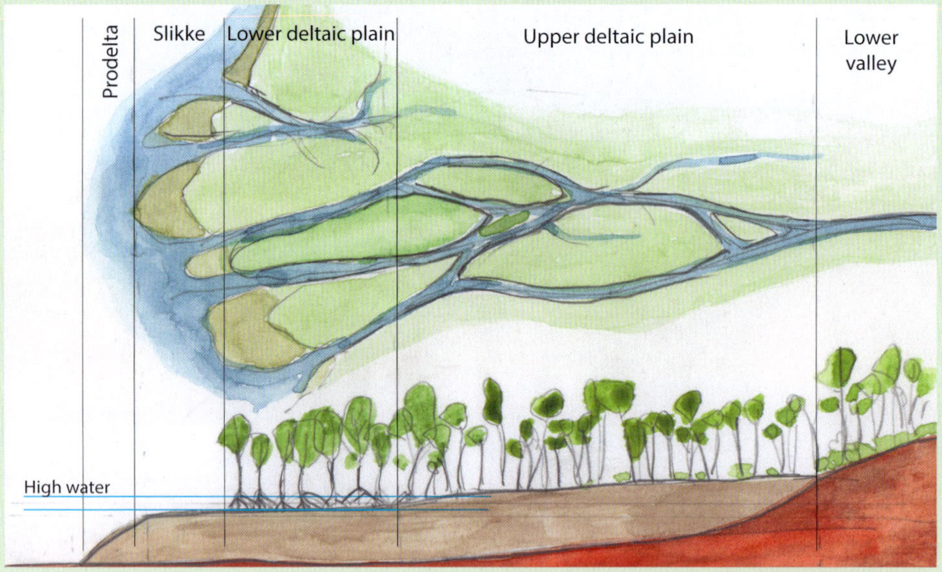

Deltas

Deltas are sedimentary structures that form when a river flows into the sea or a lake. The suspended sediments precipitate when the current slows down or stops entirely.

The structure of a delta
Although the shape of a delta may vary, most large deltas include similar structural elements. From upstream to downstream, the following can be distinguished (Figure 1.11):

(1) The alluvial valley, which is usually flooded because the formation of the delta increasing water levels upstream.
(2) The apex or root of the delta, which represents the section where the river begins splitting and depositing alluvium.
(3) The upper deltaic plain, with seasonal variations in water levels.
(4) The lower or tidal deltaic plain, with variations in water levels caused by tides.
(5) Mudflats, which are exposed at low tide.
(6) The delta front, situated below the low tide level and formed by underwater alluvium.
(7) The prodelta, formed by an abrupt sedimentary slope that progressively impinges on the continental shelf.

Types of deltas
Sedimentation is heavily influenced by coastal currents (generally not found in lakes), tides (absent from lakes), and swell (generally weak or insignificant in lakes). Vastly different types of deltas form depending on which of these three factors is predominant. Three main types can be identified:

(1) Tidal freshwater deltas (Figure 1.12) have large alluvial beds lining the channels but no deltaic tidal plains or mudflats. A typical example is the Mississippi Delta. In Gabon, this type includes the deltas of the small Ezanga River and the Rivière des Pélicans (pages 34–36).

(2) Wave-dominated deltas, which are characterized by the presence of a sandbar at the mouth of the delta (Figure 1.13). An example of this is the Rhône Delta. In Gabon, the southern Ogooué Delta is wave-dominated.

(3) Tide-dominated deltas are characterized by each river ending in a wide estuary (Figure 1.14). Examples include the Rhine Delta in Europe. In Gabon, the northern Ogooué Delta is tide-dominated.

However, most large deltas are complex and include formations and sectors corresponding to different basic types. This is the case for the Ogooué delta.

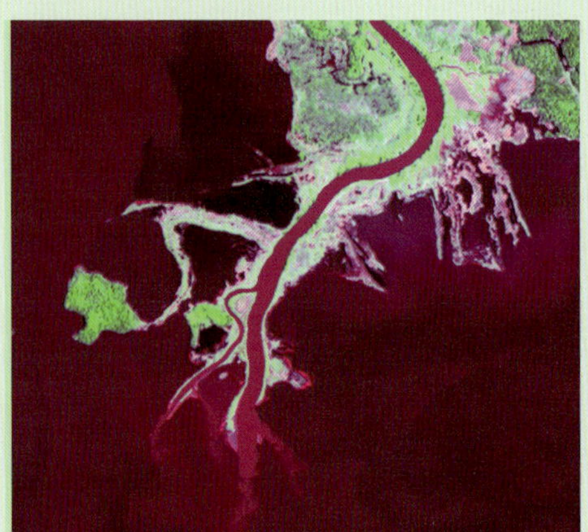

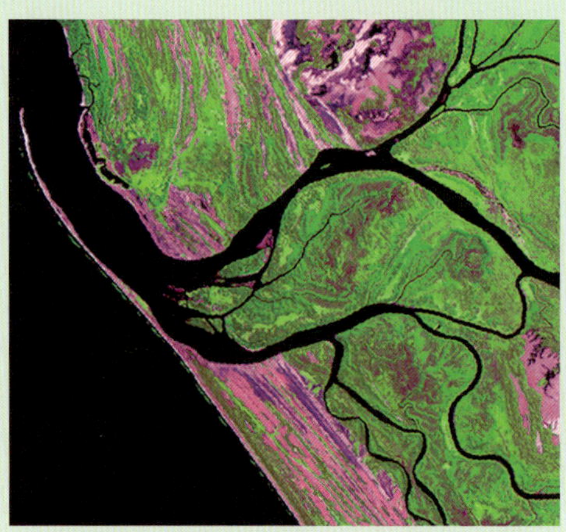

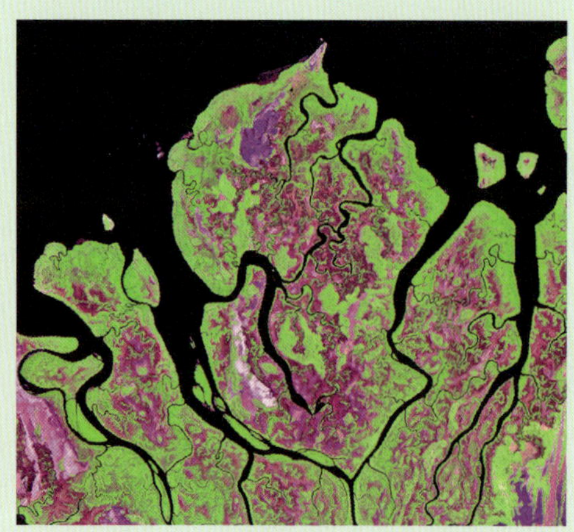

1966). In the Cape Lopez bay, salt water is limited to the area immediately along the ocean. Two surveys of the Atadali River, one of the main watercourses of the northern delta, revealed that waters 20–25 km away from Cape Lopez bay remain fresh all year round, from the surface to the riverbed (OLAM Experts, pers. comm.).

Acidity

The waters of the Ogooué have a pH of 6.83 to 7.13 on average from Ndjolé to Ngomo, with extremes from 5.8 to 8.1. They thus vary from slightly acidic to slightly basic. The pH in the southern Great Lakes (e.g., Onangué) varies from 6.6 to 8.3. The black waters of Lake Azingo are slightly acidic (Loubens, 1966).

Nutrients

Phosphate, ammonium, nitrate, and nitrite contents, the main nutrients, are very low throughout the complex (Fermon, 20015; Fermon *et al.*, 2017). The waters are thus oligotrophic.

Structure

Taking into account the considerations mentioned above, the Lower Ogooué wetlands can be divided into four main sections, each with its own structure:

(1) the **lower river valley** and its nearby wetlands upstream from Lambaréné;

(2) the **upper deltaic plain**, situated downstream from Lambaréné, including the annex lakes and inland deltas;

(3) the **lower or tidal deltaic plain**;

(4) the **coastal sand structures**, notably Mandji and Ozouri islands and their adjoining plains, along with Animba, Ompingo, and Inguéssi islands, as well as older structures such as the Koroso Plain (Figure 1.1).

The problem of the Nkomi Lagoon

The southern part of the Ogooué Delta currently flows into the Nkomi Lagoon. Thus, in this area, the Ogooué Delta and the Nkomi Lagoon have a common outlet, which is protected by an 18 km long sandbar comprised of the "Barre de l'Arbre" and "Barre des Portugais" (Figures 1.1 and 2.28).

Historically, 4,000 to 6,000 years ago, the Nkomi Lagoon's estuary was unquestionably separate from the Ogooué River (pages 45 and 46). Later, the coastal sand structures formed an enormous sandbar, which blocked the outlets of both the Nkomi Lagoon and the Ogooué River. For this reason, the delta and the lagoon are in close contact. Waters from the Djembagombé and Mpolounié rivers always flow south and end up flowing into the lagoon. At high tide, they even flow back upstream from the lagoon, where they deposit large quantities of alluvium. The shallows, which make navigation difficult across large areas during the dry season and at low tide north of Omboué, are all the result of deposits from the Ogooué River. In fact, they form the southern tip of the deltaic front, which extends into the lagoon (frame on page 24).

Furthermore, it appears that connections have existed or might still exist at times between the Lower Ogooué Delta and the basin of the Nkomi Lagoon.

According to fishermen of the Onangué lake region, there was once a connection between Lake Oguemoué and a small tributary of the Nkomi Lagoon. This connection was probably not permanent but rather was a spillway for the lake during the strongest floods and allowed fish to climb upstream from the Nkomi Lagoon directly to the Great Lakes. It would definitely have been cut through two decades ago by the construction of a raised forest track in the forestry concession surrounding the northern part of Lake Oguemoué. According to the same fishermen, multiple species that used to be frequently found in the Great Lakes have nearly disappeared since this connection was closed. During the 2014 WWF expedition, we unfortunately could not verify this information, which remains to be confirmed.

Another connection is said to exist further upstream. It would allow the waters of the Nkengué swamp to flow towards Ntchonga Bay and the Nkomi Lagoon when the Ogooué River experiences heavy flooding, probably through the Ngouboué River (Figure 1.21). According to terrain modeling for this region, this interpretation seems possible, though it also remains to be verified.

With regard to their biology, the Nkomi Lagoon and Lower Ogooué currently have a very similar if not identical fish fauna (Fermon, 2015). Despite their having distant origins, both hydrological entities, which lack any barriers that isolate them from one another, now simply represent distinct sectors of a single ecological and biogeographical entity. Furthermore, the Nkomi Lagoon has its own watershed, and as a result, its own water sources, although this is also true for all of the Great Lakes situated both to the north (Azingo, Gome) and south (Anengé, Onangué, Ezanga) of the alluvial plain, as well as the smaller marshes of the Mbiné and Abanga rivers. Moreover, the fact that a large lagoon is juxtaposed with a delta is nothing exceptional.

This phenomenon can be observed elsewhere in the world, especially those with wave-dominated deltas such as the Danube delta and the Rhône delta in Europe.

In the broadest sense of the term, the Ogooué complex thus includes the Nkomi Lagoon. It begins to the east, slightly upstream from Ndjolé, at the limit of the coastal sedimentary basin. It ends to the west at the Atlantic coast, where it includes coastal sand structures that stretch from Cape Lopez to southern Gabon. To the north, the complex includes the watersheds bordering the northern Ogooué's tributaries, notably those of the Rivière des Coupeurs, Lake Azingo, and the Mbiné marshes, and even a lower section of the Abanga River. To the south, it includes the southern edge of the Nkomi Lagoon basin and the various basins of the southern Ogooué's tributaries. Further east, it also includes the lower sections of the Ngounié River. For practical reasons, we will only treat the northern part of the Nkomi Lagoon in this book, limited by the city of Omboué. We lack sufficient data to be able to cover the rest of the lagoon and its tributaries properly.

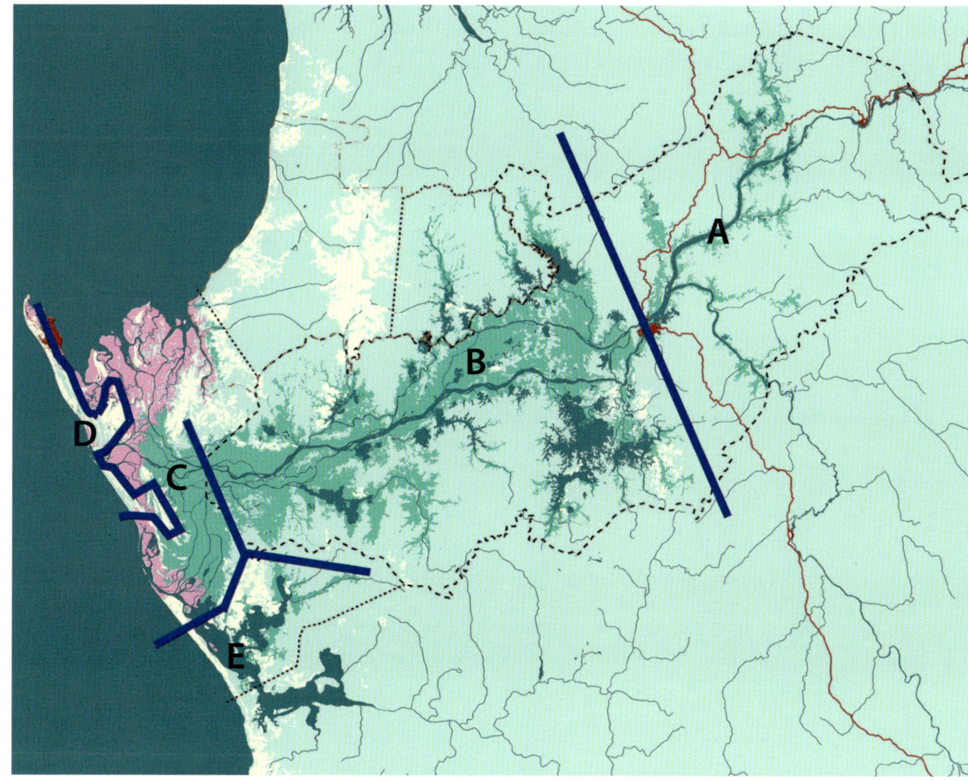

1.15. The five areas of the complex:
A. The Lower Ogooué valley, flooded by the rising river waters caused by the delta's deposits.
B. The upper deltaic plain, subject to seasonal variation in water level.
C. The lower deltaic plain, subject to daily variation in water level based on tides.
D. The coastal sand structures.
E. The lower section of the Nkomi Lagoon.

The Lower Ogooué valley

Approximately 70 km to the northeast of Lambaréné or 20 km upstream from Ndjolé, the Ogooué River enters the littoral sedimentary basin. Its single course widens progressively, measuring 1.64 km wide near the point where it meets the Ngounié (its main left bank tributary) and 2.38 km wide just upstream from

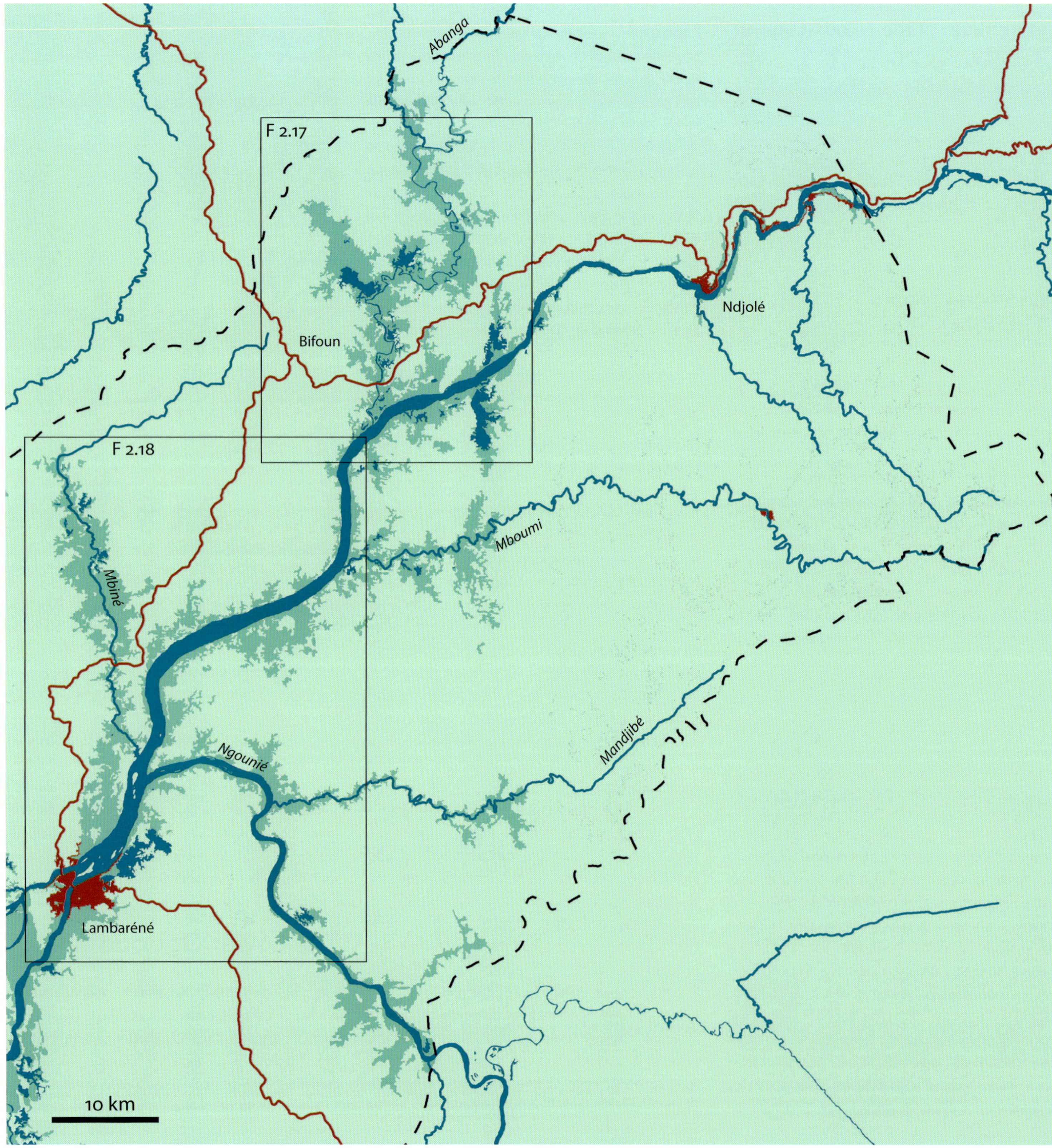

1.16. The Lower Ogooué valley, upstream from Lambaréné, with the Abanga and Mbiné rivers and various lakes.

Lambaréné (Figure 1.16). In this area, the Ogooué flows in a single channel, flanked on both sides by flood zones, which include flooded riparian forests, swamps, marshes, and lakes. The Abanga River flows in on the right bank. The Ngounié River flows in on the left bank, with its lower valley similar to that of the Ogooué, at least within the limits of the Ramsar site. The small Mbiné River also flows in on the right bank, its valley also largely inundated (Figure 1.18). Together, the humid areas in this section of the Lower Ogooué cover approximately 20,000 ha. Even though its structure is relatively simple, the Lower Ogooué has a very diverse set of habitats (Figures 1.19 and 1.20).

The river

The river reaches widths from 100–200 m to 2.35 km, and the water level is on average around 13 m at Lambaréné, 18 m above where the Abanga River flows in, and 22 m in Ndholé. The river's slope is approximately 10 cm/km. Its transverse section usually includes shoals in the center bordered by two deep channels that follow near the banks or alluvial levees (Figure 1.19). In places, these channels can reach depths of 10–12 m. In the central part, the riverbed rarely reaches depths exceeding 0.50–2 m (Fremon, 2015).

The water's physicochemical parameters were measured at its surface within the main flow of the river. According to measurements made at Ndjolé and Lambaréné in 2014, the water's pH is 6.83–6.97 on average, with higher values (7–7.5) from May to August. The average water temperature is 10.63°C–15.94°C and its transparency is 26–30 cm (Fermon, 2015). The alluvial load increases during the rainy season. Furthermore, these waters contain low levels of phosphates, ammonia, nitrites, and nitrates, which represent the main nutrients. No measurements have been made in the black waters of adjacent lakes and marshes.

Sand banks and islands

During the dry season, from the beginning of June and through to the end of September or even the beginning of October, the central part of the river's bed dries up in many places to form vast sand banks that can reach up to 2–3 m above water level. These sand banks can be divided into two zones: (1) a higher zone, which remains dry longer, comprises very white bleached sand (Figures 1.42 and 1.44); and (2) the lower zone, which emerges only at the end of the dry season, is comprised of ochre-brown to purple-brown sand with more clay (Figure 1.43).

The tops of some sand banks are partially covered in ephemeral herbaceous vegetation, mostly comprised of Poaceae, Cyperaceae, and Polygonaceae, which is nevertheless submerged or even washed away during floods.

The river also has some small islands, especially where the Ngouiné River joins it a little upstream of Lambaréné. These islands are permanent and covered in mats of Polygonaceae, *Alchornea* thickets, or sparse *Nauclea* forests. Most are recent sedimentary structures–tall sand bars on which woody species have grown and remain (Figure 1.42). Their presence suggests that flooding of the Ogooué River was much heavier in the recent past than it is today. The decrease in flooding since the 1970s (page 20) may have transformed some sand banks into permanent islands.

Alluvial levees

Both sides of the river are bordered by wide alluvial levees, usually spanning a few dozen meters and covered in flooded forests with *Uapaca heudelotii, Ceiba pentandra,* and dense undergrowth with *Alchornea cordifolia* on open areas (Figure 1.44). These forests have generally been heavily disturbed by past logging activities.

These levees are broken in places by narrow channels that drain flooded areas situated behind the levees, and which allow them to flow toward the river. Around the confluence of the Ogooué and the Abanga, these channels are very numerous and form complex networks (Figure 1.46).

Flood zones

Beyond the alluvial levees there are swamps, marshes with tall grasses, papyrus beds, and several lakes. These floodable zones vary: the width can range from 0.5–3 km (Figures 1.16 and 1.18) but are poorly defined, and their surface area varies according to the flood levels.

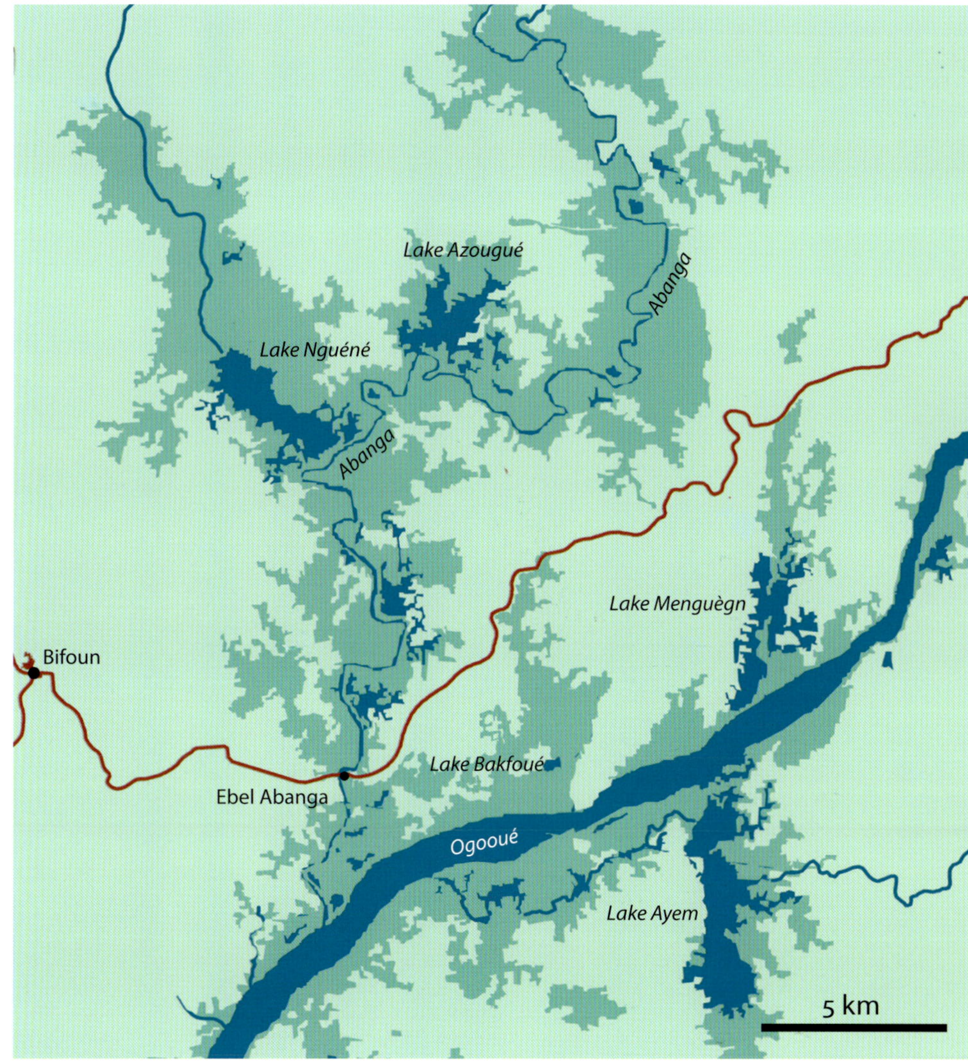

1.17. The Lower Ogooué with Lakes Menguègn, Ayem, and Bakfoué, the Abanga River and its lakes, Nguéné and Azougué, as well as the flooded and floodable areas. In red, the road from Bifoun to Ndjolé.

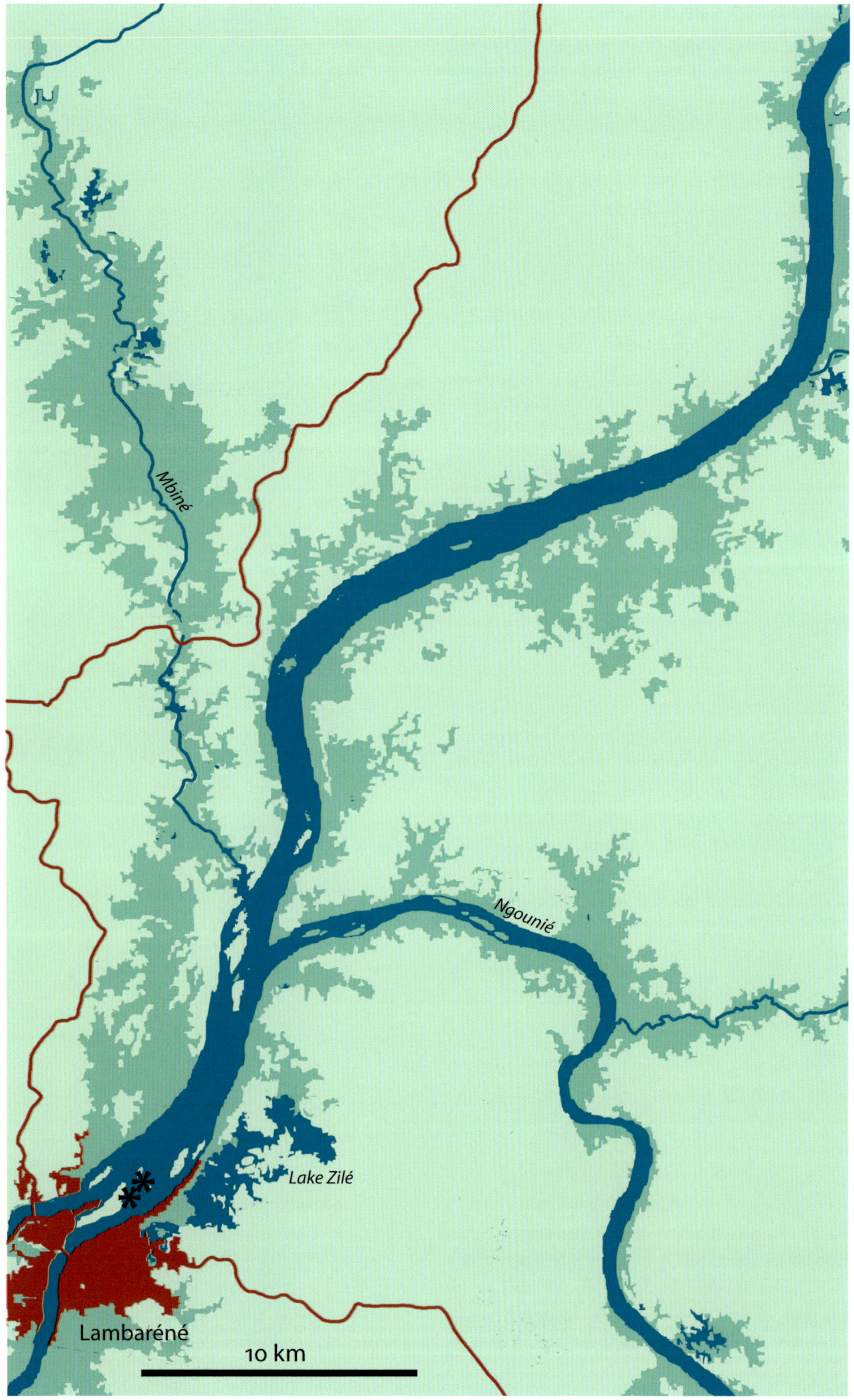

1.18. The marshes of the Mbiné River and Lake Zilé. The main colony of African cormorants, egrets, darters, and herons is indicated by asterisks (*).

In these zones, the water is black, totally devoid of alluvium and, probably, less acidic. Unfortunately, no surveys of their physico-chemical parameters have been done.

Lakes

There are three lakes on the right bank, notably lakes Manguègn and Bakfoué (Figure 1.16), and four on the left bank, the best known of which is Lake Zilé near Lambaréné (Figures 1.1 and 1.18), but the largest is Lake Ayem situated further upstream (Figure 1.17).

Lakes Manguègn, Ayem, and Zilé represent lateral valleys flooded by the Ogooué River. They present numerous islets, representing raised areas that escape flooding. Their shores facing the river are heavily fragmented and covered in dryland forest. The transition between lakes and the forest is abrupt, with only a few bays inhabited by partially floating herbaceous mats, papyrus stands, formations comprising *Ficus trichocarpa*, and swamp zones. The alluvial levees formed between the river and these lakes are mostly linear.

Herbaceous marshes

Herbaceous marshes appear to be old, dried-out lakes. Some still have nearly inaccessible, residual open-water lakes in their center, which are roughly circular and are bordered on all sides by aquatic plants. This is the case for Lake Bakfoué (Figure 1.17), which is very different from those formed when nearby valleys are flooded.

The largest grassy marsh is in the Mbiné valley, a small right-bank tributary of the Ogooué River whose flow is slowed down when the river's water level increases (Figure 1.18). Spanning 2–4.5 km in width, this marsh extends nearly 30 km from north to south and covers an area of approximately 7,000 ha. The Mbiné River flows into the Ogooué just north of the confluence of the Ogooué and the Ngounié, where it crosses an area with flooded forests. This portion is often blocked by fallen trees, but is maintained by fishermen and women who live in villages upstream on the Mbiné, mainly along the Bifoun-Lambaréné road, which crosses over the river. A large section of the marsh is situated north of this road. These marshes comprise *Echinochloa*, papyrus, and raffia groves or mixed swamp forests. Several open-water areas form small lakes (Figures 1.51–1.54).

The Abanga River

The Abanga River begins in Equatorial Guinea and flows through Lake Azougué and the Crystal Mountains. In its alluvial plain, it forms large meanders flanked by alluvial beds and vast flooded zones. It is also flanked by several lakes, the largest of which are Lakes Azougué and Nguéné, the latter fed by the small rivers Minsoulé and Nguéné, which flow into vast marshy plains

1.19. The sandbanks of the Ogooué and Lower Ngounié during the dry season (Photo: Jean P. Vande weghe).

1.20. Abanga, downstream from Ebel Abanga, with its alluvial beds and adjoining lakes (Photo: Jean P. Vande weghe).

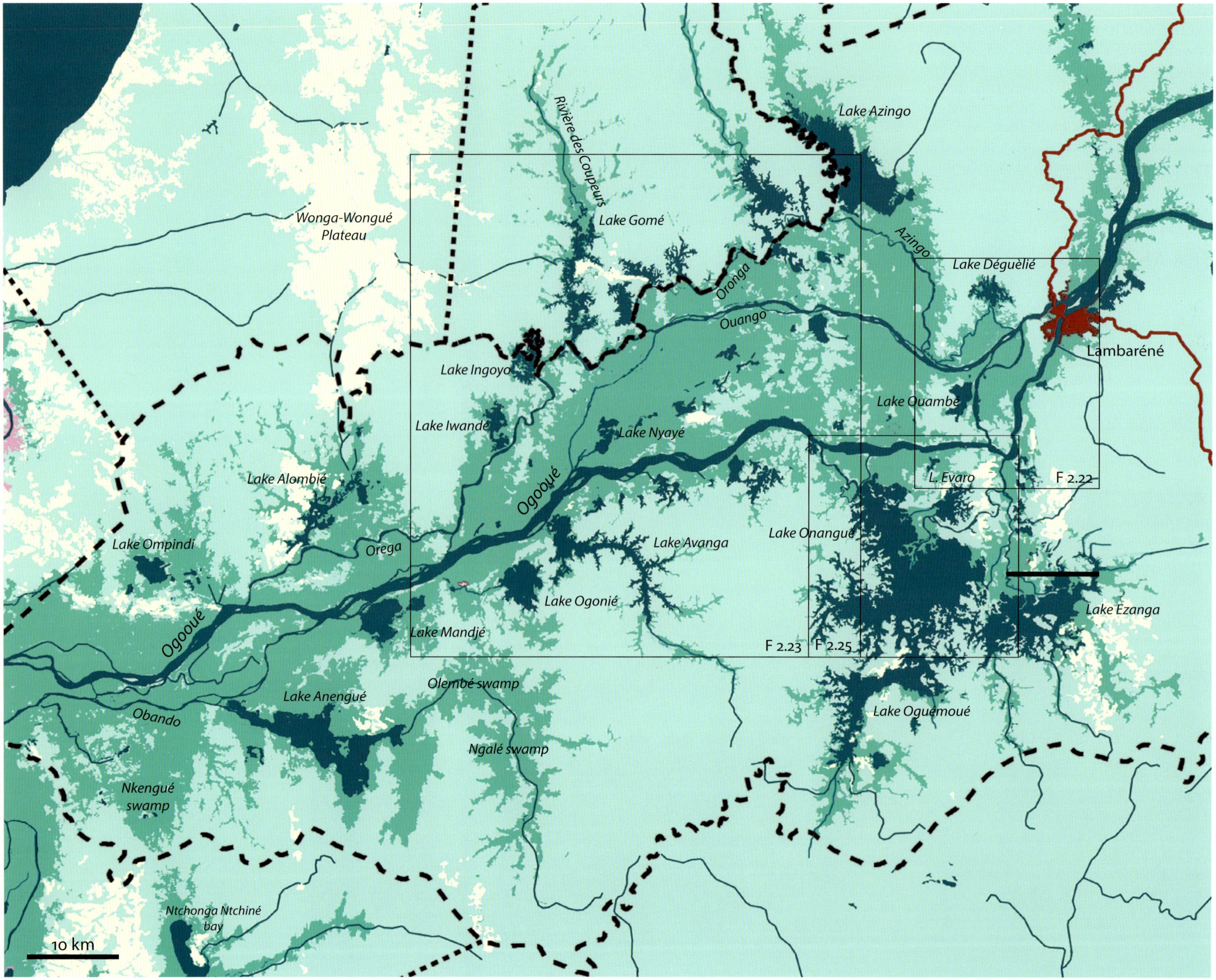
Wonga-Wongué
Plateau
Rivière des Coupeurs
Lake Gomé
Lake Azingo
Azingo
Lake Déguèlié
Oronga
Ouango
Lambaréné
Lake Ingoyo
Lake Iwandé
Lake Nyayé
Lake Ouambé
Lake Alombié
L. Evaro
F 2.22
Ogooué
Lake Ompindi
Orega
Lake Avanga
Lake Onangué
Lake Ogonié
Lake Ezanga
Ogooué
Lake Mandjé
F 2.23
F 2.25
Olembé swamp
Lake Anengué
Obando
Lake Oguémoué
Ngalé swamp
Nkengué
swamp
Ntchonga Ntchiné
bay
10 km

approximately 5-6 km long (Figure 1.50).

Upstream from Lake Nguéné, the Abanga River weaves between two alluvial beds on which stand dense but shallow riparian forests. Beyond them, open marshes and swamps, often surrounded by raffia, occur (Figures 1.47 and 1.48).

The upper deltaic plain

Near Lambaréné, which is built on hills forming part of the eastern horst chain, the Ogooué River splits in two, the southern branch being the main one (Figures 1.21 and 1.22).

These two courses cut across the hills and spread throughout a small alluvial plain whose area is 110-120 km² before crossing the eastern branch of the horst.

Beyond this, they fan out across a vast alluvial plain 110 km long and up to 30 km wide, which has very irregular outlines. It is separated in two by a pronounced bottleneck near the village of Ngoumbi (Figure 1.21). Here the plain is reduced to two channels with a combined width of 3 km. The main river course and the smaller secondary channel to the north—the Orega—both flow through low hills. Between the horst and Ngoumbi, the plain covers an area of 1,900-2,000 km², including secondary lakes and marshes. Between Ngoumbi and the lower deltaic plain, the plain covers about 1,100-1,200 km², and in all, covers 3,000-3,200 km².

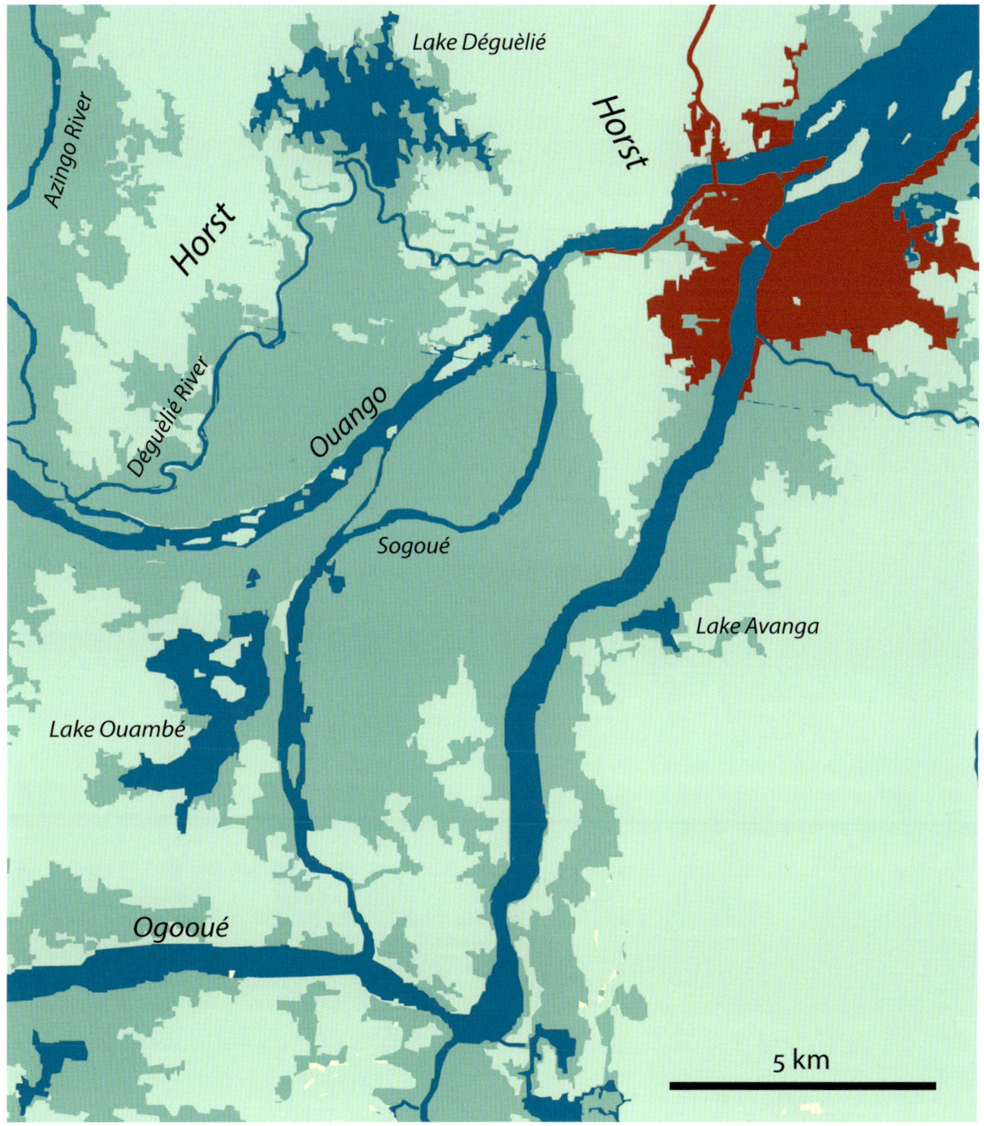

Page 30:
1.21. View of the upper deltaic plain, including the central alluvial plain, the three river channels (Ogooué, Ouango, and Oronga), and the many lakes that border the plain both to the north and south.

Page 31:
1.22. The "intrahorst" section of the upper deltaic plain, with the two branches of the Lambaréné horst and Lakes Déguèlié, Ouambé, and Avanga.

The intrahorst section

Between the two branches of the horst, the Ogooué River flows through a small plain occupying about 110-120 km² and mostly covered with grassy open, herbaceous, and swamp forests. Large areas of *Uapaca* riparian forest occur on the alluvial levees, especially along the Ouango River.

This northern river branch of the Ouango River has a small, secondary channel along its right bank, the Déguèlié River (Figure 1.22), which begins with small rapids formed by a rocky outcrop, making it difficult to access when the water level is low. It then follows the base of the horst hills, flowing into Lake Déguèlié via a mini bird's foot lacustrine delta and then continuing along the foot of the western horst hills, and terminating at the river's northern branch. Beyond these rapids, the northern channel forms a large furo* along the left bank, which joins the river's southern channel. The Ouango passes through the western Lambaréné horst range for a short distance (1 km at most). This part is about 1 km wide and has a secondary channel running in parallel along its right bank.

The river's main channel follows the foot of the coastal sedimentary basin hills before making its way through low hills of the western horst range. This southern section traverses the western branch of the horst for over 15 km. It flows through hills with ill-defined edges and whose tops are between 50 and 115 m altitude, while the river has an approximate altitude of 13 m. The banks are abrupt and at times reach up to 6 m in height. Near the center of this branch, the river bends at a right angle,

* Furos are channels that form braids between the main channels of a deltaic system.

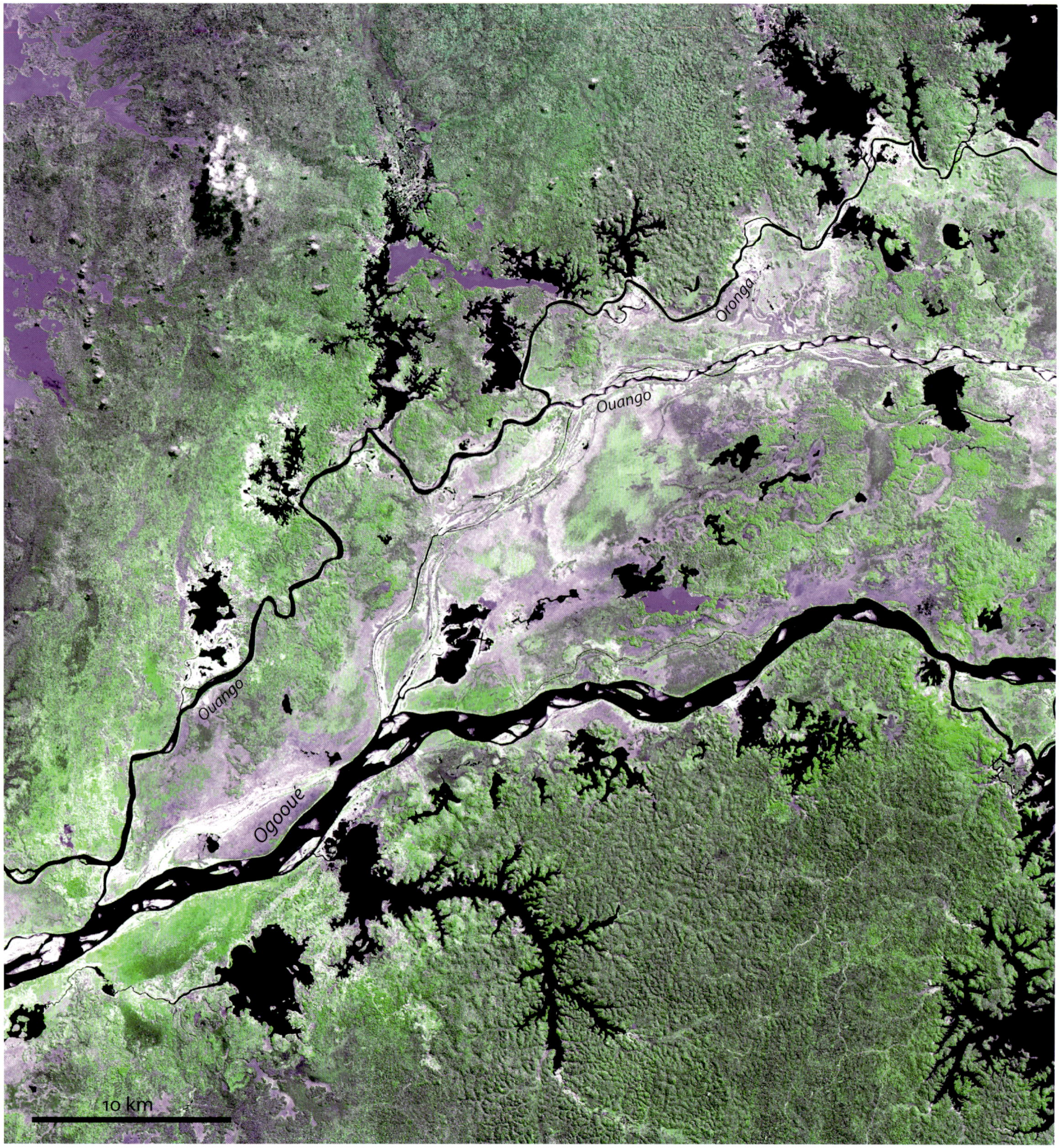

flowing straight to the west. This is the start of the Rivière des Pélicans, which feeds the lacustrine deltas of both Lake Evaro and Lake Ezanga (page 38).

The upper alluvial plain

Upon exiting the horst, both channels of the Ogooué enter a vast alluvial plain extending nearly 110 km from east to west and 30 km from north to south, which is split into two sections near the village of Ngoumbi. Across this plain, the river's surface descends from 13 m altitude to 1 or 2 m, resulting in a slope of 0.8–1.6 cm/km. The water of all three channels has high levels of alluvium and is entirely fresh.

The river

The alluvial plain is traversed by three separate branches of the river (Figure 1.21).

The main channel

Upstream from Ngoumbi, the Ogooué River proper follows the plain's southern border. Downstream, it extends across the plain toward its northern border before returning to the center, reaching a width of 1.75 km in some places. This channel has small islands covered in thickets or forest. Moreover, during the dry season, large sandbanks form, relatively isolated from one another and situated away from the riverbanks. These sandbanks are suitable for breeding birds (Figure 1.23). Navigation remains possible throughout the year, although it can be very difficult during the dry season.

The Ouango

The second-largest channel traverses the central part of the plain and measures 200–300 m in width. Around the Ngoumbi bottleneck, it turns into the Orega, but a furo connects it to the main channel. Further, a few kilometers before the end of the upper deltaic plain, it joins the main branch. There are numerous sandbanks in this part of the river, which are closely spaced but are alternately located on the right and left banks (Figure 1.23), which makes them accessible to small predators (civets, genets, and mongooses). Very few sandbanks are fully isolated by water, and thus suitable as nesting sites for birds. During August, navigation becomes completely impossible until the water level rises again at the start of the rainy season.

The northern channel

This channel starts as the Azingo River but turns into the Orongo River beyond Lake Azingo. It follows the northern edge of the plain, measuring about 100 m in width and flowing into the central channel just beyond Lake Logé, 49 km upstream from Lambaréné. This channel contains small peninsulas but no sandbanks. It is navigable throughout the year.

The furos

These channels, which connect two main branches of the river, are relatively numerous. The longest one is situated in the center of the plain (Figure 1.21); measuring 21 km in length, it connects the central channel to the main one. It is very narrow and only navigable when the water level is high.

Oxbow formations

Numerous oxbow formations dot the plain (Figure 1.3). Most of them run dry and are filled with herbaceous and shrubby vegetation, although some actively flow during floods.

The alluvial plain

This area covers an area of about 50,000 ha, including small islands of dry land to the east. It is situated at an altitude of approximately 15 m where the horst ends, but drops down to 3-6 meters in the west. The plain is divided into a few large islands by the river's channels. The central sections of these islands, which are lower than the alluvial levees that line the channel's edge, comprise herbaceous marshes with some residual lakes that are likely being filled in.

In the western region of the plain, upstream from Ngoumbi, numerous flat hills emerge above the recently deposited alluvium. These represent the tops of hills that are high enough to avoid flooding. They are generally about 35-50 m in altitude, although some reach 80 m. As with the hills of the Lambaréné horst, their outlines are sometimes hard to recognize because they are hidden by forest. These hills do not, however, belong to the Lambaréné horst. According to Chevallier *et al*. (2002), the rocks that comprise them date from the Cretaceous, mostly from the Albian stage, 100–110 million years ago.

In the eastern region of the alluvial plain, the water courses are surrounded by a mosaic of various plant formations, but dominated by forests:

- Along the edge of channels, a fringe of riparian forest with *Uapaca heudelotii* occurs; this extends across the entire area covered in the Ogooué's muddy water during floods;
- Further from water's edge, there are areas that remain wet but are shielded from the mud of the Ogooué and are covered in swamps with *Anthostema aubryanum*;
- Beyond the reach of the river's waters, notably on the hills, Atlantic coastal *terra firme* forests occur.

Toward the central and western regions of the plain, marshes are predominant. The following formations can be recognized:

- grasslands with *Echinochloa stagnina* and *Persicaria* spp.;
- papyrus stands;
- more or less open swamp forests with and without *Raphia*;
- woody thickets with *Alchornea cordifolia* and *Antidesma* on the alluvial levees, often with old *Ceiba pentandra*.

These various formations often form a complex mosaic. During the dry season, a majority of the plain dries up to one degree or another, but when heavy flooding occurs it can be flooded.

Page 32:
1.23. The central part of the alluvial plain. This satellite image clearly shows the numerous sandbanks of the Ouango River forming alternately along its right and left banks. The Ogooué sandbanks are larger but more spaced out and do not touch the riverbanks.

Page 33:
1.24. The Azingo River in the northeast of the upper deltaic plain (Photo Jean P. Vande weghe).

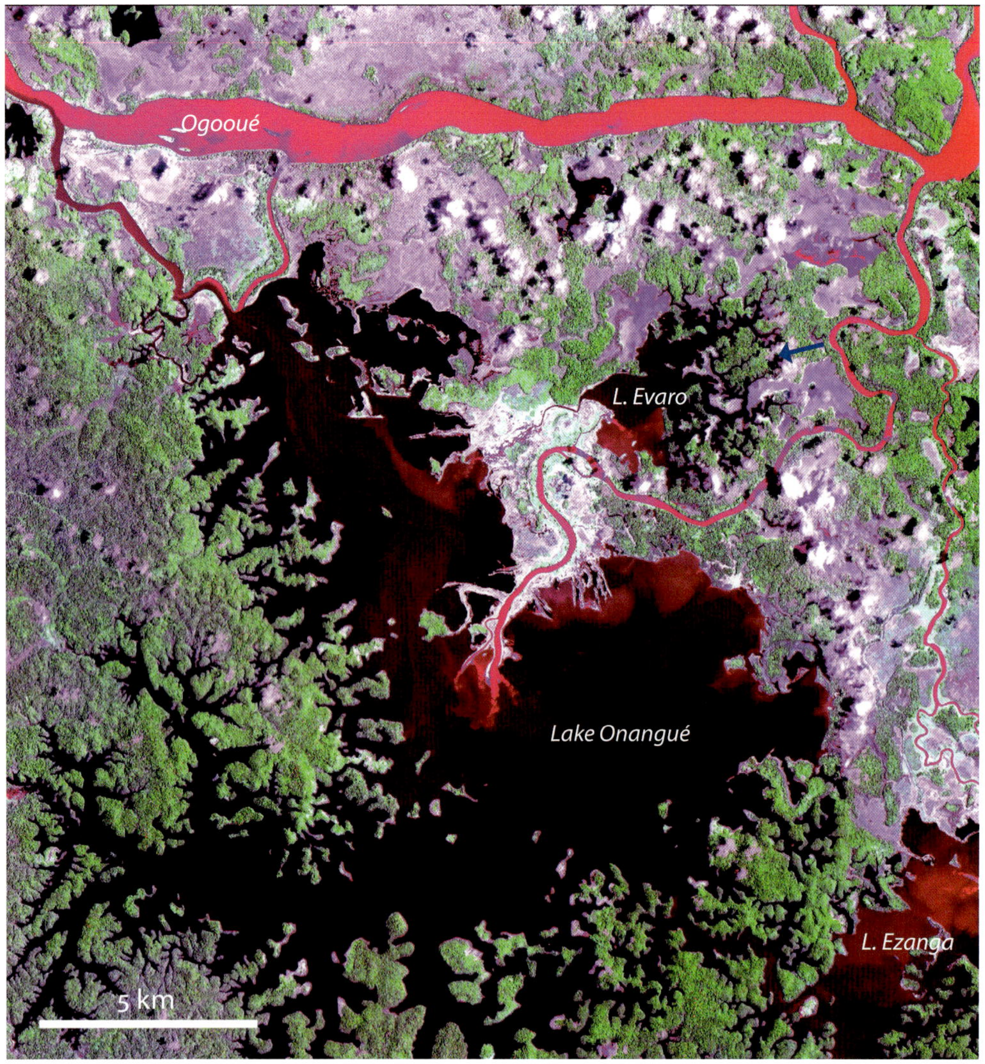

1.25. Lakes Onangué, Evaro, and Ezanga with the Rivière des Pélicans and Ezanga River, as well as their bird's feet lake deltas.

Residual lakes

In the plain's central region, where there is less alluvium because it is caught by the alluvial levees, there are two irregularly shaped lakes whose edges are more or less rounded. The largest is Lake Nyayé (Figures 1.2, 1.21, and 1.23), situated in the lowest part of the plain and entirely surrounded by tall grasslands. When the water level is high, these central lakes are accessible by canoe via small channels that connect them to either of the river's main channels. These small channels are shallow and have black water (Figure 1.9).

Lakes Ogonié, Mandjé, and Anengué are also situated in the southwestern part of the plain (Figure 1.21) and strongly resemble residual lakes such as Lake Nyayé (Figures 1.21 and 1.23). Their banks are entirely bordered by extensive marshes, dense papyrus beds, and *Ficus trichopoda* massifs, contrary to those of Lakes Onangué, Evaro, Oguemoué, and Ezanga.

Flood lakes

Further away from the main channels of the Ogooué River, in the deepest "bays," there are numerous lakes, all of about the same configuration. They are located along the edge of the plain, both north and south of the river, and in some "bays" adjoining small islands of dry land, which are present in the center of the plain. These represent valleys adjacent to the old Ogooué valley that were flooded and became filled in with alluvium, which now form a sort of dam, blocking the flow of the river.

These lakes are typical for flooded landscapes. Their banks facing the river are usually indented and generally well marked. A strip of open marshland (papyrus stands and grassy marshes with *Echinochloa*) separates these lakes from the Ogooué River, broken only by narrow channels that are often blocked during the dry season. Some small lakes only have a single channel connecting them to the river; this channel is sometimes very narrow and occasionally temporary. Other lakes lack a channel that connects them to the river, and their water reaches the river by filtering through the marsh. In general, these lakes have much less alluvium than the Ogooué River, and the water of their most distant parts is distinctly black and slightly acidic, and carries little alluvium.

A series of lakes is situated to the north of the plain, starting with Lake Azingo in the east and Lakes Gomé, Ingoyo, Iwandé, and Alombié to the west. Some of them are part of the Wonga-Wongué Reserve and the corresponding Ramsar site. These lakes are fed by the Rivière des Coupeurs, which drains most of the Wonga-Wongué Plateau's western flank, meandering tightly through the alluvial region. During floods, the valleys of these rivers cannot fully drain and become inundated (Figure 1.78), resulting in floods that extend to 10–25 km upstream from the lakes.

Lakes Onangué, Oguemoué, Ezanga, and Evaro are situated to the south, and in reality, together they form a single larger lake with extremely irregular outlines and dotted with innumerable islands representing the tops of hills that have not been submerged. Lakes Nyondé, Bagé, and Avanga are found farther to the west; they have an indented structure similar to that of the larger lakes. Open-water areas with *Nymphaea* beds are present in some of the smaller bays, between the actual shore and the papyrus stands. On land, the landscape is primarily formed of coastal forests with *Aucoumea*, which have been heavily degraded by repeated logging and more recently in some places by fires started by immigrants originating from northern Gabon (page 134).

The Great Lakes and the lacustrine deltas

Lakes Onangué, Evaro, Ezanga, and Oguemoué belong to the upper deltaic plain, with which they were originally only connected to the northwest. The northwestern extremity of Lake Onangué is in fact still connected to the Ogooué River through two channels (Figure 1.25). The Akembé River carries muddy water from the Ogooué River to this lake. The Ambila River flows from the lake, which has been partially cleared of alluvium, and brings it to the main branch of the Ogooué River near the Nogo mission. Water from Lakes Oguemoué and Ezanga feed into Lake Onangué through a series of islands with more or less abrupt relief.

The depth of Lake Ezanga ranges from 3-8 m, averaging around 3.8-6.4 m. The lake bottom is sandy and very flat starting 50 m from the banks (Loubens, 1966). The depth in the central part of Lake Onangué reaches past 10 m, even during the dry season. The greatest depth, recorded in 1911 by a marines expedition, was 17 m. Some parts of Lake Onangué are thus below sea level, which indicates that it is a very old lake and might be the extremity of the Ogooué ria (pages 45-46).

Currently, all four lakes also are fed water from the Ogooué River through the Rivière des Pélicans, which is much further up than the Akenbé River. This river, which forms a sort of "short circuit" between the main channel of the Ogooué River and Lake Onangué, begins at the mouth of the Ogooué River's main channel, 17 km from Lambaréné and between both branches of the horst. Just a bit further than 4 km from the Ogooué River, it splits to form the Ezanga River, which carries the waters of the Ogooué River to Lake Ezanga, and its main channel continues towards Lake Onangué. The Rivière des Pélicans and the Ezanga River flow south, meandering through flat hills that are most likely the end of the western horst chain (Figure 1.26).

1.26. The lake delta of the Rivière des Pélicans, with the channel feeding into Lake Evaro and the channel that drains the waters of Lake Evaro towards Lake Onangué (black arrows). During the major periods of flooding, in November or April, the Rivière des Pélicans can overflow and its waters then flow directly through the forests into Lake Evaro (white arrow).

Furthermore, these rivers form two small lacustrine deltas with multiple bird's feet lobes whose channels are bordered with heavy alluvial levees. The mouths of these deltas exhibit distinct sand accumulation. Both deltas are separated by a strip of low hills running north to south and covered in dense forest; they belong to the Lambaréné horst and are formed of Cretaceous sediments belonging to the coastal basin. The geological mapping of this sector is, unfortunately, very imprecise.

The delta of the Rivière des Pélicans

This is the most typical, and also the most well known of the deltas. It extends into the western bank of Lake Onangué for a distance of approximately 5 km, separating off a bay that has formed Lake Evaro. This delta covers approximately 1,750 ha between Lakes Evaro and Onangué (Figure 1.26). Like the others, Lake Evaro is a typical flood lake with indented borders and islands. It is fed by a small channel, bordered with tall *Ceiba pentandra* trees, which carries water from

1.27. The Ezanga River Delta in Lake Ezanga. The easternmost channel of this delta currently blocks a bay of Lake Ezanga. This was not yet the case in 1966. Alluvial beds have therefore formed over a little more than 1 km in about 40 years.

the Rivière des Pélicans (Figures 1.26 and 1.75) to Lake Onangué via another channel that flows through approximately 4 km of the delta's marshes (Figure 1.26).

During the strongest floods, in November and April, the Rivière des Pélicans overflows its banks and flows directly through the flooded forests towards Lake Evaro (Figures 1.25 and 1.26).

The Ezanga River Delta

This delta is situated to the north of Lake Ezanga and covers 3,500 ha (Figure 1.27). It contains two main channels whose alluvial levees are covered in herbaceous plants such as *Echinochloa*, *Vossia,* and *Persicaria*. The western channel flows into Lake Ezanga in a way similar to how the Rivière des Pélicans flows into Lake Onangué. The eastern channel follows the northern edge of Lake Ezanga. Since 1966, it has progressed considerably (approximately 1 km) and its alluvial levees are beginning to form a bay from the lake just as the Rivière des Pélicans has done in Lake Onangué (Figure 1.26). New islets of dry land covered in forest are forming in the middle of the delta's grassy expanses.

The lower deltaic plain

Approximately 30 km from the coast, the upper deltaic plain becomes narrower, measuring less than 10 km in width (Figure 1.21). The river's various channels come together without joining entirely. The lower deltaic plain begins at this bottleneck–which is narrower than the one in Ngoumbi–and it forms the current marine delta. The transition between these deltaic plains is very gradual and nearly imperceptible in the field. We have in fact not been able to determine how far the tides reach.

The lower deltaic plain measures 45 km at most from east to west (in the northern delta near the border between Cape Lopez Bay and Nazaré Bay) and 90 km from north to south (from Apomandé Point to the

Nkomi Lagoon), and covers approximately 190,000 ha. In fact, this plain includes three more or less active deltas with adjoining littoral sand structures. As such, this lower deltaic plain is a complex structure comprised of multiple elements of varying age.

The northern delta

Two of the Ogooué's channels, the Kojo and Atadali rivers, flow north through vast expanses of mangrove (Figures 1.28 and 1.29) to reach Cape Lopez Bay. They form a typical tidal delta.

The main channels of the delta

The Atadali River flows mostly in a straight line towards Cape Lopez Bay and runs alongside the Animba Plain. The Konjo River splits into two channels that surround a vast island, with ancient dunes in its center. The western channel also terminates at Cape Lopez Bay while the eastern channel flows into Nazaré Bay. These channels carry water from the Ogooué River. They are flanked by *Rhizophora* mangroves, which are tall alongside the rivers and lower further away. Downstream along these rivers there are pure *Rhizophora* mangroves, though further upstream they are colonized by *Pandanus* massifs, *Raphia* palms and locally *Phoenix reclinata* palms as well as numerous species of swamp or riparian forests. This freshwater vegetation forms a narrow strip that is difficult to map and reaches up to 4–13 km from Cape Lopez Bay (Figure 1.28). Moreover the Atadali and Konjo rivers are connected by multiple small furos, mainly including the Mandji and Nkomi rivers (Figure 1.28).

The tidal channels

To the west of the delta, near Mandji Island, the Doronga River drains the mangroves that are situated to the north of the Animba Plain and separated from the Atadali River

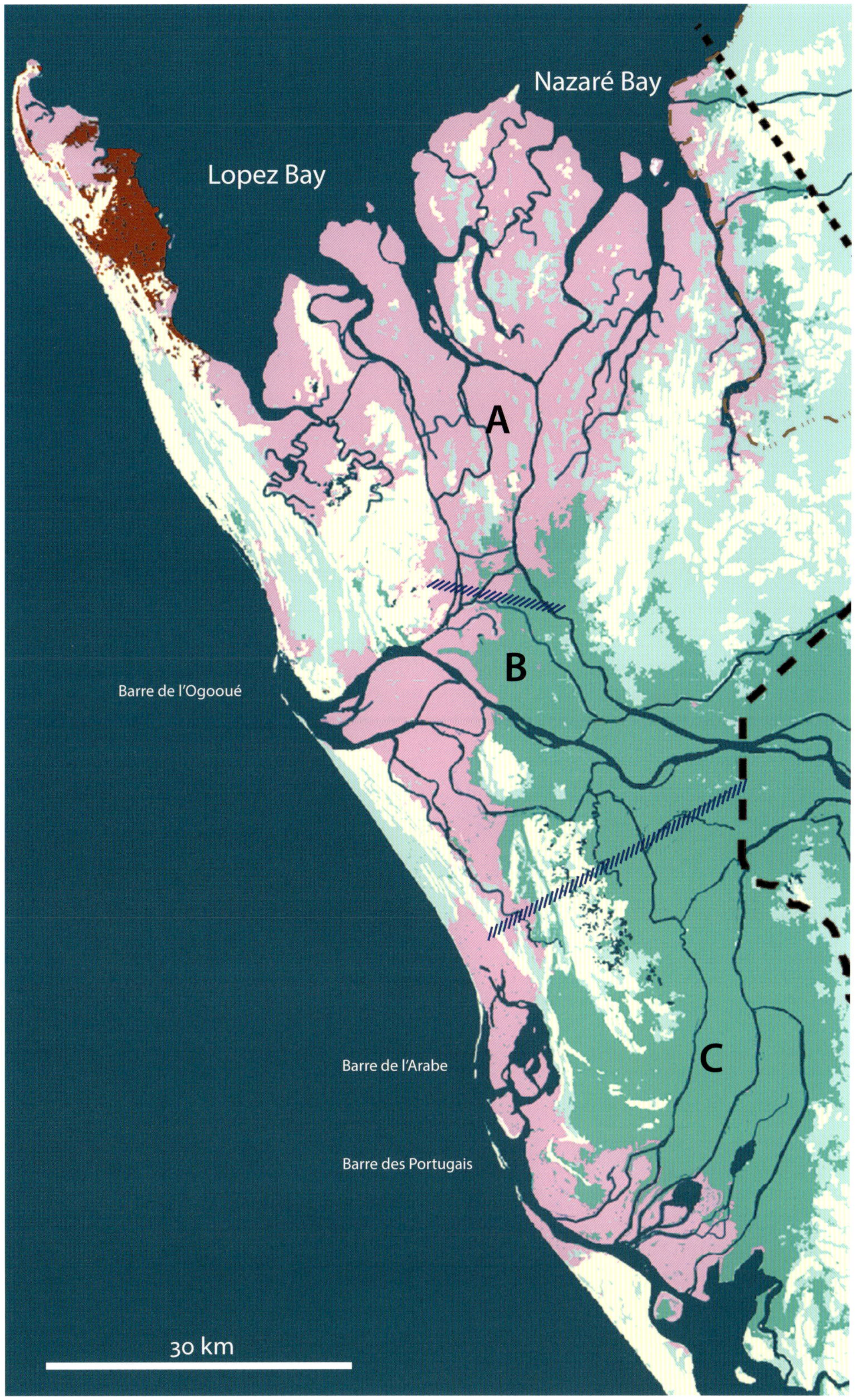

1.28. The lower, or tidal, deltaic plain with its three main sections:
A. The northern delta, an estuary delta or tidal delta.
B. The central delta, a swell delta with flared channels similar to an estuary delta, but also a significant sandbar.
C. The southern delta, a swell delta without flared channels and with strong sandbars (the "Barre des Portugais" and the "Barre de l'Arabe").

1.29. The northern delta, with its main distribution channels (Kondjo and Atadali) and its tidal rivers (Doronga, Ozima, and Gangwé). The yellow lines indicate the limit of mangrove colonization by freshwater species (especially Raphia *and* Pandanus*). High mangroves appear in green, low mangroves in pinkish brown. Savannas also appear in dark pink. Daily observations have shown that, at the asterisk (in the Atadali River), the water is completely fresh all year round, both on the surface and near the bottom (OLAM, pers. comm.). A large part of the* Rhizophora *mangroves is therefore located in brackish—or even entirely freshwater, as previously indicated by the observations of Loubens (1966).*

by vestiges of an ancient sandbar. To the east are the Ozima and Ganoué Rivers, which flow into Naeaté Bay. The first of these also drains water from the Ogooué River through small narrow rivers, but remains essentially a tidal channel. The second one is fed by some fresh water from the Mbilapé, Mpogwé, Ongomo, and Aloumbi rivers, which drain the south-western fringe of the Wonga-Wongué Reserve, although it is also mainly a tidal channel. Other tidal channels exist at the heart of the delta, notably the Oyila River, which splits Ampumandé in half.

The waters of these tidal rivers contain very little alluvium. As a result, the mangroves that grow along them are poorly developed and form a narrow curtain of medium-sized *Rhizophora* that hides vast expanses of very low *Rhizophora* reaching just 1 m high in places. Tannes are even present in the lower mangroves of the Doronga River (Figure 1.7).

The mudflats

Along the edges of Nazaré Bay and in the western estuary of the Knondjo River, mudflats reach a maximum surface area of 1,200 ha when the tide is at its lowest, according to Schepers & Martejin (1993). These mudflats are very liquid in some areas and more sandy in others. Curiously, the Knodjo River's sandy mud banks are mostly consolidated by *Crinum natans*, a freshwater Amaryllidaceae that blossoms massively in June (Figure 5.4). We have not found recent maps of the configuration of these mudflats, and it appears that, since the beginning of the 1990s (Schepers & Martejin, 1993), they have shrunk significantly—possibly as a result of a slight increase in sea level. According to fishermen and women in the region, these mudflats only emerge significantly when the tide is low and the sea is agitated. As a result, they may be visited by migratory birds, but would not be suitable for a large number of wintering bird species.

1.30. The Animba River represents the main course of the Ogooué, with islands covered with tall herbaceous vegetation and a small sandbank (Photo: Jean P. Vande weghe).

1.31. The muddy sandbanks of the Kondjo River are covered with the large Amaryllidaceae Crinum natans, *which flowers abundantly in June (Photo: Jean P. Vande weghe).*

The northern delta's dry land

In the center of some of the delta's islands, there is herbaceous coastal savanna with some shrub growth and even some dryland forests. These environments are clearly built on old coastal sand structures. There are, however, large laterite quarries at the northern tip of the Koroso Plain sandbars. This laterite could not have been formed in recent sediment and therefore probably indicates the presence of structures dating from the Tertiary on which more recent sediment has been deposited. There are also ancient outcrops near the confluence of the Kondjo and Mandji rivers. They include a ferruginous layer around 80 cm in depth.

The central delta

The river's main channel continues it course to the west before flowing into the Gulf of Guinea between Mandji Island to the north and Ozouri Island to the south (Figure 1.32). It thus traverses through the heart of the lower deltaic plain and forms the delta's central axis, where it forms. It is a narrow tidal delta or estuary delta (Lebigre, 1983), but the outlets of both its channels are protected by a narrow sandbar measuring 12–15 km long whose shape changes each year.

Up to approximately 10 km from the coast proper, the river's channels meander through swamps predominantly of *Alstonia*, *Mytragina*, *Anthostema,* and *Raphia*. The vegetation in these swamps remains sparse, but is denser in places, usually along the river banks (Figure 1.84). Alluvial levees are relatively large and sustain dense riparian forests that contain dryland species (notably *Klainedoxa*). In places, especially in the first kilometers starting from the upper alluvial plain, open marshes with *Echinochloa* and *Persicaria* as well as papyrus fields are also present.

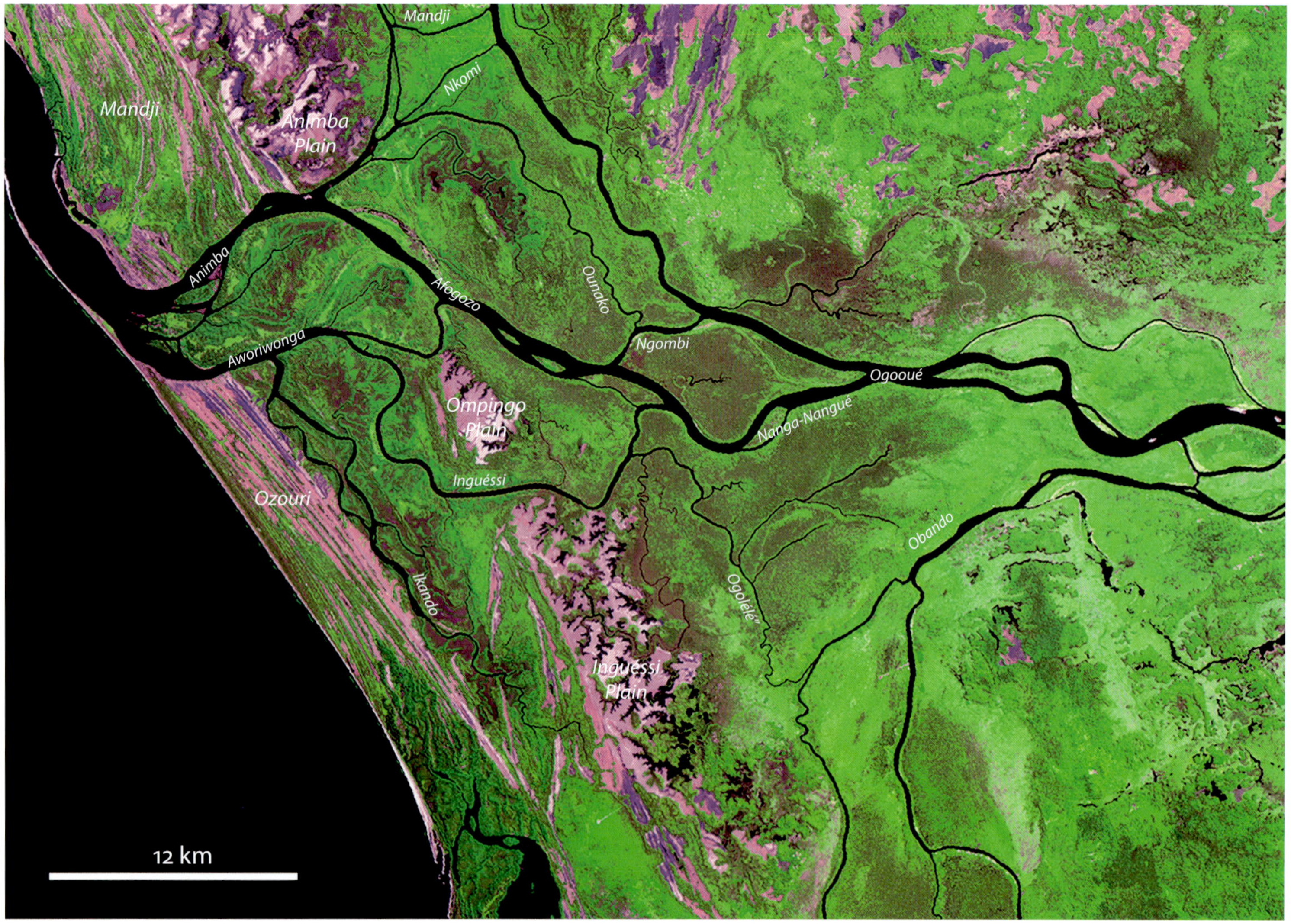

1.32. The central delta is currently the main mouth of the Ogooué. It is intermediate between a swell and a tidal delta.

1.33. A river in the central delta bordered by high mangroves but colonized by Raphia *palms (Photo: Jean P. Vande weghe).*

Near the mouth of the delta, the river splits into two nearly parallel channels (Figure 1.32). The northern channel is the main one. The narrower southern one partially carries water from the Ogooué River, but also has southern tributaries in the form of tidal channels that drain the mangrove situated between Ozouri Island and the Ompingo and Inguéssi plains. These rivers are bordered by tall *Rhizophora* mangroves often separated by a narrow band of *Raphia* (Figure 1.33).

Inside the mouth of the delta there are a few sandbanks, but there are no real mudflats (Figure 2.30).

The southern delta

One of the tributaries, the Obando River, splits off from the main channel well upstream from Lake Anengué, within the upper deltaic plain (Figures 1.32 and 1.34). It drains the waters of Lake Anengué and, once it reaches the lower deltaic plain, grows along its route south before splitting into the three rivers that form the southern delta: the Mpolounié to the west, the Adéloué in the center, and the Djembagombé to the east (Figure 1.34). These flow south-southwest nearly parallel to one another, toward the Nkomi Lagoon, through both dense and sparse swamps as well as vast *Raphia* forests. A few kilometers from the Nkomi Lagoon, this forest gives way to declining tall *Rhizophora* mangroves and an "undergrowth" of *Raphia* palms (Figure 1.102). Unlike the northern delta, the southern delta does not have any low mangroves. There are two residual lakes within the mangroves. The largest is Lake Akoumba, which is surrounded by tall *Rhizophora*.

The southern delta leads to the lower section of the Nkomi Lagoon. From there, the waters of the Ogooué River mix with those of the lagoon and flow northwest, then north toward the Gulf of Guinea. This outlet is protected by sandbars–the "Barre des Portugais" and "Barre de l'Arabe"–that have a cumulative length of 18 km, but which often change shape.

The Olendé region

All along its western edge, the southern delta is delimited by coastal sand structures: the Inguéssi Plain to the north and numerous dune ridges that are very narrow and surrounded by swamps or mangroves further to the south. These dune ridges in a way extend Ozouri Island to the edge of the Nkomi Lagoon and have been used to build the road that connects Port-Gentil and Omboué. As a result, they form a border between the delta proper and the Olendé region further west. This region contains mangroves that are not part of the southern delta. They are high mangroves that grow around a wide tidal river with several scattered islands. This region is strictly marine with pure *Rhizophora* mangroves that probably colonized vast mudflats. Low mangroves are absent, as in the southern delta.

The Nkomi Lagoon

In the northern part of the Nkomi Lagoon—the only part being taken into account here—water depth reaches 5–10 m, but vast portions are very shallow and difficult to navigate at low tide or during the dry season. These shoals are formed by sediments deposited by the Mpolounié and Djembagoumbé rivers, thus making them a part of the Ogooué deltaic complex. Nevertheless, salt water is confined to the lower parts of the outlet that border the Olendé region's mangroves.

The lagoon is separated from the ocean by a long white sandbar that extends several hundred meters to the south and is covered in dry savanna with *Loudetia* and *Melinis* that is being colonized by *Chrysobalanus* groves. The river that runs inside this sandbar is dotted with isolated stands of *Rhizophora* up to a few kilometers from Omboué. The bank opposite the lagoon (on the continental side) is bordered with dryland forests and narrow strips of flooded riparian forests. Near the Niongo Plain, savanna reaches the edge of the water. The back of Ntchonga Ntchiné Bay is bordered with vast swampy and flooded forests.

The coastal sand structures

Independently from the river alluvium that forms the upper and lower deltaic plains, coastal currents have deposited immense sand structures.

The oldest parts of these structures, currently contained within the marine delta, are clearly identifiable in satellite imagery (Figures 1.29 and 1.32). They form "fossil" dune ridges that are generally covered in herbaceous *Loudetia* savanna. The most recent formations, which are still evolving, form sandbars along the southern delta's border as well as on Mandji Island and Ozouri Island and their adjoining plains.

Sandbars

The mouth of the Ogooué River and the outlet of the Nkomi Lagoon are bordered with sandbars (barrier islands or spits) comprised of a series of sandbars separated by inlets that bend the freshwater coming from the delta toward the north, in parallel with the beach. The sandbar of the Ogooué River is 12 km long and that of the Nkomi Lagoon measures 18 km.

These sandbars are wider in places, up to 100–150 m, and covered in bare sand, pioneer vegetation with *Ipomoea pes-caprae, Canavalia rosea, Diodia* spp., *Cassytha* sp., and various Amaranthaceae and Cyperaceae (Figure 1.36). In some areas thickets have formed with *Dalbergia ecastaphyllum* (Figure 1.100). Some lower areas have been colonized by *Rhizophora*. The largest sandbars are crossed by shallow channels that empty on the side opposite the ocean.

The shape of these sandbars changes from year to year. In August 2014, the Ogooué sandbar was separated in two places, while the Nkomi sandbar was separated only once, with its entire northern half in direct contact with dry land, which explained the presence of numerous marsh mongooses, *Atilax paludinosus*, and the almost complete lack of bird colonies. The waters between the sandbars and the continent remain calm most of the time, although in August 2014, they were turbulent because of strong tides passing through the inlets at high tide.

These sandbars show traces of intermittent or occasional human presence. Until early 2014, there was a camp of West African fishermen on the Ogooué sandbar.

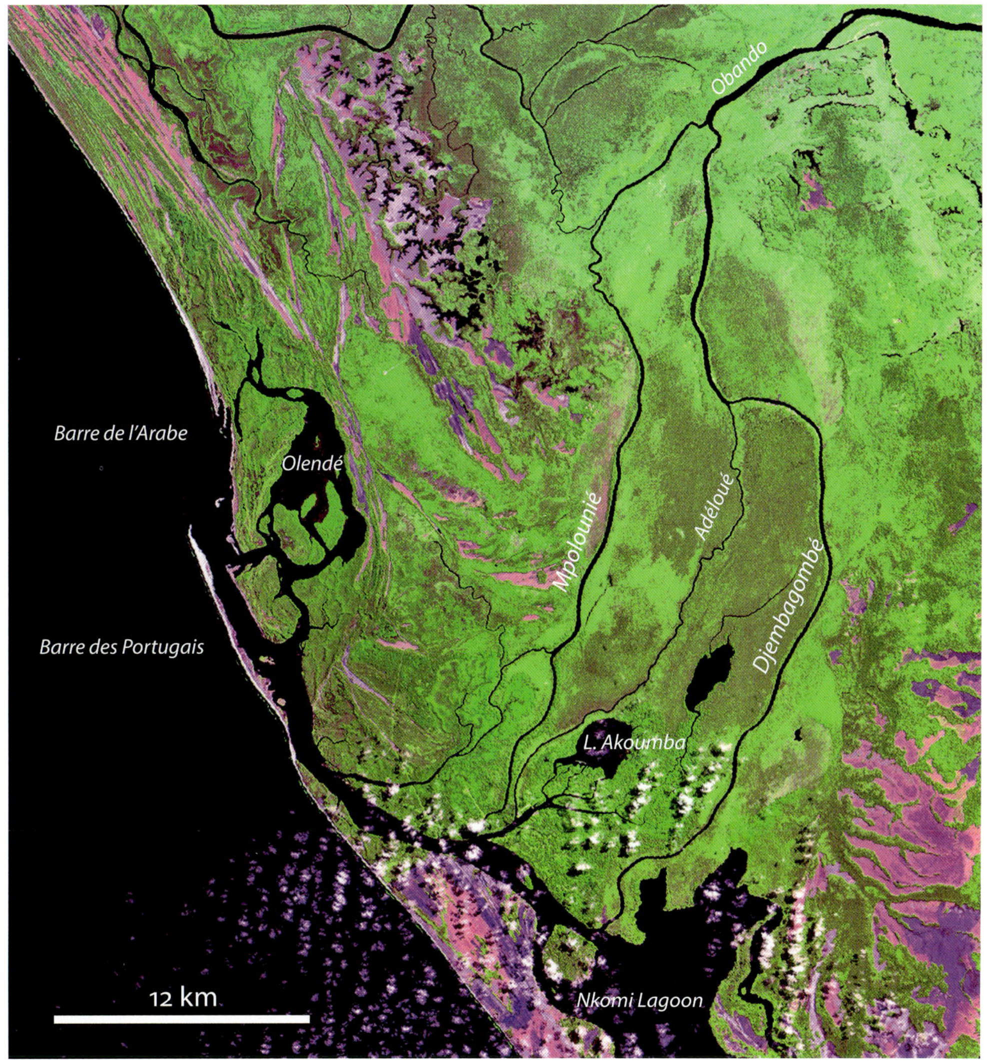

1.34. The southern delta is a typical swell delta with its three distributing rivers (the Mpolounié, Adéloué, and Djembagombé), Lake Akoumba, the Olendé Lagoon, the Nkomi Lagoon, the "Barre des Portugais," the "Barre de l'Arabe," and numerous dune ridges.

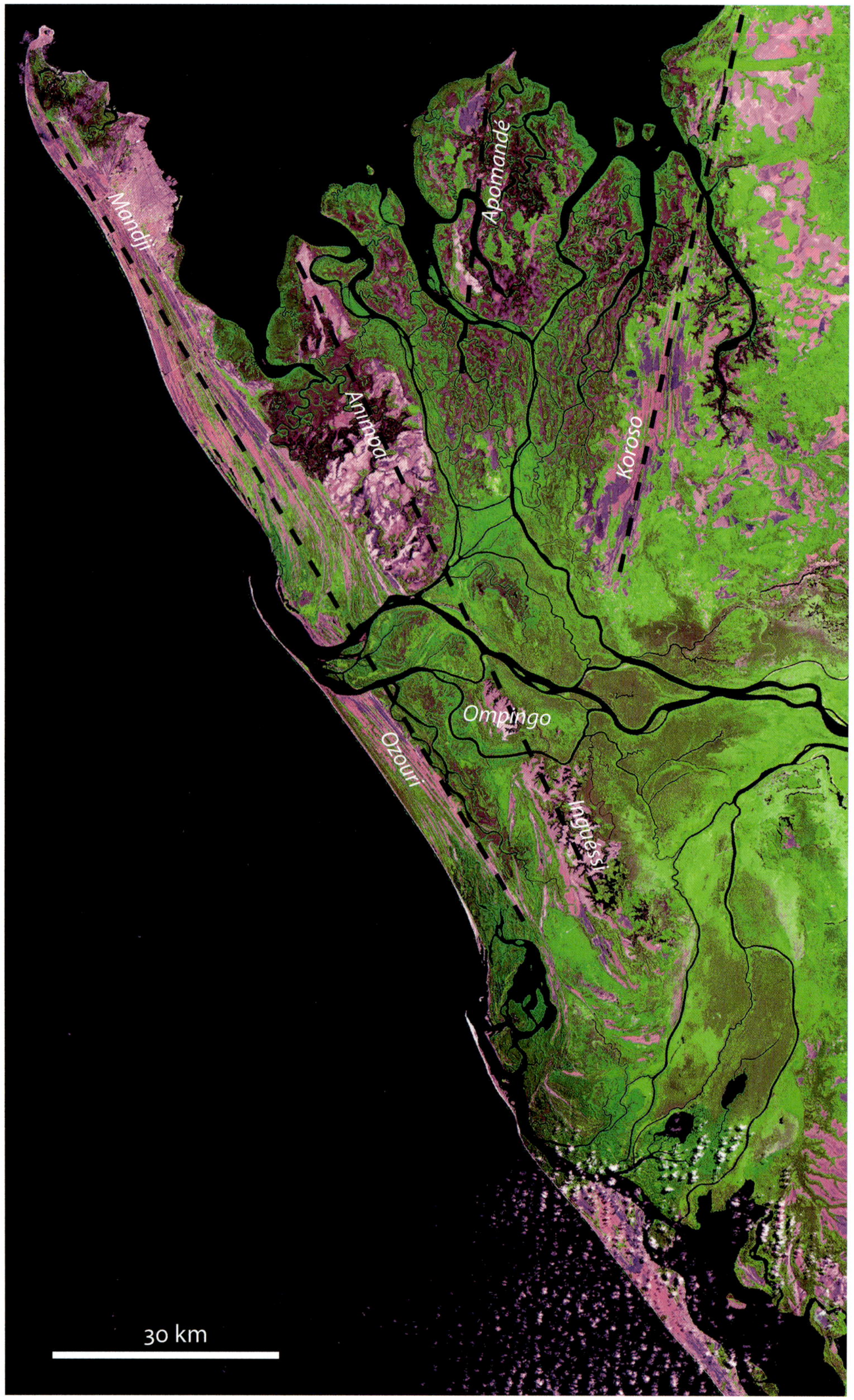

1.35. The sand structures: the Koroso Plain, the Apomandé "ridge," the Animba, Ompingo, and Inguéssi plains, and Mandji and Ozouri islands. South of the Inguéssi Plain there are many curved ridges that are part of the southern delta. The axis along which these different structures were formed has evolved over time.

Mandji and Ozouri islands

Mandji Island, which measures 48 km in length, is made up of an immense sand formation that has been deposited all along the marine delta, extending the beaches of southwestern Gabon. It contains at least 20 parallel dune ridges measuring up to 5-6 km wide in the south and covered in alternating patterns of dry herbaceous vegetation atop the ridges of old dunes and marshy vegetation or small lakes (during the rainy season) that cover the low areas between the dunes. To the west, Mandji Island is bordered by the narrow Atlantic beach.

North of Port-Gentil, the eastern edge of Mandji Island is rimmed by *Rhizophora* mangroves that grow up against narrow strips of swampy forest. *Conocarpus* and *Laguncularia* formations are found in the transition area. However, the swamp forests have largely been replaced with forests of niaouli (*Melaleuca quinquenervia*), an invasive Myrtaceae endemic to Australia and New Caledonia that was introduced after World War II in order to drain the swamps surrounding Port-Gentil.

In the southern part of the island, forests with *Chrysobalanus icaco* (Chrysobalanaceae) and *Syzygium guineense* var. *littorale* (Myrtaceae) are found, encompassing a large part of the landscape (Figures 1.95 and 1.96). They mainly occupy the low areas between dunes.

Ozouri Island extends from the mouth of the Ogooué River to the Olendé Lagoon (Figure 1.35). It is very similar to Mandji Island. Both have the same origins and were probably connected until the Ogooué River permanently separated them. Both to the north and the south of the river's mouth, the dune ridges have been bisected at a right angle by the river (Figure 1.94). Along the island's eastern edge, there are mangroves that separate it from the Inguessi Plain and whose water courses have in places eroded the dune ridges of Ozouri Island (Figure 1.94).

1.36. The "Barre des Portugais," with its vegetation colonizing the coastal sands (Photo: Jean P. Vande weghe).

The Animba Plain

The Animba Plain is situated to the east of Mandji Island. It forms a vast herbaceous expanse covering 6,000 ha, deeply indented and delineated by the northern delta mangroves to the east, specifically those bordering the Atadali River (Figure 1.35). This plain is crossed by numerous tracks left by oil prospectors.

The Ompingo and Inguéssi plains

To the east of Ozouri Island, one finds plains that closely resemble the Animba Plain and are similarly aligned.

The Ompingo Plain is totally isolated by mangroves and swamps, but can be reached by canoe via the Aworiwonga River. It covers a surface of approximately 500 ha and is flanked by an isolated dune ridge.

The Inguéssi Plain is also separated from Ozouri Island by mangroves. It covers a vast mosaic of 4,500 ha of herbaceous savanna, small lakes, mangroves, *Pandanus* thickets, swamps, and narrow strips of dryland forest (Figures 1.106–1.109) whose origin appears to be the same as that of the Animba Plain. Like the Ompingo Plain, its western border is delineated by parallel sandbars, although here they are separated by swamps. It can be reached by canoe via the Inguéssi River. During the rainy season, small lakes are formed that cover part of the herbaceous savanna.

Apomandé Point and the Koroso Plain

Coastal sandy formations are located in the center and on the eastern side of the northern delta (Figure 1.36). They represent old coastlines that had a different orientation than the more recent formations (mainly the Animba and Inguéssi), which were built along the same axis as the current coastline. The Koroso Plain has retained its parallel bands, which in fact continue farther north into the southwestern corner of the Wonga-Wongué Reserve. The rest of these areas are covered in herbaceous *Loudetia* savanna surrounded by dryland forests.

Genesis of the deltaic complex

The Ogooué is an old river; its birth is linked to that of the Atlantic Ocean. To the southwest of Port-Gentil, oil research has revealed evidence of sand deposits forming a fan that date from the end of the Cretaceous or the early Tertiary, 60–70 My ago.* These indicate the presence of an ancient estuary or delta that predates the current complex (Chevalier *et al.*, 2002). It is likely that the geological history of the coastal sedimentary basin, which is linked to the opening of the Atlantic rift, may have deeply influenced the shape of the Lower Ogooué, just as in turn, the Ogooué River influenced the disposition of sediments.

The history of the Lower Ogooué deltaic complex thus hinges on two pivotal phases. The first corresponds to

* My = millions of years

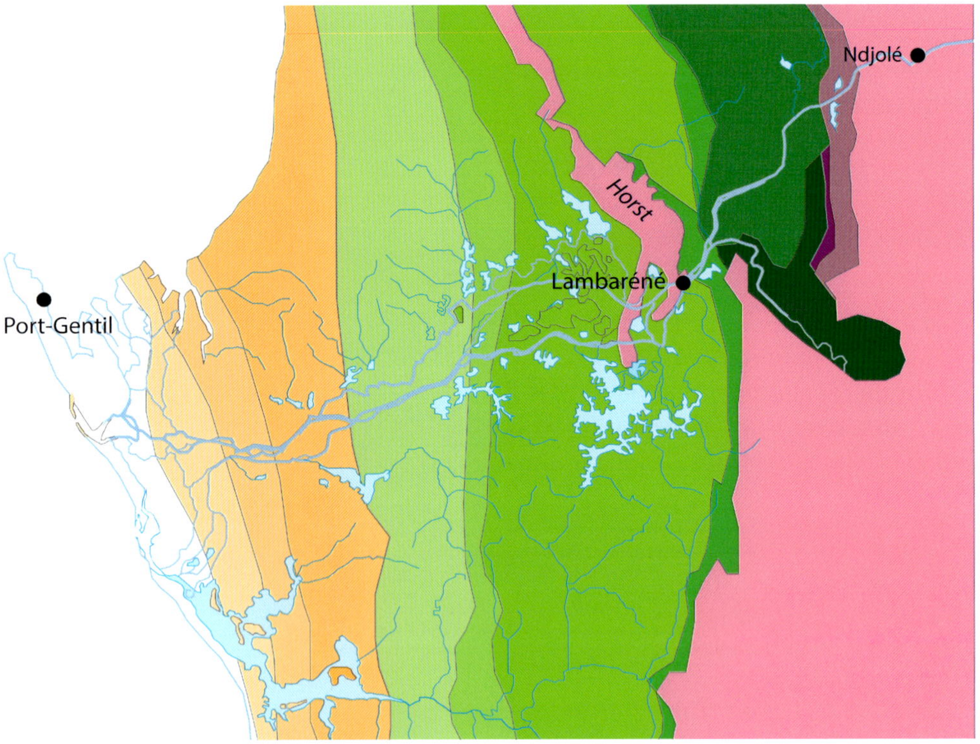

1.37. The sediments of the coastal basin, from the Permian to the Neogene. Pliocene and Pleistocene-Holocene deposits are not shown (see Figure 1.6).

- *Crystalline base (>550 My)*
- *Paleozoic sediments:*
- *Permian (290–248 My)*
- *Mesozoic sediments:*
- *Jurassic (201–145 My)*
- *Cretaceous-Neocomian (145–135 My)*
- *Cretaceous-Barremian (135–130 My)*
- *Cretaceous-Aptian (125–112 My)*
- *Cretaceous-Albian (112–99.6 My)*
- *Cretaceous-Cenomanian (99.6–93.5 My)*
- *Cretaceous-Turonian>Maastrichtian (93.5–65 My)*
- *Cenozoic sediments:*
- *Paleocene (65–56 My)*
- *Eocene (56–33.9 My)*
- *Miocene (23–7.2 My)*

the establishment of the coastal basin's "foundation," on which the current deltaic complex was built. This began 250 My* ago and ended around the end of the Tertiary, ca. 2 My ago. The second phase explains the presence and structure of the recent alluvium. It began 6,000 years ago when the Atlantic Ocean returned to its current level after a long phase of lower levels during a period with a cold and dry climate.

The coastal sedimentary basin

The basin began forming approximately 250 My ago, at the very end of the Permian, with the appearance of a rift that contributed to the fragmentation of Gondwana by separating South America and Africa.

The Lambaréné horst was formed at the very start of this Atlantic rifting. This fragment of Archean metamorphosed rock, which belonged to the Lambaréné block, broke off, was uplifted and later flattened out by erosion. Today it forms a chain of low hills that extend north to the nearby Remboué River, a tributary of the Komo Estuary. To the south, this chain divides into two branches: the eastern branch includes Azingué Island on which the city of Lambaréné sits; the flatter, western branch reaches the region of Lake Evaro.

The eastern part of the coastal basin was formed between the horst and the Crystal Mountains. It covers 10,000 km² between Cocobeach to the north and Lambaréné to the south, and comprises continental sediments that date back to the Permian, the Jurassic, and especially the first half of the Cretaceous (Figures 1.37). As such, these sediments were deposited throughout a period of 150 million years, up to about 100 million years ago. The sediments dating from the Permien and Jurassic, situated in the far east of the basin, are mainly clay and sandstone deposited in a continental depression. The sediments of the Lambaréné region are sandstone, marl, limestone, and argilites that were deposited during the Albian (Lower Cretaceous).

The western part of the coastal basin was formed between the horst and the Atlantic Ocean. It covers 500,000 km² and extends from Pointe Pongara (opposite Libreville) in the north to Mayumba in the south. Its sediments are mostly of marine origin and were deposited between the middle of the Cretaceous and the Neogene.

Between 112 and 100 million years ago, just before the middle of the Cretaceous, the Atlantic rift formed an enormous lake that extended from current Angola to western Cameroon and accumulated alluvium from nearby mountains. Thick layers of salt were deposited there approximately 110 million years ago when the lake evaporated due to drier climatic conditions. Between 112 to 90 million years ago, at the beginning of the Upper Cretaceous, this lake was slowly filled in by a sea, progressing from south to north. The layers of salt were then covered in carbonate-rich sediment, which is typical for marine sediment. However, the lake's transformation took place in several phases. Thus, after a first infringement period, water drained again from the area during the Cenomanian, approximately 95 million years ago. From the Senonian (the last 24 million years of the Cretaceous) to the Oligocene, 33 million years ago, the inland sea broadened and opened up to the north. The rift's axis shifted to the west of the Lambaréné horst.

Sedimentation continued during the Paleogene (the first half of the Tertiary). It was, however, interrupted for long periods by the ocean receding, notably during the middle and end of the Eocene and most of the Oligocene. Erosion dug deep valleys in the, then dry, continental plate, perpendicular to the coast. During periods of marine infringement, the continental plate was once again submerged and valleys were filled in through sedimentation.

A new marine infringement began throughout the Neogene, in the second half of the Tertiary, and is still ongoing today. It deposited a layer of sediment, up to 3,000 m thick in places, with outcroppings on Mandji Island, notably in Tchengé, southwest of Port-Gentil, and probably also in the center of the delta, where there is a small outcropping of ocher sand with clay and a layer of laterite at a depth of 1 m.

During the Tertiary, the entire western edge of the African continent was also lifted by tectonic movement. These movements brought about today's existing terrain, notably the Crystal Mountains, as well as the rejuvenation of the hydrographic system. They are also partly related to the deepening and widening of the Atlantic Ocean; although this was then accentuated approximately 22 million years ago by the collision between Africa and Eurasia. Since the second half of the Cretaceous, with the breakup of Gondwana, Africa had been completely isolated from the other continents and became a large island, like Australia is today.

This collision not only brought an end to the continent's isolation, the shock wave it generated also propagated beneath the Earth's crust and significantly contributed to the uplift of the entire Atlantic margin of the African plate. As a result, the Atlantic Ocean was lifted up and exposed, now forming the coastal basin's low plateau. Its surface was gouged by erosion, forming a landscape of hills. The original surface is gone except at the summits of the highest hills within the upper deltaic plain and Great Lakes.

The current delta

Although the sedimentary accretions of today's deltaic complex were conditioned by its underlying structure, the complex was primarily formed during the last 6,000 years. A detailed reconstruction of this history would require barymetric surveys, drilling, and chronological dating.

The Last Glacial Maximum

The delta's history begins during the Last Glacial Maximum, which reached its peak 18,000 years ago. At that time, the world's climate was significantly cooler and drier than it is today, and for approximately 2,000 years, the level of the Atlantic Ocean remained between 110 and 120 m below its current level (Giresse, 2008) (Figure 1.38). Around Port-Gentil, the shore line was 16.4 km further west 16,000 years ago than it is today (Figure 1.39).

Because the level of the Atlantic had been lower than it is today for about 125,000 years, the upper part of the continental plate and the adjoining land were exposed to erosion for over 100,000 years. The relief thus had more than enough time to be flattened, leaving little of the previous deltaic structure. The Lower Ogooué must have flowed through a large valley with little relief and which would probably have been covered in savanna with strips of forest along the rivers. The river formed two channels running through narrow canyons and flowed into the Gulf of Guinea far west of its current outlet. Because it likely followed an old fault, the Ogooué's water course would have been very straight.

The upper deltaic plain's formation

About 15,000 years ago, sea level started to rise again (Figure 1.38). By around 11,000 years ago it was at 80 m lower than today, and around 6,000 years ago it was just 10 m lower, reaching its current level around 4,000 years ago, although it subsequently lowered slightly before rising again, reaching 0.5-1 m higher than today about 2,000 years ago (Giresse, 1977). This rise in sea level was extremely rapid between 10,000 and 8,000 years ago (2.5 cm/year). Not only did it submerge the continental plateau—and its archaeological vestiges—but it also washed away huge amounts of soil and transported large quantities of material, which was then progressively redeposited, filling up the terrain's depressions—the sedimentation rate reached 3 m/millennium—and forming dune ridges that altered the coastline all along the entire coast of Gabon.

Between 7,500 and 7,000 years ago, sea water began to flood the valleys of coastal rivers, progressively converting them to rias. The Komo and Mouni valleys in northern Gabon have remained rias that are broadly open to the ocean. Indeed, these two rivers do not carry much alluvium, nor do the ocean currents carry large quantities of loose sand. The small coastal rivers in southern Gabon do not carry large quantities of sediment either, but as described above, ocean currents carry large amounts of sand, which is why the rias have been cut off from the ocean by large sandbars, forming the Banio, Ndugu, and Ngové lagoons.

At the level of the Ogooué, a double mechanism was engaged. As for the other coastal rivers, its low valley was transformed from 7,500-7,000 BP into a large ria which reached, 4,000 years ago, the current Onangué lake and the surroundings of the Lambaréné horst (Figure 1.39). However, unlike the other rivers, the Ogooué carried enormous quantities of alluvium which clogged the ria, while the coastal currents deposited enormous quantities of sand which transformed the ria of the Nkomi River into a lagoon and which then progressed to the north by building the sandbars of the Inguéssi and Koroso plains, as well as the islands of Ozouri and Mandji thereafter.

For the entire time that the sea level was more than 10 m lower than today, the majority of the Ogooué's alluvium went straight to the ocean. During the Last Glacial Maximum the amount of alluvium carried by the river must have been significantly lower than it is today since rainfall was lower, resulting in less erosion. With the increase in rainfall, erosion increased anew, especially in the vast regions that had been transformed into savanna. When sea level neared its current level, the river's current slowed in the flooded parts of the valley, and alluvium began to build up. A cone of sediment formed and started obstructing and plugging the ria's main channel. With the increase in sea level, the first layers of sediment were submerged and sedimentation started moving eastward until the Atlantic reached its current level. Precise dates are not available, but the upper deltaic plain likely began forming some 6,000 years ago.

Later, the Ogooué continued to deposit alluvium, raising the levels of the plain and the river to their current elevation. Near Lambaréné, the water level rose by about 13 m. The increase was less further from the upper deltaic plain's axis: 10 m near Lake Onangué, and 11 m near Lake Azingo. This phenomenon had two consequences.

First, the increased water flow of the Ogooué and its slower current upstream from Lambaréné resulted in dozens of kilometers being flooded throughout the valley. The river became wider, the enormous sandbanks present today were formed, and land on both sides of the river

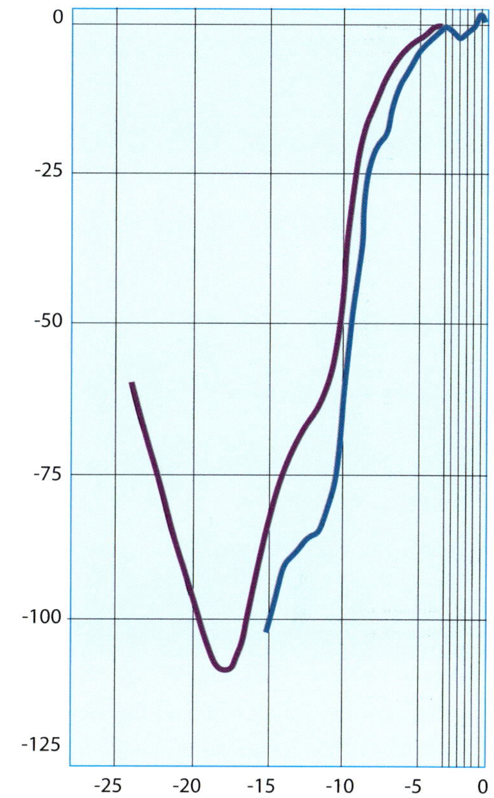

1.38. The water level in the Gulf of Guinea over the past 25,000 years, expressed in meters (vertical axis) and in thousands of years (horizontal axis). In purple, the curve for Ivory Coast; in blue, the curve for Congo. The difference between the two curves may be due to errors in the estimates, but is probably due primarily to the fact that the surface of the ocean is not absolutely flat, but rather displays "hills" and "dips" that can cause vast differences in water level over a few thousand kilometers. The two curves nevertheless show that around 18,000 BP the level of the Gulf of Guinea was 110 to 120 m lower than the current level, which was reached around 6,000 BP. Subsequently, there was a slight further drop followed by a rise in water level, reaching about 1 m higher than the current level around 2,000 BP (according to Giresse et al., 1977; Giresse, 2008).

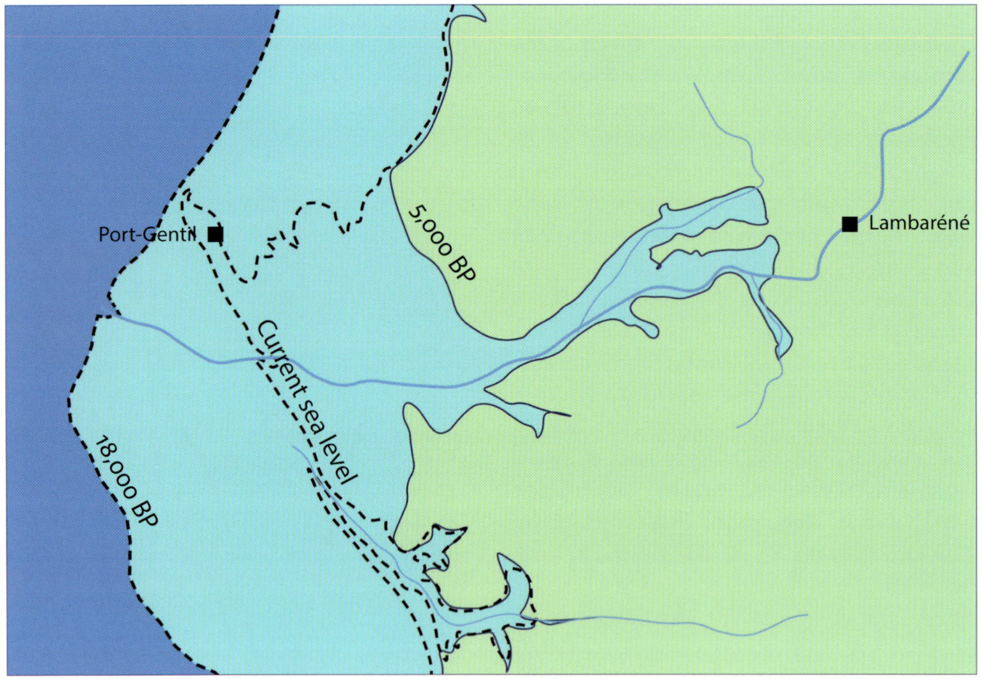

1.39. *Diagrammatic representation of the evolution of the Gabonese coast: at the Last Glacial Maximum (18,000 BP), at the maximum marine transgression 4,000 to 6,000 years ago, and today. The dotted line indicates the current coast.*

was transformed into swamps or flood lakes. The Mbiné River valley was flooded and was transformed into vast marshes.

Second, upstream from Lambaréné, alluvial accretion took place along the deltaic plain's axis and especially along the river's channels. Further from the river, alluvial accretion was minimal or even nonexistent. Lateral valleys, both north and south of the plain, likewise did not fill up. The increase in water level was nonetheless sufficient to flood vast portions of these valleys, forming flooded landscapes with heavily indented outlines, long narrow bays, and numerous islands representing elevated areas that escaped flooding (Figures 1.72–1.74).

When the water level is at its lowest (toward the end of September or even the beginning of October), the dry beaches along the edges of Lake Onangué reveal layers of laterite, but laterite does not form underwater!

The waters of the Ogooué reached roughly their current height rather recently (1,000 years at the most) and overflowed south in the area situated between two branches of the horst. They then formed the Rivière des Pélicans and Ezanga River, which wind across the plateau between low hills before flowing into Lake Onangué and Lake Ezanga. This river thus diverts part of the alluvium from the Ogooué, depositing it in the two small inland river deltas. The Rivière des Pélicans delta split a bay away from Lake Onangué, which now forms Lake Evaro. This delta is currently progressing toward the western shore of Lake Onangué and might eventually divide the lake in two. Nevertheless, this separation will not be total because the waters of Lakes Ezanga, Onangué, and Oguemoué will remain connected to Lake Ogooyé around Ngomo Mission.

Coastal sand deposits

The lower deltaic plain must have started forming approximately 4,000 years ago. Its shape, however, has been heavily influenced by the sand deposits carried in by ocean currents. Indeed, while the Ogooué's upper deltaic plain was being formed, the coastal currents continued to deposit sand, almost certainly where there were tertiary outcrops. The sand was deposited by the Atlantic swell as parallel dune ridges. Following the delta's transformation and the formation of the coastline, certain sand deposits were probably eroded while others were being formed. As a result, it is currently difficult to reconstruct the evolution of the coastline precisely, although some structures have been dated approximately (Giresse, 1997). The oldest is the Koroso Plain dune ridge, which continues in a straight line up to the western corner of the Wonga-Wongué Reserve, covering 42 km. This dune ridge was probably formed between 6,000 and 4,000 years ago. The sandbars present on either side of the Oyila River came next, between the two channels of the Kondjo River and up to Apomandé Point.

A third series of sand deposits appear to have formed between 4,000 and 2,000 years ago. This includes the dune ridges that separate the Atadali and Otenga rivers as well as the Animba, Ompingo and Inguéssi plains, which are all aligned. Finally, the most recent sand deposits comprise Mandji and Ozouri islands, which date back to the last two millennia. The oldest dune ridges, which are still visible via satellite imagery, are now engulfed by the delta, far from the coast. The orientation of these dune ridges is different from more recent sand deposits, which suggests the marine delta once had a very different configuration than it does today.

The lower deltaic plain

The evolution of the lower deltaic plain is difficult to trace back through time, although it is clear that all three deltas date from very different periods, as indicated by their vegetation.

The central part of this tidal plain, which undergoes freshwater flooding, contains typical freshwater features such as swamps, narrow riparian forests, and more or less open, vast marshes. In many cases, these open marshes have undergrowth of *Cyperus papyrus* and *Alchornea cordifolia*, along with trees of varying density (e.g., *Raphia*, *Mitragyna*, *Alstonia*), a portion of which are dead or dying. This could indicate that the marshes represent a regressive phase following the historical presence of more dense forest. This phenomenon could be the result of an increase in water level, diminished nutrient availability, or a combination thereof. Regardless of the explanation, the center of the lower deltaic plain is currently changing.

The southern delta appears to be the oldest. Its mangroves are limited to a strip no wider than 6 km and are being colonized by *Raphia* groves. In fact, in many places, larger *Rhizophora* trees are disappearing (Figure 1.103). Only in the area of Olendé, which resembles a tidal stream, do large, pure mangroves remain. Along the Mpolounié and Djembagombé rivers, mangroves are being actively colonized by *Raphia*, up to 1–2 km from Nkomi Lagoon. With the appearance of the Nkomi Lagoon sandbar, the flow of the southern delta, which was probably initially tidal in nature, was slowed and become a swell delta. Currently, only the larger rivers still flow in the direction of the Nkomi Lagoon, while smaller ones are nearly stopped today and are in the process of being colonized by *Raphia* groves.

More recently, the waters of the Ogooué seem to have been divided in the north and begun flowing into the Cape Lopez Bay, forming a typical tidal delta with numerous tidal streams.

Inside the delta, mangroves extend to 20-25 km inside the plain. They cover a wide area, but are mainly low mangroves. There are fringes of high mangroves along the river banks, but they are being colonized by species from the freshwater swamps that form a narrow band along the river's edge (page 100). Only the tidal rivers have escaped this colonization. It is clear that the Ogooué carries seeds from the riparian forests and initiates the colonization process of the mangroves.

Finally, less than 2,000 years ago, the Ogooué divided Mandji-Ozouri Island in half and flowed back into its previous channel, straight to the ocean. Aerial photos clearly show how both of the Ogooué's main channels cut across the multitude of dune ridges on Ozouri Island at a right angle, abruptly eroding them (Figure 1.94).

This suggests that the central delta is the most recent of the three, although it also shows abundant signs of colonization by freshwater species. Pure mangroves only reach 4-5 km into the delta. Further inland toward the plain, smaller rivers, which are bordered by large *Rhizophora*, are all being colonized by *Raphia*. It has been 1,000 years since the water rose back to its current level. Along the shoreline, new dune ridges have been deposited. The tip of Mandji Island has rapidly progressed to the north, forming a succession of hooks on which the city of Port-Gentil was built, and the barrier islands at the mouth of the deltas were also extended in a hook-like manner. The shape of the southern delta stabilized and the progression of the northern delta has slowed. At a certain point, the northern delta even regressed, as shown by the relict islands now situated beyond the delta's front: Apari Island faces the mouth of Kondjo River, and Kobi, Neviga, and Angodia islands are located inside Nazaré Bay.

Acknowledgments

The author thanks Gaëlle Viennois (IRD), who prepared most of the maps. He is also grateful to the countless people who helped him explore the Ogooué delta from 1999 to 2017, especially Michael Fay, who provided many opportunities to fly over the delta, and Lee White (previously in charge of WCS Gabon) who tasked him with the evaluation of this exceptional site from 2000 to 2002, and as a result stirred his interest for this unique complex of wetlands.

Bibliography

BAHAA-EL-DIN, S.M., 1999. *Directory of important bird areas in Egypt*. BirdLife International. Palm Press, Cairo.

CHEVALLIER, L, J.F. MAKANGA & R.J. THOMAS (Compilateurs.), 2002. *Carte géologique de la République Gabonaise : notice explicative*. Council for Geoscience, Pretoria.

FERMON, Y., 2013. *Caractérisation ichtyologique des bassins versants gabonais*. WWF, Libreville.

FERMON Y., 2015. Caractérisation ichtyologique dans le massif forestier d'Evaro, Bas Ogooué, Gabon. WWF Gabon-Aimara. 132 p.

FERMON, Y., L. CHIRIO, A. LAMBOJ & J. SULLIVAN, 2017. Les poissons. Pp. 222-249 in : T. STÉVART & J.P. VANDE WEGHE. *Les milieux humides du bas Ogooué : le delta oublié*. ANPN, Libreville.

GIRESSE, P., 1969. Carte sédimentologique des fonds sous-marins du delta de l'Ogooué. *Cahiers Océanographiques*, 10: 965-994.

GIRESSE, P. & G. KOUYOUMONTZAKIS, 1973. Cartographie sédimentologique des plateaux continentaux du sud du Gabon, du Congo, du Cabinda et du Zaire. *Cahiers de l'ORSTOM, Série Géologie*, 5(2): 235-257.

GIRESSE, P., G. KOUYOUMANTZAKIS & G. MOGUEDET, 1977. La fin de la Transgression Holocène sur les Littoraux Atlantiques d'Afrique Equatoriale et Australe (Gabon, Congo, Angola, São Tomé, Annobon). Rôles Eustatiques et Néotectoniques. *Association Sénégalaise pour l'Etude du Quaternaire de l'Ouest Africain, Bulletin de liaison*, 50: 59-82.

GIRESSE, P., 2008. Tropical and Sub-Tropical West Africa-Marine and Continental Changes during the Late Quaternary. *Developments in Quaternary Science*. Elsevier, Amsterdam.

GUILLERM, J.-M., 1975. Variations saisonnières des transports côtiers dans le Sud-Est du golfe de Guinée. *Bull. U.O.F.*, VII (1-2).

HISARD, P. & A. MORLIÈRE, 1973. La terminaison du contre courant équatorial subsuperficiel atlantique (courant de Lomonosov) dans le golfe de Guinée. *Cah. ORSTOM, sér. Océanogr.*, vol. XI, 4: 455-464.

KLINGE, H. & K. FURCH, 1991. Towards the classification of Amazonian floodplains and their forests by means of biogeochemical criteria of river water and forest biomass. *Interciencia*, 16: 196-201.

KOLESNIKOV, A.G., S.G. BOGUSLAVSKI, G.N. KUKLIN, V.A. SHIREY & V.G. KIRYOUKIN, 1971. Lomonosov Current in the Gulf of Guinea. *Oceanology*, 11: 311-315 (traduction américaine).

LEBIGRE, J.M., 1983. *Le littoral du Gabon : aspects géomorphologiques et biologiques*. Ministère de l'éducation nationale, Institut pédagogique national, Libreville.

LÉRIQUE, J., 1983 Hydrographie, Hydrologie. Pp. 14-17 in: J. BARRET. *Géographie et cartographie du Gabon*. Ministère de l'Education Nationale de la République Gabonaise, Libreville.

LOUBENS, G., 1966. *Biologie de* Polydactylus quadrifilis *dans le Bas Ogooué (Gabon). Peuplement en poissons des eaux saumâtres et son origine*. Thèses de la faculté des sciences, Université de Paris.

MAHÉ, G., J. LERIQUE & J.-C. OLIVRY, 1990. Le fleuve Ogooué au Gabon. *Hydrol. continent.*, 5 (2): 105-124.

MBEGA, J.-D., 2004. Biodiversité des poissons du bassin inférieur de l'Ogooué (Gabon). Volume 1 et Volume 2. Thèse de Doctorat en Sciences Faculté des Sciences. Université de Namur. 614 p.

ONDO ASSOUMOU, E., 2006. *Dynamique des paysages végétaux du littoral centre-ouest du Gabon autour de Port-Gentil : approche spatiale et analyse des données de terrain*. Thèse de l'université Paul-Valéry, Montpellier III.

SCHEPERS, F.J. & E.C.L. MARTEIJN, 1993. *Coastal Waterbirds in Gabon, winter 1992*. Zeist, WIWO.

WAUTHY, B., 1977. Révision de la classification des eaux de surface du golfe de Guinée (Berrit 1961). *Cahiers ORSTOM, sér. Océanogr*. XV, 3, 279-295.

WORBES, M., 1997. The forest ecosystem of the floodplains. Pp. 222-264 in: W.J. JUNK (ed.). *The Central Amazon Floodplain, Ecology of a Pulsing System*. Springer-Verlag, Berlin, Heidelberg.

Appendix: The delta by image

1. The Middle Ogooué

1.40. The central section of the Ogooué River flows over a bed of ancient rock, exposed when water levels are low, as can be seen in Lopé National Park, just upstream from the Okanda gates (Photo: Jean P. Vande weghe).

1.41. Downstream from the Okanda gates, the Ogooué River flows between hills covered with herbaceous savannas (Photo: Jean P. Vande weghe).

2. The lower river valley

1.42. In the coastal sedimentary basin, the Ogooué flows over sand that it has carried down from throughout its entire watershed, although mainly from the Batéké Plateau. During the dry season, this sand appears in the form of large bare sandbanks. An island covered in thickets with Alchornea, *as well as sparse trees, stands 2 km upstream from the large island on which the town of Lambaréné is built; this island hosts a large colony of herons and egrets (Photo: Jean P. Vande weghe).*

1.43. Near the confluence of the Ogooué and the Ngounié, the vast sandbanks take two forms. The highest sandbanks, exposed starting in June, are made of white sands; they attract certain species of nesting birds. The lower sandbanks, which are not exposed until August, are made of dark sand; no birds nest there (Photo Jean P. Vande weghe).

Appendix: The delta by image

Page 50:
1.44. Upstream from Lambaréné, the banks of the Ogooué are covered with dense thickets upon which there are large kapok trees, Ceiba pentandra *(Photo: Jean P. Vande weghe).*

Page 51:

1.45. A flooded landscape, formed by a mosaic of floodable forests, swampy forests, and open waters, extends back from the shore, itself formed by an alluvial bed (Photo: Jean P. Vande weghe).

Page 52:
1.46. At the confluence of the Ogooué and the Abanga River, which carry brown water heavily laden with alluvium,they are also joined by a third channel,which drains black water from the marshes located behind the alluvial levees (Photo: Jean P. Vande weghe).

Page 53:
1.47. The Abanga River, which essentially drains the Crystal Mountains, flows between two alluvial beds on which floodable riparian forests are found. Further back there are swamp forests and areas with relatively stagnant black water. Small raffia groves are visible at the edge of the riparian forest (Photo: Jean P. Vande weghe).

Page 54:
1.48. The Abanga River, lined up with its alluvial beds, annex lakes, and vast expanses of marshland (Photo: Jean P. Vande weghe).

Page 55:
1.49. The confluence of the Abanga River and Lake Nguéné, where black and brown waters mix (Photo: Jean P. Vande weghe).

Appendix: The delta by image

1.50. A vast marshy plain stretches out where the river that feeds Lake Nguéné meets the lake; it is largely flooded during the high waters of November-December and March-April (Photo: Jean P. Vande weghe).

Appendix: The delta by image

1.51. The lower part of the Mbiné valley forms a vast marsh covered with papyrus stands and herbaceous mats with Echinochloa*, and is bordered by raffia trees. The forests are partly dryland forests, partly floodable forests (Photo: Jean P. Vande weghe).*

Appendix: The delta by image

Page 60:

1.52. The Mbiné meanders through a vast swamp of papyrus stands and herbaceous mats with Echinochloa. *Dryland forests cover the tops of the hills, which have escaped flooding (Photo: Jean P. Vande weghe).*

Page 61:
1.53. The furthest "bays" of the Mbiné River have not been blocked by alluvium and now form lakes, partly covered with floating vegetation (Photo: Jean P. Vande weghe).

Page 62:

1.54. Some small, lateral valleys of the Mbiné valley are flooded, but they are in the process of being blocked by papyrus beds and floating meadows with Echinochloa. *The tracks of an elephant crossing the valley can be seen slightly left of center in the image (Photo: Jean P. Vande weghe).*

Page 63:
1.55. Other small valleys are home to black water lakes from which solitary trees emerge (Photo: Jean P. Vande weghe).

3. The upper deltaic plain

1.56. At Lambaréné, the Ogooué splits into two main channels that then cross the two branches of the horst (a strip of Precambrian rocks, detached and raised during the formation of the Atlantic rift). A large furo, the Sogoué, branches off from the northern course, the Ouango, to join the southern channel, which forms the Ogooué proper. The image shows Lake Ouambé and, in the foreground, the Sogoué. The forested hills covered with okoumé are part of the horst. On the edge of the Sogoué there are thickets of Alchornea *and large kapok trees,* Ceiba pentandra *(Photo: Jean P. Vande weghe).*

Appendix: The delta by image

Page 66:
1.57. Beyond the horst, the channels of the Ogooué stretch out across the upper deltaic plain. The image shows the Azingo River, which runs along the northeast edge of the plain to reach Lake Azingo (Photo: Jean P. Vande weghe).

Page 67:
1.58. When the water level is low, small channels drain black water from the marshes to the channels of the Ogooué. During major floods, the current reverses direction and the same channels empty the brown waters of the Azingo River into the heart of the marshy plain, where they deposit their alluvium and participate in clogging the marshes (Photo: Jean P. Vande weghe).

1.59. Between the main channel of the Ogooué and the Ouango lies a vast marshy plain with Lake Nyayé at its center. This lake, which is frequented by manatees, is surrounded by Echinochloa *meadows and papyrus beds from which emerge dense arboreal and shrubby islets. The herbaceous vegetation is etched with the tracks of large mammals (buffalo and elephants) (Photo: Jean P. Vande weghe).*

Appendix: The delta by image

1.60. This image shows the channels that drain the black waters of Lake Nyayé into the Ogooué. Beyond that, a long and very narrow furo, the Ntondi, links the Ouango (the northern channel) to the Ogooué; it is only accessible by canoe during major floods. Dense swamp forests extend beyond the Echinochloa *meadows and papyrus beds (Photo: Jean P. Vande weghe).*

Appendix: The delta by image

1.61. The great central plain extends to the west, beyond the Ntondi channel. In this section, it is crossed by an old, closed channel of the Ogooué whose course is indicated by ancient alluvial beds covered with shrub and tree formations (Photo: Jean P. Vande weghe).

1.62. Further west, the plain and its past channels become wider. Tracks of large mammals are omnipresent (Photo: Jean P. Vande weghe).

Appendix: The delta by image

Page 76:
1.63. In its western part, the upper deltaic plain is largely covered with dense swamp forests (Photo: Jean P. Vande weghe).

Page 77:
1.64. This marshy forest cover becomes less dense in some parts, giving way to papyrus beds. The trees with white trunks are Alstonia congensis *(Photo: Jean P. Vande weghe).*

Page 78:
1.65. A swamp forest with numerous Alstonia congensis *(Photo: Jean P. Vande weghe).*

1.66. A marshy forest with many different species, in particular Symphonia globulifera *(trees covered in red flowers) (Photo: Jean P. Vande weghe).*

Page 79:
1.67. A swamp forest with numerous raffia palm trees (Photo: Jean P. Vande weghe).

1.68. A dense swamp forest, including many Anthostema aubryanum *(trees with a crown forming concentric rings) in the area of Lake Gomé (Photo: Jean P. Vande weghe).*

Appendix: The delta by image

Page 80:

1.69. The Nkengué marsh extends south of the Obano channel downstream from Lake Anengué. It is a vast, flooded plain with herbaceous vegetation and crossed by innumerable animal tracks—where humans fail, large mammals survive (Photo: Jean P. Vande weghe).

Page 81:
1.70. The Nkengué marsh is drained by irregular channels and dotted with dense circular stands of Pandanus *(Photo: Jean P. Vande weghe).*

Appendix: The delta by image

Page 82:
1.71. The Rivière des Pélicans with Lake Evaro in the distance. Between the two lies an herbaceous savanna crossed by a small river (Photo: Jean P. Vande weghe).

Page 83:
1.72. Lake Evaro with its deeply indented banks represents a typical flood landscape. Lake Onangué is visible in the distance, separated from Lake Evaro by the lake delta of the Rivière des Pélicans (Photo: Jean P. Vande weghe).

Appendix: The delta by image

Pages 84 and 85:
1.73 and 1.74. The steep and deeply indented banks of Lake Evaro and Lake Onangué testify to flooding (Photos: Jean P. Vande weghe).

Appendix: The delta by image

1.75. The Rivière des Pélicans is bordered by dense thickets from which large, widely spaced Ceiba emerge (Photo: Jean P. Vande weghe).

Appendix: The delta by image

Page 88:
1.76. The delta of the Rivière des Pélicans is a typical lake delta, with long channels bordered by heavy alluvial beds, which extend far into Lake Onangué (Photo: Jean P. Vande weghe).

Page 89:
1.77. The Rivière des Pélicans, with its alluvial beds surmounted by tall Ceiba trees. The open waters behind the alluvial beds are protected by tall grass mats (Echinochloa *and* Vossia) *and serve as refuges for hippopotamuses (Photo: Jean P. Vande weghe).*

1.78. The Rivière des Coupeurs, in the foreground, feeds into Lake Gomé. During major floods, the forests behind its alluvial beds are inundated (Photo: Jean P. Vande weghe).

1.79. A flooded area behind the alluvial beds of the Rivière des Coupeurs (Photo: Jean P. Vande weghe).

Appendix: The delta by image

1.80. Lake Gomé is located north of the upper deltaic plain. Like the Great Lakes of the south, it forms a flooded landscape. In the distance, the conical hills that emerge from the plain attest to the presence of an old erosion surface from the Tertiary (Photo: Jean P. Vande weghe).

1.81. The southern part of Lake Gomé. Unlike the southern Great Lakes, its shores are fairly flat and bordered by floating herbaceous mats with Echinochloa *(Photo: Jean P. Vande weghe).*

1.82. The shores of Lake Logé, located just east of Lake Gomé, are bordered by extensive papyrus beds (Photo: Jean P. Vande weghe).

1.83. The outlet of Lake Gomé drains black waters from the lake into the brown waters of the Ouango (Photo: Jean P. Vande weghe).

4. The lower deltaic plain and coastal sand formations

1.84. The main channel of the Ogooué is covered with dense swamp forests where it enters the lower deltaic plain, but a floodable riparian forest grows on the alluvial beds, with more varied and larger trees. In particular, these include many species typically found in the dryland forests of the interior of the country. A fringe forms at the edge of the water with raffia or floating herbaceous beds with Echinochloa*. The ecosystem of this plain is no longer subject to seasonal floods, like those of the upper deltaic plain, but rather to daily tides, which are more or less pronounced depending on the lunar cycle. Flood forests and sandbanks are becoming scarce and giving way to swamp forests—which are becoming dominant—and also to mudflats (Photo: Jean P. Vande weghe).*

1.85. The small rivers in the center of the lower deltaic plain do not have alluvial beds. They are bordered by marshy forests, and at the water's edge, there are often clumps of Nauclea pobeguinii, *whose horizontal roots are densely interlacing (Photo: Jean P. Vande weghe).*

Appendix: The delta by image

1.86. The central part of the lower deltaic plain is covered with swamp forests and floodable forests. The difference in water level between high and low tide rarely exceeds 50 cm, although during white water tides it can reach over 1 m (Photo: Jean P. Vande weghe).

Page 98:
1.87. The westernmost arm of the Ogooué's northern delta is the Atadali River. Here it is photographed near Agélégé Island. The Kondjo River is visible in the background. The rivers are bordered by large, high Rhizophora *mangroves, which hide large sections of low mangrove (Photo: Jean P. Vande weghe).*

Page 99:
1.88. Between the Atadali River, Mandji Island, and the Animba Plain stretch the tortuous channels of the Doranga River, a typical tidal river that only carries the salt water of Cape Lopez Bay, as well as some fresh water from the Animba Plain during the rainy season. The meanders of this river are surrounded by low mangroves with some higher formations here and there. Its water does not carry alluvium from the river; it is therefore clear and salty. The pale, straight lines that mark the landscape are old tracks opened for oil exploration (Photos: Jean P. Vande weghe).

1.89. The narrow strip of Rhizophora *that borders the Atadali River near Mondorobé grows in entirely freshwater. Samples made over a period of one year, both above and below the surface, never detected any salinity. A large part of the mangrove of the lower deltaic plain thus grows in freshwater, and* Rhizophora *is in no way an indicator of salinity (Photo: Jean P. Vande weghe).*

1.90. The mangroves that thrive in waters loaded with alluvium are lush, and the Rhizophora trees can reach over 30 m in height (Photo: Jean P. Vande weghe).

Page 102:
1.91. The Ikando River, which runs along the Ozouri Island, is a tidal river bordered by pure mangrove. The high mangroves are, however, much wider than those of the northern delta. Ozouri Island is visible in the distance, along with the construction of the Port-Gentil-Omboué road (Photo: Jean P. Vande weghe).

Page 103:
1.92. Further upstream, in the central part of the lower deltaic plain, the mangroves are colonized by Raphia. *Some small rivers are completely overgrown with these palm trees (Photo: Jean P. Vande weghe).*

Page 104:
1.93. The central mouth of the Ogooué, with the sandbar that diverts the waters of the river to the north in the foreground. The Animba River, which forms the main arm of the Ogooué, is visible beyond. This image dates from 2006, long before the construction of the bridge that links Mandji Island to the north and Ozouri to the south (Photo: Jean P. Vande weghe).

Page 105:
1.94. The northern part of Ozouri Island has clearly been eroded by the Aworiwonga River, the southern branch of the central mouth of the Ogooué: the parallel dune ridges have been cut off at a right angle. On the right is a meander of the Ikando River, which has also "bitten" into the sand formations of Ozouri Island (Photo: Jean P. Vande weghe).

Appendix: The delta by image

Page 106:
1.95. The southern part of Mandji Island is heavily forested. A strip of coastal forest is visible on the image, flanked on both sides by high mangroves (Photo: Jean P. Vande weghe).

Page 107:
1.96. The southern part of Mandji Island, with its parallel dune ridges covered with dense forests interspersed with herbaceous vegetation. The central mouth of the Ogooué is visible in the distance. This image shows that the dune ridges currently being deposited do not have the same orientation as the older ones that form the island (Photo: Jean P. Vande weghe).

Page 108:
1.97. The "Barre des Portugais" (Photo: Jean P. Vande weghe).

1.98. On its eastern flank the "Barre des Portugais" includes young mangroves (Photo: Jean P. Vande weghe).

Page 109:
1.99. Other parts of the "Barre des Portugais" are colonized by herbaceous plants and shrub vegetation with Dalbergia ecastaphyllum *(Fabaceae) (Photo: Jean P. Vande weghe).*

1.100. The southern delta, with the Djembagombé River bordered by large swamp forests and, in the distance, high residual mangroves. The water is fresh in this part of the lower deltaic plain (Photo: Jean P. Vande weghe).

Page 112:
1.101. The southern delta, with the Nkomi Lagoon on the left. On the edge of the lagoon, the mangroves are almost intact. To the right, closer to the Djembagombé River, the Rhizophora *thin out and are gradually being replaced by a* Raphia *grove (Photo: Jean P. Vande weghe).*

Page 113:
1.102. Between the Djembagombé and Adéloué rivers, there is a dense stand of high Rhizophora *with dense* Raphia *"undergrowth" (Photo: Jean P. Vande weghe).*

Appendix: The delta by image

Page 114:
1.103. The confluence of the Mpolounié and Nkendji rivers, with shrinking mangroves and Raphia groves (Photo: Jean P. Vande weghe).

Page 115:

1.104. The Djembagombé River bordered by its marshy forests. On the left the Nkomi Lagoon, bordered by tall, dying mangroves (Photo: Jean P. Vande weghe).

Appendix: The delta by image

1.105. A vast open, marshy plain stretches between the Mpolounié River and the Inguéssi Plain, covered with large grasses and papyrus beds; it is also bordered by marshy forests. Close examination shows that the swamp forests are in decline and that there are dying trees in the open areas. This image, like many others, points to the fact that mangroves are being replaced by raffia trees, which in turn are replaced by mixed swamp forests before making way for open marshes. In the open areas, there are countless large mammal tracks (buffalo, sitatungas, forest hogs?) (Photo: Jean P. Vande weghe).

Page 118:
1.106. The herbaceous formations that stretch between the Inguéssi Plain and the Ozouri Plain are colonized by Pandanus *(Photo: Jean P. Vande weghe).*

Page 119:

1.107. The Inguéssi Plain forms a complex patchwork of small lakes, dense Pandanus *stands, floodable forests, swamp forests, dryland forests, dry savannas, and wet savannas. Parts of the plain still show parallel structures (Figure 1.109), but elsewhere they have disappeared. Animal tracks extend along the water's edge near small lakes (Photo: Jean P. Vande weghe).*

1.108. Some small lakes are covered with floating herbaceous mats (Photo: Jean P. Vande weghe).

1.109. A small group of buffalo seeking refuge in riparian thickets. In the background, dense Pandanus *stands (Photo: Jean P. Vande weghe).*

1.110. *Some parts of the Inguéssi Plain still have parallel sand structures, although they are largely covered by coastal forests (Photo: Jean P. Vande weghe).*

Page 124:
1.111. The Olendé sector does not receive water from the Ogooué and is covered in pure mangroves with Rhizophora*; it is like a tidal river (Photo: Jean P. Vande weghe).*

5. The lagoon region

Page 125:
1.112. To the east of the southern delta lies the Ntchonga Ntchiné Plain, mainly covered with dry savannas. It represents a structure similar to the Wonga-Wongué Plateau, although smaller and lower; its sands also seem to date from the Pliocene. The Nkomi Lagoon is visible in the foreground (Photo: Jean P. Vande weghe).

2.1. A "camp" in one of the main rivers of the northern delta (Photo: Jean P. Vande weghe, June 2016).

2.2. Isolated dwellings, constructed along the Ouango River, downstream of Lake Gomé (Photo: Jean P. Vande weghe, June 2011).

2.3. A loose "agglomeration" on the shore of one of the lakes in the northern part of the upper deltaic plain (Photo: Jean P. Vande weghe, March 2017).

2.4. Isolated houses along a channel that connects the Obando River to the Ogooué, at the transition between the upper deltaic plain and the lower deltaic plain (Photo: Jean P. Vande weghe, April 2006).

2. Human populations

Jean P. Vande weghe & Ehoarn Bidault

Contrary to the two large deltas of the Nile and Niger (page 20), the Ogooué Delta has a relatively small human population. According to the 2013 survey (government data), the population of Moyen-Ogooué and Ogooué-Maritime provinces, which encompass the Ramsar site, are 69,287 and 157,562 inhabitants, respectively. Since the city of Lambaréné has about 25,000 inhabitants and Port-Gentil has close to 137,000 people, the population density in rural areas is just 2.3 inhabitants/km² in Ogooué-Maritime and 0.89 inhabitants/km² in Moyen-Ogooué. In fact, Ogooué-Maritime province is sparsely populated throughout, although pressures resulting from the inhabitants of Port-Gentil extend dozens of kilometers beyond the city limits, notably reaching to the Nkomi Lagoon region and beyond. The rural population is also very dispersed since the villages of the Lower Ogooué region have not been grouped together as they are in the rest of Gabon—a point that will be revisited below.

Despite its small human population, the Lower Ogooué region has been inhabited for "a long time," and while many landscapes might seem to be intact, the impact of these populations is not negligible since they have lived mostly off of available natural resources (fishing, hunting). The density of mammals and birds is in fact low—one could say that it is just as low as the human population. The region was probably subjected to human influence for a long time, and the resulting impact has been exacerbated throughout the 20th century with the development of industrial logging and the oil industry, the construction of new cities, and the introduction of firearms, outboard motors, and nylon nets, to name just the main factors.

Previous populations

The various archeological discoveries in Gabon have been summarized by Clist (1995). More recently, some new discoveries have been made, but they have not fundamentally changed our understanding of the region's cultural chronology, which can be summarized as follows. The oldest vestiges that have been found in this region to date are from the Middle Stone Age (Figure 2.5). They are Lupembian tools found in Ndjolé, which indicate the presence of hunter-gatherers in the region between 40,000 and 12,000 BP (Clist, 1995). Tools from the Late Stone Age have been found on the edges of the Nkomi and Iguéla lagoons, Those from Iguéla are dated from 3,680 and 6,300 BP (Clist, 1995). The Ogooué region was clearly traversed by groups of hunter-gatherers living in very low density.

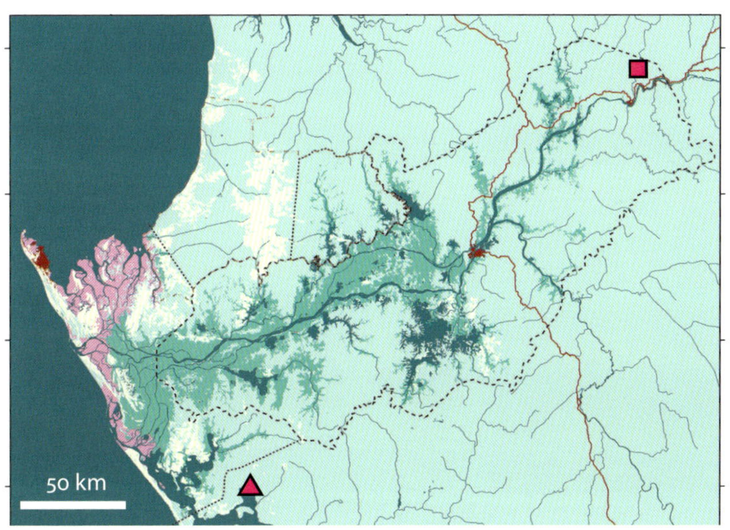

2.5. Locations of artifacts from the Middle Stone Age (■) and the Late Stone Age (▲).

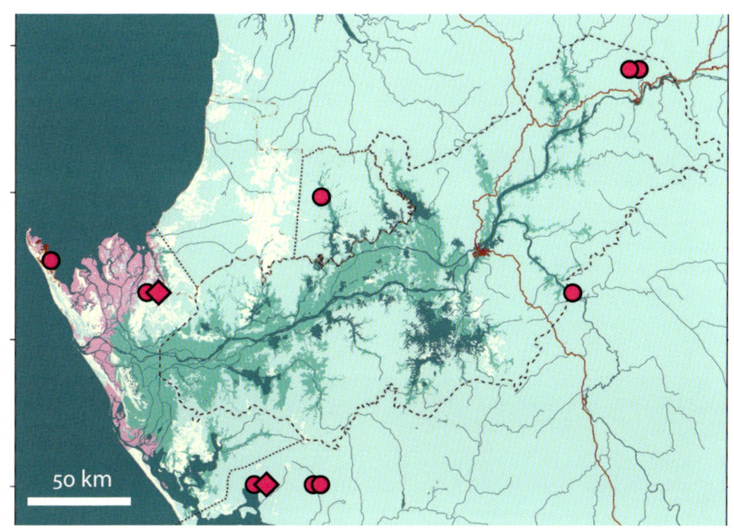

2.6. Archeological finds of Neolithic artifacts (●) and the Early Iron Age (◆).

Discovery of old agricultural structures in the Lower Ogooué's humid savanna

by Richard Oslisy

The combined study of aerial photography taken using airplanes and drones alongside high-resolution satellite images has resulted in the discovery and cataloging of numerous traces of ancient agricultural structures in the Lower Ogooué's humid savanna. The aerial survey work initiated in 2015 led to discoveries from Pongara National Park to the Ogooué Delta, west of Envonga village, in the savanna at the confluence of the Ogooué and Agouma rivers south of Lambaréné and the savanna north of Tchenguifala village on Lake Anengué.

The specificity of these agrarian structures comes from their vast numbers, covering an area of 10 km², their recurring presence in humid savanna areas, and their diversity in shapes: ridges and furrows measuring up to 500 meters long, mounds, rectangular and circular raised beds of varying sizes, connected by access trails, irrigation channels, etc. These represent raised beds erected to protect crops from floods in areas subject to continuous or seasonal flooding. The layering of soil extracted during the excavation of the surrounding water channels increased its sustainability. During their construction, these channels were dug out in between or next to the fields and were filled with water during the rainy season, simplifying irrigation during the dry season. The decomposition of aquatic plants in these channels created fertile organic mud that could be used periodically to renew the raised beds and enrich the soil with nutrients.

Raised bed crops were already known to exist in several South American countries: the Llanos de Mojos in Bolivia, around Lake Titicaca in Bolivia and Peru, at the mouth of the Guayas River in Equator, in the Bogota highland savanna and the San Jorge River on the coast of Colombia, the Llanos de Apure and the Middle Orinoco in Venezuela, in areas along the coast of Guyana and Suriname, and on Marjó island in Brazil. Most of these structures have been dated to the first millennium of our era, but some, including the oldest, were built as early as 100 BC.

In Central Africa, in his famous 1966 book entitled "De l'Atlantique au fleuve Congo," the geographer Gilles Sautter indicated the presence of raised housing and agriculture to the north of Brazzaville (Republic of Congo) in the humid savanna of the "Land of Three Rivers" (Likouala-Musaka, Alima Nkeni), where the Likouba and Likouala ethnic groups currently cultivate manioc. There are a few other references to these structures throughout the literature, but little is known about their origin.

In Gabon, these structures were not known until the first aerial flights began in 2011 with the help of Mike Fay, during which we were intrigued by rows of structures in the wetlands. The zone with the largest area is situated inside the Ogooué Delta, east of the village of Envonga, where we found quadrangular and circular mounds sometimes connected by paths or aligned ridges of varying lengths (Figure 2.7). The second-largest area comprises an alignment of ridges found in the humid savanna of Ndiawé to the northeast of Lake Anengué (Figure 2.8). The third-largest area is found south of Lambaréné, not far from the village of Matadi; there we found numerous mounds, some of which aligned perfectly perpendicular to the main body of water (Figure 2.9).

Since 2015, we have conducted a pair of campaigns in the humid savanna along the Gabonese coast, focusing both on aerial reconnaissance to locate these structures using drones, and excavations executed by digging trenches to reveal the deposit stratigraphy and provide carbon samples used for radiocarbon dating. Using an auger, we sampled this organic deposit along a relatively longitudinal transect at depths of 0–20 cm and 20–40 cm in order to collect plant remains, seeds, and other phytoliths. This will allow us to obtain information on the types and species of plants that have been cultivated in the past.

The first dating results for the sites along the coast of Gabon indicate two periods centered around 2,200–1,900 BP and 900–500 BP. Within the context of all the dates relating to these same areas, these results coincide with Neolithic and later Metal Age populations that were present during these same periods.

This raised bed agriculture practiced in humid savanna declined in favor of slash-and-burn agriculture in forested areas, which leads to three fundamental questions:

- When did this shift in agrarian system take place?
- What were the climatic or anthropic factors that contributed to this shift?
- Which plant species were cultivated and/or domesticated?

Although no data are currently available regarding ancient occupation of the raised bed fields in Congo, the small, limited literature shows that the banks of some Congolese rivers were inhabited by Neolithic populations around 2,500–2,000 BP and later by Metal Age populations. These Congolese valleys have barely been the subject of archaeological research, and much remains to be done!

One might think the agrarian structures of both Gabon and Congo were the result of the first Neolithic populations that farmed using stone tools such as polished hoes and axes, with which it would be difficult to cut down trees for slash-and-burn agriculture. Analysis of the soil extracted by augers from the ridges in Pongara National Park in Gabon showed numerous small splinters and shards of silex. These were certainly the result of impacts along the edges of tillage implements made of flint, caused during weeding and harrowing operations.

According to this hypothesis, the Neolithic populations (3,000–2,000 BP) erected the first of these raised-bed fields in the flooded lower-lowlands of the savanna. As for the Metal Age populations (starting from 1,800 BP), they are thought to have been the first to farm forested land thanks to their mastery of iron tools, which would have been more robust and efficient. Thus, they could have continued to use wetland raised-bed fields for a while, progressively transitioning, depending on regional context, to slash-and-burn agriculture.

2.7

SAUTTER, G., 1966. *De l'Atlantique au fleuve Congo ; une géographie de sous peuplement (République Congolaise et République Gabonaise)*. Paris, La Haye, Mouton.

DENEVAN, W.M., 2001. *Cultivated landscapes of native Amazonia and the Andes*. Oxford University Press.

MCKEY, D., S. ROSTAIN, J. IRIARTE, B. GLASER, J.J. BIRK, I. HOLST & D. RENARD, 2010. Pre-Columbian agricultural landscapes, ecosystem engineers, and self-organized patchiness in Amazonia. *Proc. Nat. Acad. Sci. of the United States of America*, 107: 7823–7828.

2.8

2.9

Tools dating from the early Neolithic were found at Mbilapé, not far from the Gangwé River, the westernmost river of the northern delta. They date from approximately 2,400 BP. Others were found in Ikengué along the banks of the Nkomi Lagoon (Figure 2.6). They date from the same period. Late Neolithic tools were found in Mbilapé, Ikengué, Ofoubou, Ntchengué (Mandji Island), and Ndjolé (Figure 2.6). All these tools are a testimony to the first wave of Neolithic migration coming to Gabon along the coastal region.

Vestiges from the ancient Iron Age were found in Mbilapé and Ikengué, but more discoveries would certainly have been made if more thorough investigations had been led.

Traces of raised-bed agriculture were recently found in the humid savanna situated between Komo and the Lower Ogooué. In the region we are focusing on, some were found in the Koroso Plain, at Lake Mpaga, at Lake Anengué and near the Rivière des Pélicans (highlight box, pages 128 and 129). These vestiges indicate that a population which practiced intensive agriculture lived in this region approximately 2,000 years ago and was still present 1,000-500 years ago. We know nothing about this population except that part of the vestiges is currently covered in forest, which indicates that they lived in an environment less forested than it currently is.

Current populations

The oldest information available regarding the current populations in the area are from Van Linschoten (1594), De Marees (1602), Van den Broecke (1705), and Dapper (1667), Dutchmen who never visited the coasts of the Gulf of Guinea but collected accounts from sailors working for the East India Company who stopped over in Cape Lopez Bay on their way to the Far East. Their boats stopped in Cape Lopez Bay because there were

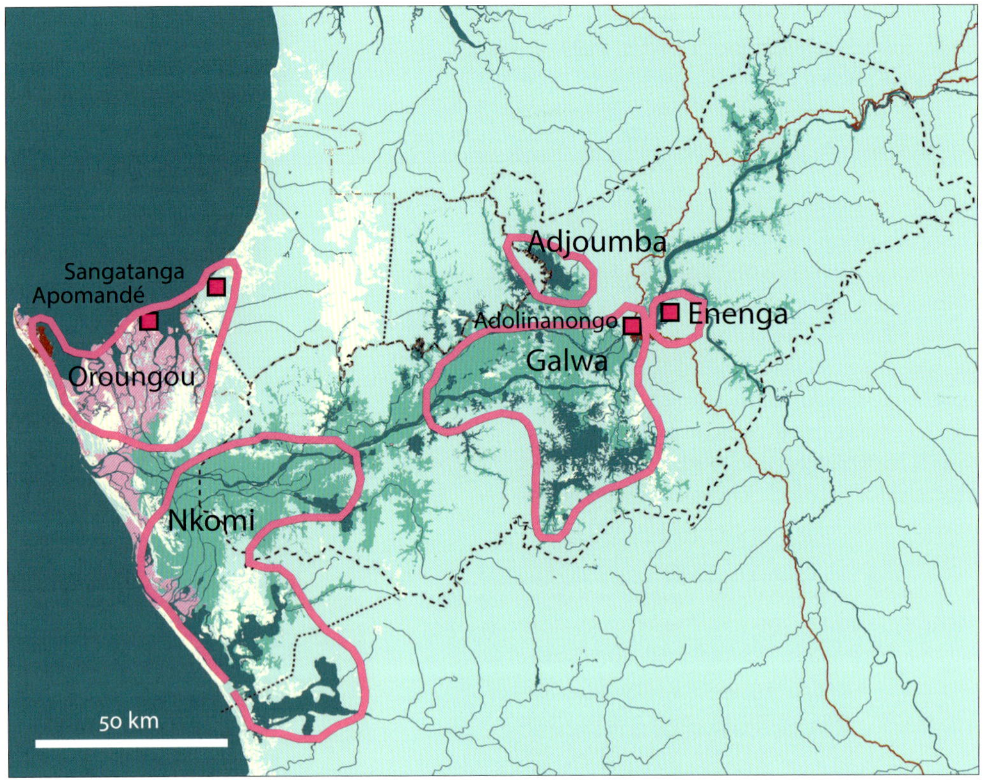

2.10. Schematic indication of the area occupied by Myené peoples in the Lower Ogooué region: the Oroungou, Nkomi, Adjoumba, Galwa, and Enenga (based on Merlet, 1990). The red lines that delimit the "territories" occupied by the different ethnic groups clearly do not correspond to borders in the Western sense; at most, they delimit an area of dominant influence that fluctuated over time.

no dangerous sandbars to cross in order to land. Sandbanks and currents could nevertheless be dangerous for large sailboats of that time when the wind was low. The remains of the *Mauritius*, which sunk near Port-Gentil on March 19, 1609, and was discovered in 1985, illustrates this danger (Vande weghe, 2007). Sailors made use of this stopover to buy smoked meat (elephant, "wild beef," and antelope). This began the "bushmeat hunting and trade." They also stocked up on fresh water and fished using small canoes. In just half a day and with a single net, they caught enough fish to last the rest of the trip back to the Netherlands.

According to accounts from these sailors, the population of Cape Lopez Bay in the 1600s and the beginning of the 1700s seems to have belonged to the Vili ethnic group, or at least their language and culture was heavily influenced by those of the Kingdom of Loango. Then came changes, apparently caused by migration from the center of the country (Tsongo country), the lower Ngounié region, and the regions generally south of the Ogooué (Merlet, 1990).

In their contact with Europe and its much-sought-after products, these coastal populations progressively differentiated from inland populations as they distanced themselves from the declining populations of Loango. They also underwent a strong "myénization" under the growing influence of the Mpongwé from the Komo Estuary (in the Libreville region). Regardless of the history of these peoples, in the 19th century, when the French progressively took over the region that would later become Gabon, the Lower Ogooué region was occupied by the Myéné ethnic group (Figure 2.10).

The Oroungou essentially occupied the northern delta and Mandji Island. Their royal residence, Apomandé, was built on the sandbar south of Fètiche Point in 1790 (Figure 2.10). They had encampments on Mandji Island and probably also along the continental coast, southwest of the present-day Wonga-Wongué Reserve. Having engaged heavily with the Atlantic slave trade and fearing reprisal from the French and English marine forces, they moved their royal residence toward Sangatanga along the coast of the Wonga-Wongué Reserve at the beginning of the 19th century. There they could hide from ships patrolling along the coast behind thickets and coastal forests (Raponda-Walker, 1960; Merlet, 1990). The Nkomi had settled around the Fernan Vaz Lagoon and their territory encompassed the southern delta. The Galwa lived mostly in the eastern part of the upper deltaic plain and around the Great Lakes. Their domain extended to Achouka to the west and their royal residence, Adolinanogo, was situated approximately where the Schweitzer Hospital in Lambaréné currently stands. Finally, the Enenga lived on the shores of Lake Zilé and the Adjoumba around Lake Azingo.

The forests situated away from the river, and notably in the valley of the Rivière des Coupeurs and along the Bokoué River (a tributary of the Komo River), were inhabited by Kellé, an ethnical group related to the Kota.

Starting in 1868, Fang people came from the north, pushing the Galwa to south of the Ogooué, and advanced up to the Great Lakes region, where they built numerous villages. Today they are in fact the majority ethnic group in this region.

Starting at the end of the 19th century, especially due to the increase in logging activities, a wide diversity of populations from outside the region moved into urban centers and forest encampments.

Colonization

In 1854, two American missionaries, William Walker and Ira Preston, traveled up the Ogooué. Soon after, Paul Belloni Du Chaillu visited Sangatanga and the northern delta, where he observed a sea turtle hunt. He returned to the country between 1858 and 1865 and visited the Fernan Vaz Lagoon (Nkomi Lagoon). He then departed on an expedition to Lake Anengué, where he saw many huge Nile crocodiles. Starting in 1860, numerous British, German, Portugese, Dutch, Spanish, and French factories were established all along the coast of future Gabon.

The French navy, which had established its presence in Libreville in 1838, was specifically interested in the Ogooué, which was then an important trade route, and thought it might provide access to the Congo River basin. As early as July 1862, Lieutenant Serval and the naval surgeon Griffon du Bellay visited the northern delta region, but their ship was blocked by sandbanks and they were forced to continue their voyage by canoe. Thus, they discovered Onangué Lake and the sacred island of the Galwa, which housed an enormous colony of aquatic birds. They returned to the region over land in December and reached Lake Zilé, where the Enenga lived. The English salesman Robert Bruce Walter also reached the Adolinanongo region by land in 1865 and Ndjolé in 1866. In 1867, a French navy ship reached the confluence of the Ogooué and the Ngounié, and an accord was reached between the French and the Enenga. In 1868 the French navy navigated downstream to Fernan Vaz and signed an accord with the Nkomi. The Marquis of Compiègne and the naturalist Alfred Marche arrived in 1873, followed by Admiral Couriault du Quilio, who signed an accord with the Galwa the same year and one with the Oroungou a bit later, in which the French obtained ownership of Mandji Island.

In just 11 years, the Lower Ogooué was "conquered" by Europeans and the privileges and taboos that previously regulated and limited traveling along the river were abolished. Toward 1883, Pierre Savorgnan de Brazza established the Ndjolé military outpost in order to ensure security along the river and, starting in 1875, open it to trade. Commerce then greatly increased. In 1886 Gabon, which had remained the territory of the French navy, became a French colony. Lambaréné became an administrative outpost under the jurisdiction of Ndjolé. Numerous factories were set up along the shores of Lakes Ezanga, Oguémoué, and Onangué. Toward 1915 they numbered more than 30.

Following the arrival of traders and administrative staff, missionaries moved in. The Kangwé protestant mission, currently encompassed within the limits of Lambaréné, was built in 1874 by American missionaries. The first Catholic mission of Azengé Island (Lambaréné) was built in 1881. The Sainte-Anne mission was built on the Nkomi Lagoon (Figure 2.11) in 1897. The Catholic mission in Ndjolé was constructed in 1897 on the left bank of the river, and the Ngomo mission (Figure 2.12) was built at the confluence of the Ambila River (the spillway of Lake Onangué) and the Ogooué in 1898 by the Alsacian missionary Ernest Haug, who also contributed significantly to our ichthyological knowledge of the delta (page 224).

Logging activities grew starting in the 1880s, mainly for okoumé, *Aucoumea klaineana*. It was centered around Lambaréné. Toward 1920, this resulted in the "okoumé fever" (Pourtier, 1989). Farmers abandoned their fields and rushed to logging camps; famine ensued. The global economic crisis of 1929 reined in these forestry activities, plunging the Lower Ogooué region into doldrums. After World War II, bulldozers were introduced which, along with the general economic recovery, gave fresh impetus to logging activities. Since the forests of the coastal basin were nearly "depleted," logging activities moved inland. Nevertheless, the city of Lambaréné remained important since it was a necessary passage between the north and south of the country, although the bridges that now connect both banks of the river were only built in 1975.

Mandji Island, on the other hand, developed much later. Cape Lopez Bay had been a stopover for European merchant

2.11. The Sainte-Anne mission on the Nkomi Lagoon, built in 1887 (Photo: Jean P. Vande weghe).

2.12. The Ngomo mission, established in 1898 (Photo: Jean P. Vande weghe).

2.13. At the edge of Lake Evaro, a tree cut down with a chainsaw in order to harvest a little bit of honey (Photo: Jean P. Vande weghe, July 2006).

vessels and a refuge for pirates who used the delta's rivers to hide. In 1770, Irish settlers had set up a factory on the island. At the very start of the 19th century, the Portuguese were still taking slaves aboard at Mandji Island. It was later abandoned and around 1860 it only had temporary hunting or fishing encampments. It was purchased in 1873 by the French navy in a treaty signed with the Oroungou and, in 1880, Savorgnan de Brazza had warehouses built on the island in preparation for his 1879-1882 expedition, which led him to the Batéké Plateau and to the discovery of the Brazzaville region. A post office was set up on the island in 1889 as well as a customs office in 1894, and merchants then progressively set up shops. An English factory was even established in Yombé in the farthest part of the upper deltaic plain, not far from Anengué.

Toward 1900, the urban area that had formed on Mandji Island started to take the form of city and was named Port-Gentil. In 1915 it became the administrative seat of the Oroungou territory. Starting in 1950, it became the main port for okoumé export and, starting in 1956, it also became the center of oil exploitation (Figure 2.22).

The current situation

Geographic distribution of populations

The Lower Ogooué Ramsar site encompassed approximately 250 villages, campsites, and construction sites in 2003 (Le Duc Yeno, 2015). These "urban areas" were mostly situated along the Libreville-Lambaréné-Mouila national road, all along the Ogooué River main channels and around numerous lakes. The highest population density was to the east of the Great Lakes. Aerial surveys in 2012 and 2017 nevertheless indicated that many people lived in isolation, more or less hidden among the forests along the lakes north of the upper deltaic plain (Figures 2.1 and 2.2).

Outside of rural populations, the combined population of the cities of Lambaréné, Ndjolé, and Port-Gentil was approximately 167,000 inhabitants.

Lifestyle

A socioeconomic analysis of the rural populations of the western part of the upper deltaic plain was conducted by WCS as part of a project for strengthening capacity for managing national parks and biodiversity (Loundou *et al.*, 2013). The following is a summary of the key points:

- The majority of villages predate the independance of Gabon but encampments are generally recent.
- Settlement locations were chosen based on access to the Ogooué (the main communications channel), past employment opportunities among the then numerous logging enterprises, and the abundance of natural resources (fish, soil fertility, etc.)
- With a few exceptions (e.g., Gomé, Dakar), most villages have a very small population, whereas floating or seasonal population numbers are very high.
- Villagers depend mostly, whether directly or indirectly, on natural resources to ensure their survival through procuring food, fuel, fiber, medicinal plants, and construction materials. Commerce in these resources is often a household's only income source, especially among the poorest people.
- Villages have a diversity of basic infrastructure: school buildings are sparse, health infrastructure is rarer still, there is no clean water infrastructure, and the villages are not connected to a public electric grid.*
- No villages have a market, and the few shops that exist sell products at much higher prices than in urban centers—how could they do otherwise?
- Social cohesion is strong in a majority of villages, a minority of which show community initiatives, and there are matrimonial bonds between a majority of villages.†
- Employment opportunities are rare and idleness is frequent, especially in villages that previously were associated with logging camps.
- Fishing is the main source of income for a majority of households; it is practiced throughout the lakes neighboring the Ogooué. Open access to fishing areas, the increasing number of fishermen, current fishing methods, and the type and quality of tools used by each individual all threaten the sustainability of this practice.
- Agricultural plots are relatively small. The majority of fields are close to homes, the banks of lakes, and the Ogooué. Most of their production is dedicated to direct consumption.
- Few households practice trapping; rifle hunting is mainly a means for

* Solar panels had been installed in three villages: only one worked in Issangui, two in Akoun-Nlam, and all were out of use in Olamba, where the batteries supposed to store energy had been stolen. According to the local inhabitants, the reason these installations do not function is nevertheless due to a lack of maintenance from the company that installed them. Some households were equipped with personal generators. Some shared electricity with their neighbors, but contributions to collective fuel purchases remained the main issue.

† This cohesion is sometimes more theoretical than practical; indeed, many families live in isolation following disputes. Despite this, many residents complain about not having access to schools or mobile phone networks!

procuring protein for the household, and the surplus is sold or traded within the villages.* Nonetheless, some logging companies supposedly practice commercial hunting, notably at Lake Oguémoué; as anywhere else, "outsiders" are blamed for this problem!

- Harvest of nontimber forest products is a widespread commercial or subsistence activity within the study area.
- Mechanized logging, though not common, has been observed at the lakes of Canton Sud and along the Ogooué.
- Forested areas provide most of the materials for housing construction and building canoes.
- The majority of households within the study area consider themselves to be living in poverty.

Environmental impacts

Fishing

Long ago, fishing was practiced using traps and nets made of plant fibers (Du Chaillu, 1863, 1868; Raponda-Walker, 1960; Raponda-Walker & Sillians, 1995). With the arrival of nylon nets after World War II, it became the main subsistence activity for a large portion of the population. According to the socioeconomic study, it represents the main activity for 58% of the households interviewed.

Fishing is done in all lakes and on the Ogooué, especially at the mouths of rivers, swamps, and reed beds–which are fish spawning grounds (Loundou *et al.*, 2013). During flooding or when the water recedes, fish dams are used in the Ogooué's channels. In some villages, women empty streams and ponds during the dry season. Single-hook angling or longlining (in the southern delta and the Nkomi Lagoon) have also been observed.† Bamboo fish traps are used to capture shrimp along the Ogooué.

Two very destructive fishing techniques were introduced to the region by foreign fishermen. Tapage ("beating" in English) involves surrounding a body of water with gill nets and then driving the fish into them by hitting the surface of the water with poles or paddles. Iroka is similar to tapage, but the pole is replaced by an outboard motor. The noise produced drives fish, especially "carp," toward the nets, increasing yield. These fishing techniques are used exclusively at night, which causes conflicts between the local population and fishermen mostly coming from Lambaréné.

In some areas fishing is practiced by both sexes and, aside from a few fishermen and women of West African origin, notably from Mali and Nigeria, most are from Gabon. In any case, in the lower deltaic plain, we have found camps used exclusively by West Africans.

Within the entire wetlands complex, there are an estimated 4,400 fishermen and women and more than 2,200 canoes (page 238).

Fish are stored using refrigeration, salting, and smoking. Part of the catch is eaten by the village members and part is sold. The main actors at the northern lakes are vendors from Cameroon based in Lamabaréné or Libreville, who often arrive with their basic necessities (rice, condiments, tobacco, etc.) and alcohol, which they sell or trade with the fishermen and women.

Overall, fishing is no longer sustainable. Fishermen and women accept this themselves, and more than 74% of respondents noted a significant reduction in fish stocks since the 1990s (Loundou *et al.*, 2013). The causes they cited were:

- free access to fishing grounds and unemployment, which have caused an increase in the number of fishermen and women,
- the use of banned or overly long nets (several kilometers),
- the introduction of destructive fishing techniques,
- oil dumping by boats belonging to oil companies.

Not only has the quantity of fish caught diminished, but the species being caught have changed also. The giant African threadfin, *Polydactylus quadrifilis,* which represented 60% of the tonnage caught in the 1960s (Loubens, 1966), now only represents 20% of the catch. The size of tilapia caught has diminished greatly, and a large portion of the catch now comprises "sans nom" *Heterotis niloticus*, introduced in the 1950s (page 231). Beyond this, "secondary" fishing is important: fishermen and women capture many slender-snouted crocodiles (OELO, 2012–2016).

In fact, fishing activities in the Lower Ogooué are caught in a vicious circle that will be destructive in the long run: the intensification and diversification of fishing activities has led to a reduction in fish stock, which, in turn, drives more intense fishing.

2.14. Fishing in one of the small lake deltas (Photo: OELO).

* Evaluating the scale of hunting activities through a socioeconomic survey is very difficult since many rural people know that they do not respect the law, and their answers are therefore largely biased. Only very indirect and informal approaches facilitate obtaining information on this very sensitive subject.

† Longlines baited with spoiled meat are used to catch crocodiles and softshell turtles. On several occasions, fishermen have been observed capturing 10 Aubry's flapshells at a time.

2.15. Residents of a nearby village burned the forest on the shores of Lake Evaro (Photo: Jean P. Vande weghe, October 2006).

2.16. Manatee hunting is prohibited but still openly practiced (Photo: OELO).

OELO by Heather Arrowood

The Organisation Ecotouristique du Lac Oguémoué (Ecotourism Organization of Lake Oguémoué, OELO) was created in November 2010 when a small group of Lake Oguémoué residents decided to protect their lake and the surrounding environment for future generations. The first OELO project was to build a site for ecotourism, Tsam Tsam, in order to create a revenue source for the local community through biodiversity protection. In four years, OELO's mission has expanded to encompass protecting biodiversity and the ecosystem services it offers, raising environmental awareness in local communities, encouraging youth to manage their environment, and offering help for inhabitants of the Lower Ogooué with the objective of generating revenue through sustainable use of environmental resources.

As for the introduction of exotic species, we know little about its real consequences, but it is probably part of this vicious circle and represents a last, destructive attempt to compensate for the disappearance of indigenous species.

Agriculture

Subsistence slash-and-burn agriculture is officially the second most important activity of the Lower Ogooué population. The variety of products cultivated is, however, limited to maniok, plantain bananas, sweet potatoes, and chillies. The fields are small and less than 3 km away from housing. Destruction of the fields, especially by elephants, is said to be the main reason for their small size (Latour, 2005; Loudou *et al.*, 2013).

With a few exceptions, almost all of the production is used for local consumption. However, along the Ogooué, women send a large portion of their production to Port-Gentil, where they can sell it for better prices. A boat regularly collects these products.

All together, there are few signs of present and past activity. Indications of ridge agriculture are an exception, although they are found in the savanna and this practice has not been used for centuries (highlight box, page 128). Many, if not most, households scattered throughout the Lower Ogooué complex do not have agricultural plots that are visible from the sky, besides a few oil palms and sometimes a few banana trees (Figures 2.1, 2.2, and 2.4). The socio-economic study cataloged the agricultural territories of the upper deltaic plain, but it is difficult to determine how this relates to reality.

Populations in the Great Lakes region and near Lake Déguélié burn large forest blocks (multiple hectares), apparently more as a means of marking their territory than for real agricultural needs (Figure 2.15). This is the only part of Gabon where this phenomenon has been observed.

Hunting for meat

Though very few inhabitants are willing to admit they hunt, this practice is very present throughout the Lower Ogooué. Cartridge shells are in fact more frequently found in forests than animal tracks.

In the past, hunting was done using more "traditional" means (harpoons, spears, and traps made of natural materials). Prior to independence, rifles were rare, and often only people of significant social standing owned them. Nevertheless, the Kellé, elephant-hunting specialists, had already moved large amounts of ivory to Cape Lopez Bay.

Today, hunting is omnipresent in the sense that people shoot at anything, anywhere, and in all seasons. Hunting laws are disregarded, and fully protected animals are freely sold on markets or even in the restaurants of Port-Gentil and

Lambaréné, especially water chevrotain (*Hyemoschus aquaticus*) and "caiman." However, since February 2015, the mayor of Lambaréné has taken measures to ban the sale of any protected species, though this does not prevent hunters from killing them.

Some fishermen and women combine hunting and fishing, and as previously mentioned, employees of logging company are said to indulge in commercial hunting.

Differentiating between subsistence and commercial hunting is, however, difficult. Strict commercial hunting certainly exists, and there are specialists who hunt exclusively for commerce, although selling game is more often a question of opportunity. Hunters need cash, and when there is an opportunity, they sell their catch–which is in fact often the only thing they can sell.

To estimate the true magnitude of hunting and the number of kills would be extremely difficult since many hunters know they are acting outside the law and hide their activities, especially since some regulatory measures have been taken.

Thus, the quantity of game for sale in Lambaréné markets (Table 2.1) does not necessarily reflect either hunting pressure or current animal populations. Game can be sent to other markets or sold surreptitiously. Sometimes, game sold in Lambaréné was hunted elsewhere in the country. No study has been conducted in Port-Gentil, even though it represents a much larger market than Lambaréné and offers game from the Wonga-Wongué Reserve, the area around the Nkomi Lagoon, and the edges of Loango National Park.

In short, a reduction in the amount of game being sold may be the result of reduced hunting just as much as it may be the result of reduced animal populations or sales shifting to clandestine circuits. Furthermore, when game populations become very low, many hunters abandon their trade, which is no longer profitable. Some actually admit to this and explain that the game has now "gone too far away." These same individuals also assert that they cannot practice agriculture because too many animals destroy their crops! An analysis of hunting comes in direct conflict with unspoken truths and clear contradictions.

Nevertheless, multiple observations indicate that hunting pressure is or has been extremely high in the near past:

(1) The amount of game for sale on the Lambaréné market alone was truly large in 2005 (Table 3.1), especially for slender-snouted crocodiles (locally referred to as "caiman" or "false gharial"), dwarf crocodiles, Aubry's flapshell turtles, guenons, water chevrotains, red forest hogs, sitatunga, and various duikers, especially when taking into account the low densities of observed individual in the field for these species (Latour, 2005). From 2012 to 2016, some numbers diminished considerably. Might there in fact be a real decrease in most animal populations? As mentioned above, it is difficult to know exactly what this decrease in numbers is caused by. Besides, it is also difficult to compare observations from 2005 with those from 2012 and 2016 because the control methods used might not have been the same.

Animals sold directly to restaurants must also be added to those sold in markets. Game being sold in restaurants does not involve the same variety of species, but the quantities observed are also very large, especially the slender-snouted crocodile, Aubry's flapshell turtle, brush-tailed porcupine, red forest hog, and duikers. The statistical drop, especially the disappearance of some entirely protected species, is in large part due to the 2015 decision by the mayor of Lambaréné to ban all sales of protected game (entirely or partially) within the city limits.

(2) While the numbers for Lambaréné are a cause for concern, the question remains regarding Port-Gentil. This city is not only larger than Lambaréné but also has inhabitants working in the oil industry with much more significant financial means.

(3) The density of large mammals throughout the zone of study is very low when compared to other regions of Gabon. The study conducted in the region of Lakes Onangué, Evaro, Ezanga, and Oguémoué by Latour (2005) shows that if some of the studied sections of forest have higher animal crossing levels than others, it is not rare to travel for kilometers without crossing any signs of animal presence. The various species of duikers are especially rare. Even the blue duiker (*Philantomba monticola*) and the red river hog (*Potamochoerus porcus*), which still have dense populations

Table 2.1. Number of animals found in the Isaac market in Lambaréné (columns 1–3) and in Lambaréné restaurants (columns 4 and 5), according to Louembet (2005) and OELO (2015, 2016). The first values indicate the number of animals observed, and the second, in parenthesis, the extrapolated number for one year. The lines in red indicate fully protected species.

	April 18 – June 4, 2005 Marché Isaac (Lambaréné) (Louembet, 2005)	March 23, 2012 – March 9, 2013 Marché Isaac (Lambaréné) (OELO, 2013)	September 18 – March 15, 2015 Marché Isaac (Lambaréné) (OELO, 2015)	Sept. 16, 2014 –March 14, 2015 Restaurants Lambaréné (OELO, 2015)	Sept. 4, 2015 – March 10, 2016 Restaurants Lambaréné (OELO, 2016)
Days of observation	57	27	53	74	11
Long-snouted crocodile	63 (393)	21 (283)	31 (213)	146 (720)	4 (133)
Dwarf crocodile	121 (755)	12 (162)	20 (138)	3 (15)	0
Nile crocodile	7 (43)	0	0	0	0
Aubry's softshell turtle	43 (268)	9 (121)	1 (7)	3 (15)	2 (66)
Forest monitor	11 (68)	0	0	0	0
Kinyxis erosa	29 (181)	0	0	0	0
Seba's python	39 (243)	2 (27)	1 (7)	0	0
Forest elephant	6 (37)	1 (13)	0	0	0
Manatee	0	9 (13)	0	0	0
Unidentified monkey	93 (580)	11 (149)	5 (34)	32 (158)	1 (33)
Collared mangabey	0	2 (27)	0	0	0
Black colobus	0	3 (40)	0	0	0
Forest porcupine	33 (206)	0	11 (76)	33 (163)	4 (132)
Hippopotamus	2 (12)	1 (13)	0	0	0
Water chevrotain	47 (293)	4 (40)	0	0	0
Red forest hog	109 (680)	16 (216)	34 (234)	0	6 (199)
Dwarf buffalo	6 (37)	3 (40)	0	4 (15)	0
Sitatunga	43 (268)	22 (297)	37 (255)	0	3 (99)
Unidentified duiker	69 (430)	10 (135)	7 (48)	18 (89)	1 (33)
African civet	13 (81)	0	0	0	0
Total	734 (4,584)	125 (1,690)	147 (1,012)	440 (2,170)	21 (697)

2.17 and 2.18. Protected animals on sale at the bushmeat market in Lambaréné, including slender-snouted crocodiles, which have not been prohibited from sale since February 2015 (Photos: OELO).

elsewhere in Gabon, are very rare in the Great Lakes region. Small primates and forest elephants are also less numerous than in the rest of Gabon, except for the region found east of Lake Oguémoué. This same study shows that elephant tracks increase at distances greater than 15 km from Ogooué villages (Latour, 2005), which indicates that, regardless of what all the villagers assert during studies, hunting is truly taking place. With regard to elephants, only Salonga National Park in DRC, which has seen excessive poaching for years, has an elephant density lower than that of the Ogooué Great Lakes region (Latour, 2005).

2.19. Damage caused by logging is not limited to the construction of tracks. Damage caused to the forest cover is much less spectacular but more insidious. These tracks not only open the canopy, they also affect the undergrowth. Furthermore, these activities are very selective and among the exploited species, many do not regenerate or do so with great difficulty. In the long term, logging activities therefore impoverish the forest. Finally, logging is inevitably accompanied by hunting, which is not selective at all. In some concessions, hunting in fact compensates for the very low wages of the workforce (Photo: Ehoarn Bidault).

(4) Large birds are generally very shy. The pink-backed pelican (*Pelecanus rufescens*), the African fish eagle (*Haliaeetus vocifer*), and the western osprey (*Pandion haliaetus*) all fly away when approached from afar. Multiple bird species have nevertheless nearly or entirely disappeared in the last few years: the saddle-billed stork (*Ephippiorhynchus senegalensis*) hasn't been seen since 2006 (page 284), and Hartlaub's duck (*Pteronetta hartlaubii*) has become very rare (page 280). Colonies of herons and egrets (Ardeidae) in the Lambaréné region have lost 50–75% of their numbers since 1999 without changes to their natural habitat. The population density of large aquatic birds, notably the African fish eagle, is also extremely low. Even the grey-rumped swallow (*Pseudhirundo griseopyga*), which nests on sand banks, has diminished considerably, probably as a result of children gathering their eggs and progeny (Vande weghe & Christy, 2015; Christy & Vande weghe, 2017).

(5) Forest hogs, sitatunga, and elephants are exclusively nocturnal, which is also proof of hunting pressure.

(6) Finally, a March 2017 aerial survey has shown that in areas that are difficult to access for hunters or that simply cannot be reached, and which biologists cannot reach either, the population density of large mammals is still relatively high (pages 74, 80, and 81). Numerous tracks through the vegetation confirm this.

Besides village hunting, city inhabitants sometimes hunt for pleasure. At the beginning of 2016, pink-backed pelicans were massacred in Nazaré Bay by inhabitants of Port-Gentil during a weekend outing. Sometimes these hunting activities are organized to obtain meat for a wedding feast. The most sought-after prey are hippopotamus and manatee.

Ivory hunting

Hunting for ivory is a very old activity in the region and was previously very significant. Based on the quantities of ivory exported from Gabon in the 1860s, Du Chaillu (1868) feared at the time for the survival of elephants. In the last quarter of the 20th century, this practice was mentioned by Wilks (1990) in the report for a study conducted by Conoco Onango. Today it still exists, but is even harder to evaluate than game hunting. A few years ago (Pradel, pers. comm.) a significant number of elephants were killed inside the Wonga-Wongué Reserve by a band of hunters from around Gomé Lake.

Logging

Traces of logging activity, past or present, are omnipresent in the Lower Ogooué complex (Figure 3.19), and a large part of the forests are licensed for exploitation (Figure 3.20). This region was in fact one of the first to be exploited in Gabon, starting at the end of the 19th century, due to the proximity of navigable rivers, which made transporting logs easy (Pourtier, 1989; Vande weghe, 2012).

The entire Great Lakes area shows traces of past and present logging activities. Abandoned machinery (bulldozers, tractors) can be found in forests, and there are probably no forested areas situated on dry land without traces of logging exploitation, old or new, with the exception of a few extremely small areas, such as the hills rising above the alluvial plain. In contrast, very few flood forests or swamps have been impacted by logging.

The forest cover is more or less all secondary vegetation, and most of the tall okoumé (*Aucoumea klaineana*) trees were felled long ago. In truth, few tree species are affected by this threat, except for *Diospyros subargentea* and *Diospyros cleistantha*, but many species of grasses and shrubs from the undergrowth, epiphytes and lianas, are indirectly threatened. Although clear-cut logging is not common in Gabon, when a tree is felled for its wood, the resulting disturbance has an impact on surrounding species, and skidding trails or other large trails are cleared to extract the logs. As a result, populations of rare species can be indirectly affected, notably *Palisota albo-anthera, Drypetes verrucosa*, and *Oncoba ngounyensis*.

The opening of forest tracks can have a local impact on the valley bottoms they cross by changing their water regime or by lifting soil that will then be washed away toward rivers. These indirect effects may have local consequences on populations of rare species specific to these environments, which are usually unimpacted by logging activities themselves, such as *Veyretella hetaerioides, Salacia diplasia,* and *Scaphopetalum mannii* (Bidault *et al.*, 2017).

Skidding trails and roads, even old ones, facilitate human access to deeper parts of the forest as well as the transport of hunted animals over long distances. For the purposes of a protected area, it would thus be preferable to monitor these access routes, to exclude industrial logging activities, and to limit and regulate small "family" logging activities, which are often equipped with vehicles or motorized tools that are not legally permitted under "household logging permits" (Latour, 2005).

Industrial agriculture

In the region north of Lambaréné there are plantations of rubber (*Hevea brasiliensis*) and oil palm (*Elaeis guineensis*). However, they cover a very small area, accounting for less than 1% of the entire complex. Their establishment requires clearing forests, but the main danger they pose resides in the use of pesticides, which inevitably end up flowing into the Ogooué system.

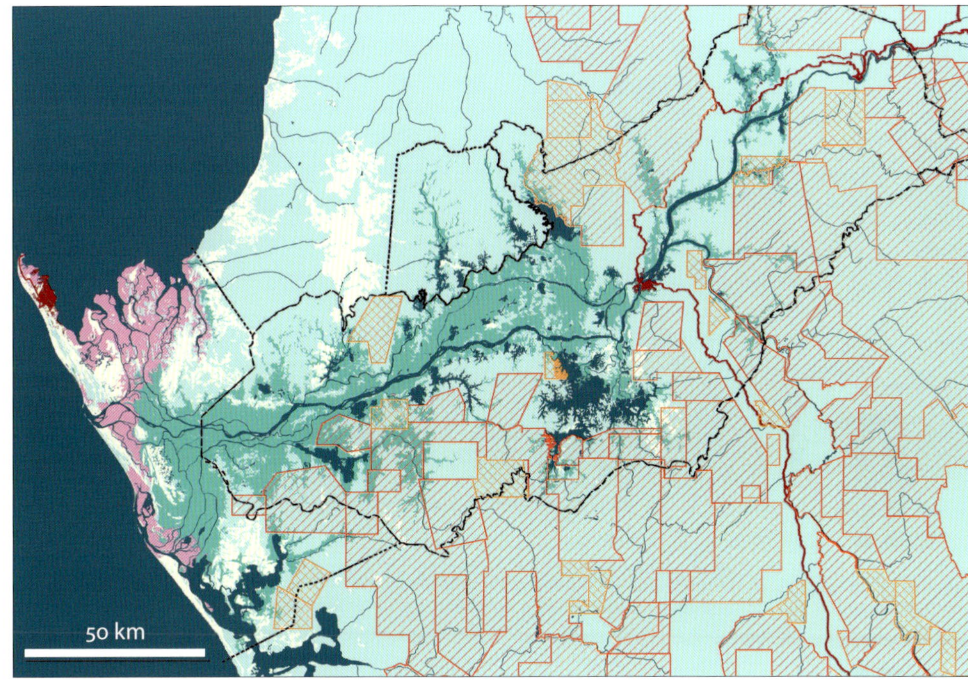

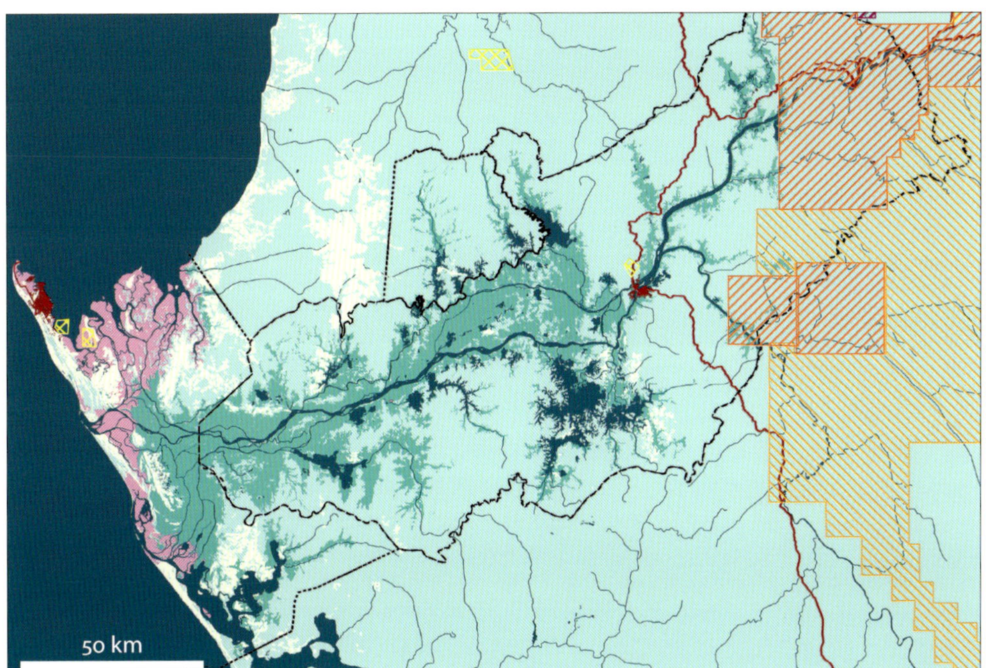

2.20. Most of the forests of the Lower Ogooué region are used for industrial logging.

Permits without proper development

Provisional agreements for development, exploitation and transformation

Logging concession managed for sustainable management

2.21. The far eastern part of the current Ramsar site is largely allocated to mining activities. In fact, these activities are almost exclusive to areas outside of the coastal sedimentary basin.

Mining exploration permits

Concessions and operating permits

Mining prospecting authorization

Quarrying permits

2.22. Industrial crops east of Lambaréné (Photo: Jean P. Vande weghe).

The oil industry

This represents a major economic factor in the Lower Ogooué, and is concentrated around Port-Gentil and Lake Ezanga. The damage caused by this industry is varied and can be long lasting. The seismological tracks that disfigure the landscape surrounding Port-Gentil and part of the northern delta are still visible over 30 years after having been cleared. The most important direct threats this creates are oil pollution, toxic waste* pollution, and gas flares. Oil pollution is caused by leakage, which can affect large areas and multiple cases of which have been reported at Lake Ezanga. Gas flares are extremely deadly for fauna. Although oil companies forbid access to these sites, employees have reported burnt insects at the foot of some gas flares, forming piles dozens of centimeters thick that also include a few dead birds and bats.

The oil industry also threatens rare and significant plant species, such as *Diospyros subargentea* in dryland forests, as well as numerous species restricted to coastal habitats present notably on Mandji Island, such as *Brachystelma letestui, Plectranthus* sp. nov., *Vangueriella dowsettiae* sp. nov., *Simirestis klaineana*, and *Guyonia arenaria* (Bidault *et al.*, 2017).

Finally, the oil industry attracts a large external human population to meet its labor needs, which results in an unavoidable increase in hunting and fishing.

* Highly toxic products generated by boreholes can be released into waterways. An instance of this was observed in the southwestern part of the Wonga-Wongué Reserve, and the products were discharged into the Gangwé River.

The mining industry

The mining industry is not well developed in the Lower Ogooué, but there are nevertheless mining projects that also pose threats (Figure 2.21). Thus, in Mabounié, rare species from dryland forests and flooded forests at the bottom of valleys and alongside rivers are threatened by mining infrastructure and trails cleared for their use. Furthermore, some species are only known to exist in the Mabounié mining concession, and as a result are particularly vulnerable and exposed to any activity that might take place within it. These plants include, among others, *Sabicea golgothae* and *Sabicea mabouniensis* (Bidault *et al.*, 2017).

Road construction

Aside from the national road connecting Libreville, Lambaréné, and Mouila and a few access routes for industrial plantations or logging concessions, the Ogooué remains nearly free of roads. Toward the end of the colonial period, there was a project to build a road to connect Port-Gentil to Libreville through the Animba Plain, the northern delta, and the Wonga-Wongué Reserve. However, it was never built. In 2014, a project to build a road connecting Port-Gentil to Omboué and then Yombi was begun, and is currently being finished (Figure 2.23). Ultimately, a 93 km long road will connect Port-Gentil to Omboué.

The path this road takes traverses various Ogooué estuaries (central and southern delta) and numerous unique habitats present on Mandji Island and Ozouri Island, which contain many rare and endangered plant species, notably *Brachystelma letestui, Vangueriella dowsettiae* sp. nov., *Guyonia arenaria*, and *Simirestis klaineana* (pages 152–154). This road is a major threat to the flora of this part of the Lower Ogooué, above and beyond its direct footprint, as it will encourage populations to settle all along its Port-Gentil-Omboué section. As a result, it will contribute to the transformation of forests into secondary growth in areas along its path, which contain habitats unique to Gabon and are already heavily degraded on Mandji Island but relatively well preserved on Ozouri Island. Furthermore, this road traverses vast wetlands, yet the roadwork has led to the construction of dikes and other artificial impoundment structures that disturb the water regime of the wetlands and, as a result, have adverse effects on the fauna and flora they contain. A few years ago, a forest road built south of Lake Oguémoué blocked the lake's outlet (page 25). Measures must be taken regarding fauna to avoid increased hunting. Yet even before the completion of the Port-Gentil-Omboué road, game from the edges of Loango National Park has already been exported towards Port-Gentil.

Of course, these roads help reconnect isolated populations, but their construction should be accompanied by impact studies. In truth, however, no such studies are done or they are of poor quality. Financed by those who will conduct the roadwork, they are often biased from the start. Finally, during the construction phase, environmental rules are followed as little as possible. Soil that must be cleared is often simply pushed into the forest or savanna, blocking creeks, quarries are left open, and oil leakage is frequent.

River transportation

Apart from pollution, river transportation does not seem to have adverse effects (Figure 2.25), but it is possible that manatees may be hurt by it. Furthermore, such transportation can interfere with fishing.

Cities

Urban areas result in significant pollution since most of the sewage ends up entering the aquatic ecosystem or in aquifers.

2.23. The bridge being constructed on the Ogooué; it connects Mandji Island to Ozouri Island (Photo: Ehoarn Bidault).

2.24. The port and industrial sections of Port-Gentil (Photo: Jean P. Vande weghe).

2.25. River transport is quite substantial on some of the river's channels (Photo: Jean P. Vande weghe).

This includes detergents, a wide variety of toxic chemicals, plastic bags, and used oils. The outskirts of Port-Gentil are notable for their piles of trash, wrecked cars, and more. Unplanned urbanization and "uncontrolled" construction disfigure landscapes and destroy important habitats (Figures 2.24 and 2.25). For example, the type population of *Brachystelma letestui* (page 150), found in Cape Lopez Bay, was destroyed by the growth of the Port-Gentil urban area since, regardless of our efforts, we were unable to relocate it.

Exotic invasive plant species

At a global scale, and especially in tropical regions, invasive exotic species represent a serious threat to both flora and fauna. In the Lower Ogooué, the most abundant invasive exotic species are *Pistia stratiotes, Chromolaena odorata, Scoparia dulcis, Stachytarpheta jamaicensis, Mimosa pudica*, and *Ageratum conyzoides*. They do not impact all the flora in the same way, nor with the same intensity. *Ageratum conyzoides, Mimosa pudica, Stachytarpheta jamaicensis,* and *Scoparia dulcis* were introduced a long time ago and are now widespread throughout Africa; they are mostly found in degraded areas and along trails, where they replace native species. As such, they threaten two species native to the Lower Ogooué that inhabit roadsides directly (*Sabicea mabouniensis* and *Sabicea golgothae*), but they are rarely found in natural habitats.

Other invasive exotic species are much more virulent and represent major threats for both flora and fauna, in particular *Pistia stratiotes*, which is found in large numbers in the wetlands of the Lower Ogooué, where its rapid growth enables it to colonize vast stretches of water. Although few significant species exist only in submerged habitats, *Pistia stratiotes* can cover wide areas and is thus a real threat to this flora.* Finally, *Chromolaena odorata* is a large, partly woody grass that grows rapidly and can colonize large areas of dry land, to the detriment of native species, to the point of forming large, monospecific populations. This very aggressive species is a great threat for the flora in general, including significant species.

Some invasive species were introduced intentionally. This is the case of the weeping paperbark (*Melaleuca leucadendra*), a tree belonging to the family Myrtaceae, which originates from Australia and Indonesia. It was introduced to Mandji Island after World War II as a means of drying out the wetlands, and it is now invading the back mangroves.

* The beautiful but sinister water hyacinth, *Eichhornia crassipes*—whose ecology is similar to that of *Pistia stratiotes*—is present in Gabon, especially in certain gardens in Libreville, but it has not yet been observed in the Ogooué Delta. If it were to spread to this region, the damage could be colossal. However, as is most often the case with invasive plants, as long as no damage has been caused, nothing will be done. Once the damage becomes noticeable, nothing can be done other than at great cost.

Climate change

Situated at low altitude and reaching the Atlantic Ocean, the Ogooué Delta forms a network of wetland habitats with vegetation that is heavily dependent on salinity (very locally), water quality, and water depth. Climate change, through rising sea levels, represents a serious threat to the water balance of the entire Ogooué Delta. If this balance were to be disturbed, its fauna and flora would be particularly impacted, starting with the coastal habitats and those connected to rivers.

Already, the Nazaré Bay mudflats seem to be disappearing, which negatively affects palearctic wader populations.

As for vegetation, the littoral forests with *Fegimanra africana*, where significant species are found, such as *Vangueriella dowsettiae* sp. nov. and *Plectranthus* sp. nov., could be reached by rising water levels. The white sand littoral savanna, where the rare and endangered *Brachystelma letestui* occurs exclusively, would also be heavily affected.

Along the river, it is the flood forests and levees with *Alchornea*—which border the river inland—that could be threatened by changes in water salinity caused by rising sea levels. As a result, significant species such as *Cissus leemansii* could be heavily impacted.

Future threats

Beyond these current and very real problems, there are potential threats that might bring even greater harm to the deltaic complex.

In the past, several projects were proposed to build a dam across the Middle Ogooué, any of which could one day be reactivated. As part of mining projects, specifically the Maboumine project, partial deviation of the Ogooué using canals has been proposed as a way to allow ships to reach the mining site upstream from Lambaréné throughout the year. This project now seems to have been abandoned, but it could become active again at any moment. It is highly probable that such a "deviation" would affect the entire complex. The smallest of the three African deltas would thus be destroyed and could end up being cited in school books as yet another victory of Man over Nature.

Acknowledgments

The authors thank Heather Arrowood who, as part of the OELO association, allowed the use of certain photographs. They also thank Richard Oslisly, who expanded this chapter with a highlight box about archaeological findings regarding raised bed agriculture.

Bibliography

BIDAULT, E, O. LACHENAUD & T. STÉVART, 2017. La flore. Pp. 143-171 in T. STÉVART & J.P. VANDE WEGHE (éds.). *Le delta de l'Ogooué*. ANPN, Libreville.

CHRISTY, P. & J.P. VANDE WEGHE, 2017. Les oiseaux. Pp. 267-303 in: T. STÉVART & J.P. VANDE WEGHE (éds.). *Le delta de l'Ogooué*. ANPN, Libreville.

CLIST, B., 1995. *Gabon; 100000 ans d'Histoire*. Centre Culturel Français Saint Exupéry - Sépia, Libreville.

DAPPER, O., 1667. *Description de l'Afrique*. Amsterdam.

DE MAREES, P., 1602. *Description et récit historique du riche royaume d'or de Guinée*. Amsterdam.

DU CHAILLU, P.B., (1863) 1997. *Voyages et aventures en Afrique Equatoriale*. Sépia-Centre Culturel Français Saint-Exupéry, Libreville.

DU CHAILLU, P.B., (1868) 2002. *L'Afrique Sauvage*. Ed. du Luto, Libreville.

FERMON, Y., L. CHIRIO, A. LAMBOJ, J.-F. LIWOUWOU & J. SULLIVAN, 2017. Les poissons. Pp. 223-249 in T. STÉVART & J.P. VANDE WEGHE (éds.). *Le delta de l'Ogooué*. ANPN, Libreville.

LATOUR, S., 2005. *Inventaire de faune autour des lacs du delta de l'Ogooué*. WCS, Libreville.

LE DUC YENO, S., 2015. *Atlas cartographique du Site Ramsar du Bas Ogooué*. MBG-WCS-AMAPIRD-WWF, Libreville.

LOUBENS, G., 1966. *Biologie de* Polydactylus quadrifilis *dans le bas Ogooué (Gabon). Peuplement en poisson des eaux saumâtres et son origine.* Thèses de la faculté des Sciences de l'université de Paris.

LOUNDOU, P., M. STARKEY, W. MBOMBE TSONGUE & B. BIMBISSA MINANGA, 2013. *Evaluation socio-économique du complexe Evaro- Wonga-Wongué*. WCS, Libreville.

LOUEMBET, MBOUMBA S., 2005. *Approche d'étude pour une connaissance de l'exploitation d'un mammifère aquatique: le lamantin* (Trichechus senegalensis) *dans la région des lacs du sud de l'Ogooué*. ENEF-WWF, Libreville.

LOUEMBET, MBOUMBA S., 2006. *Enquête d'évaluation de la chasse aux lamantins et de la grand chasse, pour une stratégie de conservation au massif forestier Ebel Abanga*. WWF, Libreville.

MERLET, A., 1990. *Légendes et histoire des Myéné de l'Ogooué*. Centre Culturel Français Saint-Exupéry-Sépia. Libreville et Paris.

OELO, 2012-2016. *Rapports de la vente illégale de viande de brousse*. Libreville.

POURTIER, R., 1989. *Le Gabon. Tome I. Espace, Histoire, Société*. L'Harmattan, Paris.

RAPONDA-WALKER, A., 1960. *Notes d'Histoire du Gabon*. Brazzaville.

RAPONDA-WALKER, A. & R. SILLANS, 1995. *Les plantes utiles du Gabon*. Fondation Raponda-Walker-Sépia-Centre Culturel Français Saint Exupéry, Libreville.

VAN DEN BROECKE, P., 1705. *Voyages de Van Den Broecke au Cap vert, en Angola et aux Indes Orientales en 1605*. Amsterdam.

VANDE WEGHE, J.P., , 2007. *Loango, Mayumba et le bas Ogooué*. ANPN. Libreville.

VANDE WEGHE, J.P., 2012. *Wonga Wongué*. ANPN, Libreville.

VANDE WEGHE, J.P. & P. CHRISTY, 2015. *Les oiseaux du complexe de milieux humides du bas Ogooué*. Rapport inédit. WWF, Libreville.

VAN LINSCHOTEN, J.H., (1594) 1619. *Histoire de la navigation de Jean Hughes de Linschot Hollandois*. Amsterdam.

WILKS, C., 1990. *La Conservation des Ecosystèmes forestiers du Gabon*. UICN, Gland.

2.26. There are many uncontrolled constructions on Mandji Island. Some even build in floodable zones (Photo: Jean P. Vande weghe).

2.27. There are also many unauthorized dumps around Port-Gentil (Photo: Jean P. Vande weghe).

3. The flora

Ehoarn Bidault, Olivier Lachenaud & Tariq Stévart

History of botanical collecting

More than 10,600 georeferenced botanical specimens from the Lower Ogooué are currently contained in various databases. These specimens were collected during four main periods: 1870 to 1950, 1951 to 1990, 1991 to 2005, and 2006 to 2016.

1870–1950

The history of botanical exploration in the Ramsar site region of the Lower Ogooué reaches back to the first European exploration of Gabon. Explorers first approached Gabon from the coast, preferring to settle in sheltered areas such as bays and the mouths of rivers. Henri Hua (1900), Auguste Chevalier (1917), and Raponda-Walker & Sillans (1961) provide detailed accounts of the various botanical and nonbotanical expeditions that were conducted in Gabon at this time.

Aside from a few specimens collected by Griffon du Bellay in 1874, no botanical collections were made in the Middle Ogooué until François-Romain Thollon, between 1883 and 1895 (Chevalier, 1917). From 1891 to 1892, J. Dybowski and C. Chalot brought back 80 specimens collected in various parts of Gabon, and especially in the Lower Ogooué. Other explorers passed through the region, such as M.A. Jolly, who navigated up the Ogooué up to Ndjolé in 1891, but most of their collections were made in the region of Libreville.

Then Breton Francis Fleury (1882–1919), then secretary of the permanent office for the Ministry of Colonies, collected under Auguste Chevalier's numbers. In all, he collected 1,059 specimens in Gabon (from numbers 26,104 to 27,163). While Chevalier was visiting the banks of the Komo and Rembоué rivers, which are both outside of the Ramsar site, Fleury visited sites situated between Cape Lopez Bay and Lambaréné, as well as all the way to Fernan Vaz. Notably, he visited the Ogooué Great Lakes region. As a result, he is responsible for the discovery of *Synsepalum fleuryanum* and *Chrysophyllum ogoouense*, both endemic to the Ogooué River basin.

During this period, 589 collections—now georeferenced—were made by the first European explorers in Gabon (Figure 3.2). These collections come from across the entire Lower Ogooué, and especially near Lake Ezanga, Port-Gentil, Lambaréné, and between Lambaréné and Ndjolé.

It should also be noted that while the Lower Ogooué provided the main access route inland for explorers, it was mostly ignored by the first botanists who visited Gabon. There was some collecting done in the region, but it pales compared to the efforts of numerous botanists who worked in the Libreville region, where Father Klaine alone made over 3,500 collections. Furthermore, the most prolific botanist to have collected in Gabon, Georges Le Testu, never visited the lower section of the Ogooué and only collected a few

3.1. Ctenolophon *forest (Photo: Ehoarn Bidault).*

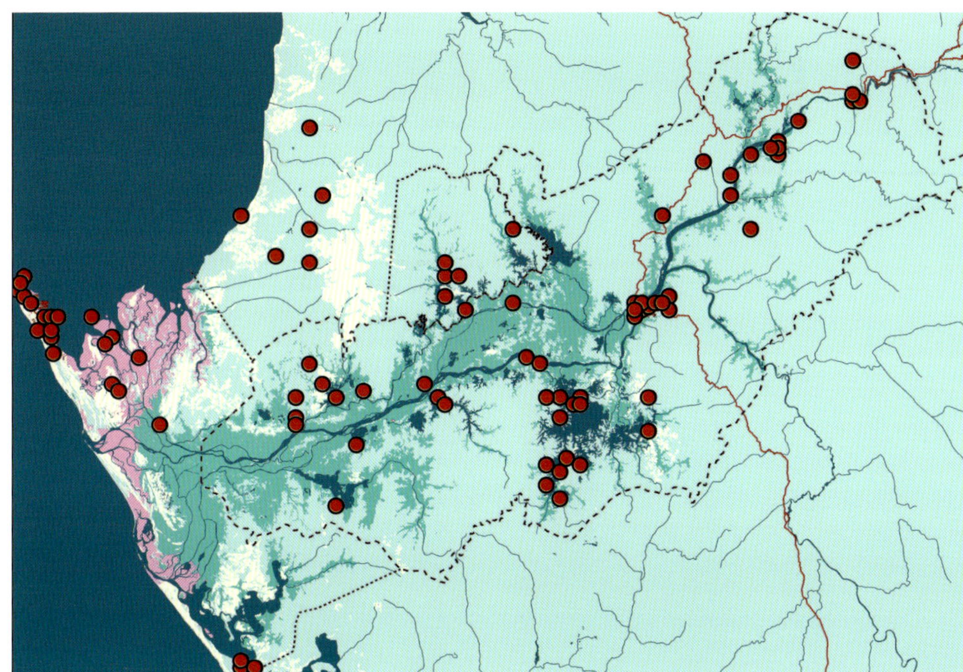

3.2. Distribution of botanical collections from the period of 1870–1950.

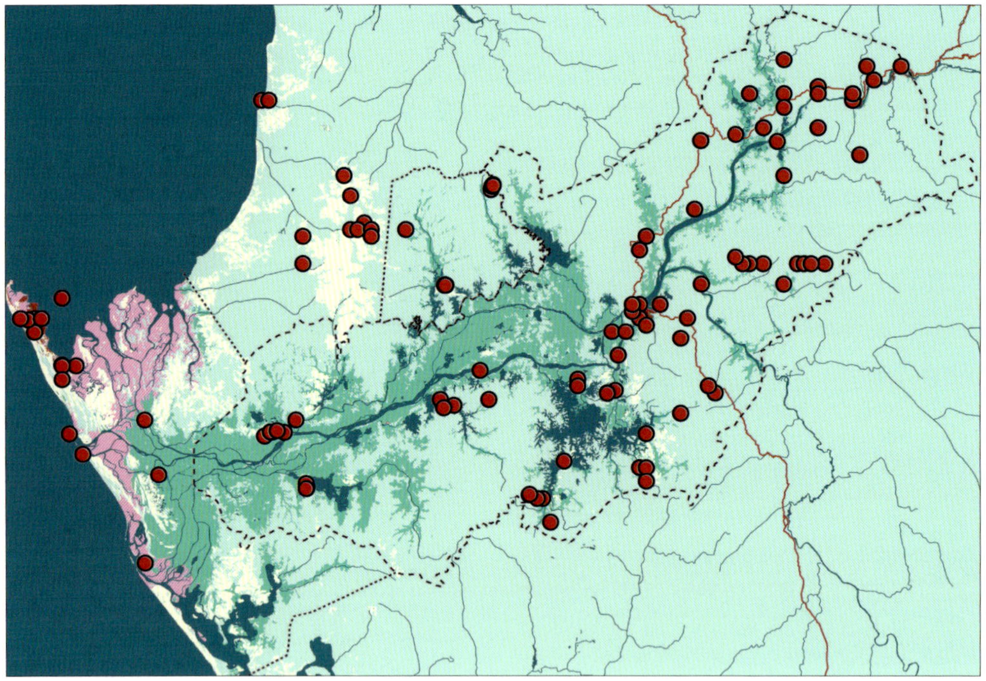

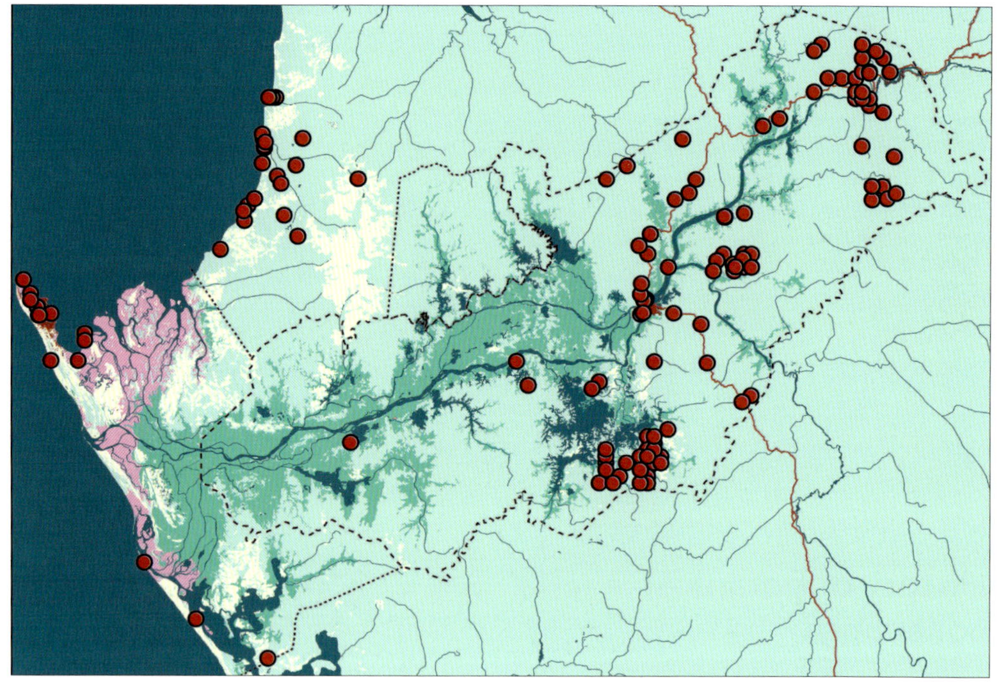

3.3. Distribution of botanical collections for the period of 1951–1990.

3.4. Distribution of botanical collections for the period of 1991–2005.

specimens around Port-Gentil (Raynal, 1968). The resulting lack of data can be attributed to the inaccessibility of the area, aside from the easily navigable river.

1951–1990

Throughout the second half of the 20th century, until 1990, 1,893 georeferenced botanical collections were made in the Lower Ogooué (Figure 3.3). During this period, Nicolas Hallé, who was especially active in Gabon as both a collector and as coordinator of the Flora of Gabon series, made over 900 collections in the Ogooué region during the first half of the 1960s. At the end of the decade, Frans Berteler, M. van Raalte, J.J.F.E. de Wilde, and A.J.M. Leeuwenberg were the first Dutch botanists to explore Lower Ogooué, later joined by the French botanist J.J. Floret. These inventories focused mainly on the dry land forests of the Lower Ogooué valleys between Lambaréné and Ndjolé, as well as the area around Port-Gentil. Some collections were also made south of Lakes Onangué and Ezanga, and, closer to the river, not far from Lakes Ayanga, Anengué and Ompindi.

1991–2005

During this 15-year span, 1,828 geo-referenced collections were made in the Lower Ogooué. This corresponds to a period of intensive collection by Dutch botanists from the Wageningen University herbarium, including F.L. van Nek, A.M. Louis, and J.J. Wieringa. Others to collect in the area include C. Wilks and G. McPherson, who were the first to collect in Gabon for the Missouri Botanical Garden (MBG) since A. Gentry, who did not work in the Lower Ogooué. During the period, Yves Issembé was the first Gabonese botanist to collect plants systematically in the Lower Ogooué. This work was done mostly in the dryland forests of the Lower Ogooué valleys, around Ndjolé, as well as east of Lambaréné and around Lake Ezanga (Figure 3.4).

The entire lower course of the river was, however, largely neglected during this period, and sampling for the entire area is still very low, which remained true until very recently, as described by Nicoll & Langrand (1986), Wilks (1990), and Doumenge *et al.* (2003a & 2003b). Although these authors emphasize the region's great potential for conservation, due to the diversity of habitats it presents, none of them provides a floristic list to support their claims, nor do they make a case for meaningful botanical exploration efforts in the area. The collections map published in 2005 in the Plants of Gabon Checklist (Sosef *et al.* 2005) shows a notable absence of collections from the Lower Ogooué, between Lambaréné and Port-Gentil. This is partly explained by difficult field conditions, which orient most of botanists toward more accessible dryland forests.

2006–2016

Starting in 2006, the MBG began working in the Lower Ogooué. Their activities began to expand in scope in 2008 with the work of A. Boupoya near Lake Evaro. Starting in 2012, extensive inventories were conducted in the Mabounié area, southeast of Lambaréné, at the edge of the Ramsar site. Over a period of two years, more than 2,500 collections were made in this area, representing more than 1,000 taxa.

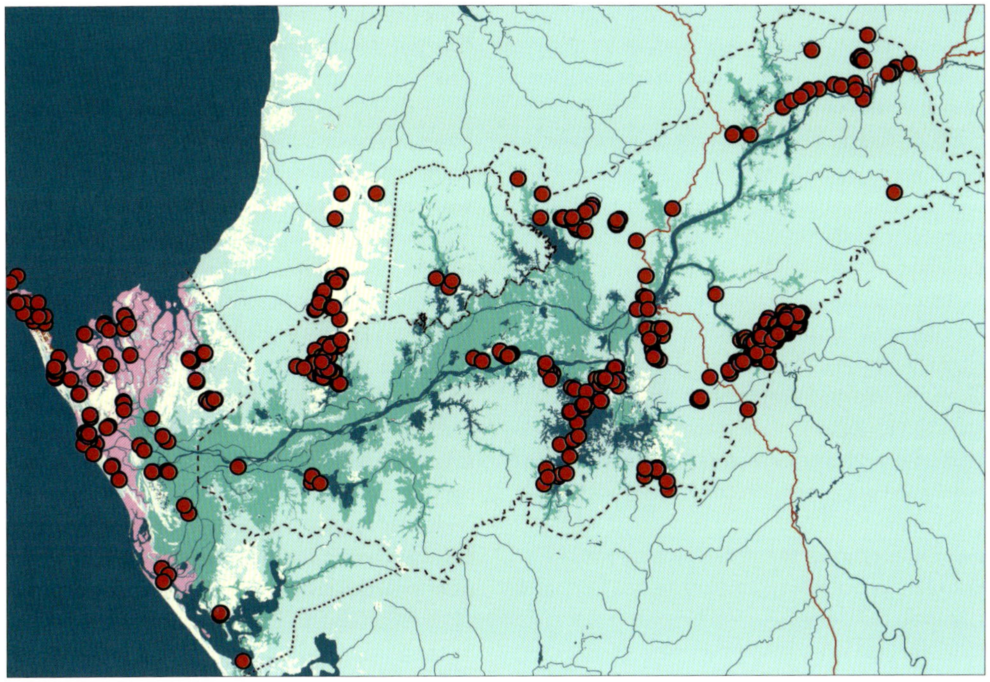

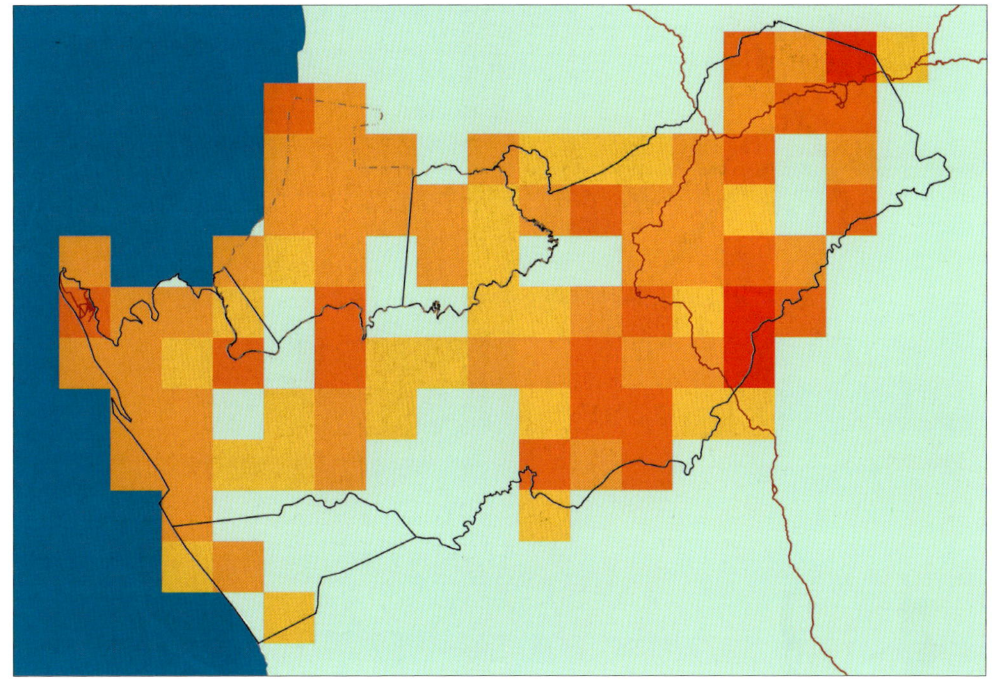

Between 2013 and 2016, major collecting expeditions were undertaken around Lakes Onangué, Ezanga, Azingo and Alombié. For the first time, the area's wetlands, marshes, and swamps were systematically inventoried. In 2015 and 2016, the lower part of the Ogooué Delta was catalogued by the MBG in order to overcome the lack of botanical data. Some inventory work was also carried out in the Fernan Vaz Lagoon as well as the Lower Ogooué valley. Throughout this 10-year period, 6,310 botanical collections were made, representing more than half of the known georeferenced specimens from the Lower Ogooué dating back to the start of botanical exploration (Figure 3.5).

Despite these intensive inventories carried out over the past 10 years, many areas in the Lower Ogooué remain devoid of botanical collections. Figure 3.6 shows that the sampling effort within the Ramsar site has been very heterogeneous. Dryland habitats have been the most explored, as demonstrated by the three grid cells in which more than 500 collections were made, near Mabounié, southeast of Lambaréné, and in the vicinity of Ndjolé. The flora of these areas is now considered to be well documented.

Finally, the flora of large sections of Lower Ogooué has been insufficiently explored, as shown by the grid cells where fewer than 100 collections have been made. These large empty sections of the map include the entire region separating Lake Ezanga and Lake Alombié, which comprises large marshes and flooded habitats making it difficult to access, even by boat. However, the northeastern part of the lagoon is also markedly devoid of collections, even though it contains large sections of dryland forest.

This situation should be placed in perspective with the species accumulation curve (Figure 3.7), which does not currently show signs of decreasing: each new expedition in the area continues to yield

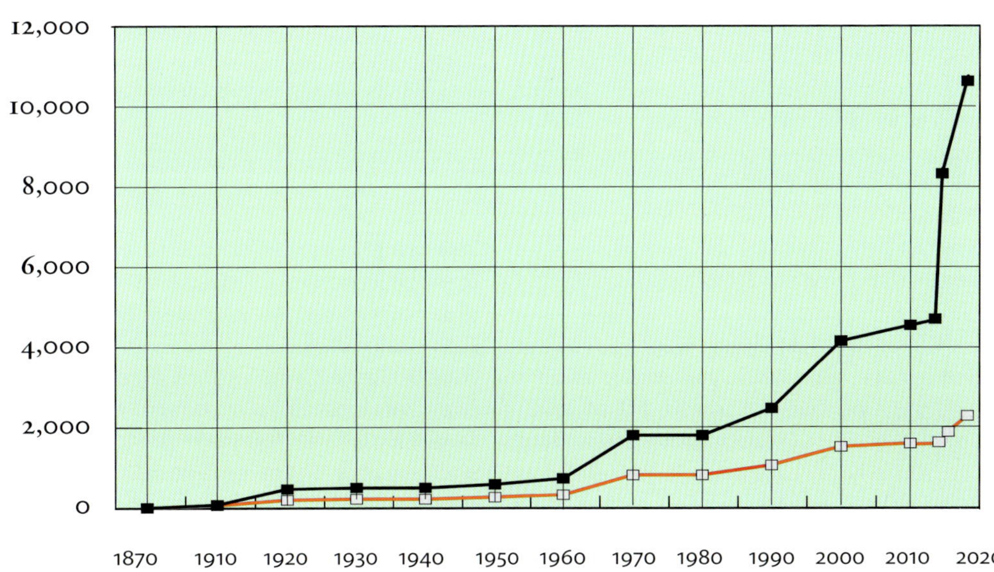

3.5. Distribution of botanical collections for the period of 2006–2016.

3.6. Number of botanical collections per 15 square km.

- *1–10*
- *11–100*
- *101–500*
- *501–2,701*

3.7. Curves representing the number of collections (▬) and taxa (▬) accumulated between 1870 and 2016.

new insights regarding the flora. Botanical exploration in Lower Ogooué, which has been on the rise recently, thus still has a lot to offer.

Table 3.1. The 10 families with the most taxa (species and taxa at lower ranks, as well as genera) in the current Lower Ogooué Ramsar site, as well as within the limits of the proposed Ramsar site

	Current Ramsar site		Proposed Ramsar site	
Family	Number of genera	Number of taxa	Number of genera	Number of taxa
Rubiaceae	65	287	66	302
Fabaceae	91	184	98	203
Cyperaceae	7	53	10	93
Orchidaceae	23	57	26	66
Apocynaceae	27	60	27	64
Annonaceae	21	59	22	63
Poaceae	27	36	37	58
Euphorbiaceae	34	47	35	56
Acanthaceae	22	40	23	41
Malvaceae	22	37	23	30
Other families	418	1,016	449	1,107
TOTAL	757	1,876	816	2,092

The flora of the Ramsar site and its extensions

To date, the flora of the Lower Ogooué Ramsar site—as defined by the latest proposed outlines—is estimated to comprise 2,092 taxa (2,174 when the Wonga-Wongué Ramsar site is also included), belonging to 816 genera and 180 families. The flora of Gabon is currently estimated at 6,000 taxa, which means the Lower Ogooué has 35%—more than a third—of the country's known flora. The Lower Ogooué Ramsar site alone—as currently defined, without the proposed extensions—has 1,876 of these taxa (790 genera and 165 families) (Table 3.1).

Within the Port-Gentil extension of the Ramsar site, 380 taxa (in 288 genera and 112 families) have been observed, 193 of which are not known to exist in the current Ramsar site. Similarly, the Lake Gomé extension hosts 85 taxa (in 76 genera and 34 families), 16 of which are not currently known to exist in the Ramsar site, and the Fernan Vaz extension includes approximately 50 taxa (in 47 genera and 28 families), 20 of which have not been found in the Ramsar site.

It is worth mentioning that, among the three Ramsar site extensions, the one at Port-Gentil has the largest number of plant taxa not found elsewhere in Lower Ogooué (180, or almost 50% of the area's known flora), at least partly because the area contains a number of unique habitats, which do not have equivalents elsewhere within Lower Ogooué.

The Wonga-Wongué Ramsar site, as defined by its current borders, currently hosts 358 taxa in 272 genera and 97 families, 80 of which are not present elsewhere in the Lower Ogooué.

Table 1 presents the 10 families with most taxa within both the Ramsar site's current and planned borders.

Due to its substantial size (1,967,000 ha) and habitat diversity, the flora of Lower Ogooué is very diverse. Among the 10 families with most taxa, both within its current and planned borders, Rubiaceae and Fabaceae are the two largest. Both have, on average, two to three times more taxa than any of the other families in the list. This coincides with the situation observed throughout Gabon and is a result of the fact that both families are present in all types of environments. Fabaceae include many trees and shrubs found in dryland forests, but also many herbaceous taxa and vines present in marshes and savanna. Meanwhile, Rubiaceae include few large trees but encompasses a significant portion of the shrubs, suffrutex, and herbaceous undergrowth; Rubiaceae also include typical savanna herbs.

The other larger families in the list are also widely represented in forests, in the form of medium-sized shrubs and trees (Euphorbiaceae, Annonaceae, Malvaceae, Apocynaceae), herbaceous plants (Acanthaceae), vines (Apocynaceae), and even epiphytes (Orchidaceae). These families are also found in the grasslands and marshes of Lower Ogooué, although represented by different genera and species than those found in forests. However, except for orchids, they are less diverse in these environments than in forests.

Two large families in Lower Ogooué, Poaceae and Cyperaceae, are less diverse within the current Ramsar borders than they would be under the proposed extensions. The new borders encompass the maritime part of the delta, which surrounds Port-Gentil and in which there is much more savanna than in other parts of Lower Ogooué. Furthermore, the flora of these coastal savannas is not found elsewhere in the Ramsar site; many of the species found in these areas are in fact restricted to this habitat. Poaceae and Cyperaceae thus contribute more to species diversity within Lower Ogooué when the new limits are considered.

The flora of structural units

The flora of the five structural zones found in Lower Ogooué is represented in Table 3.2 and described below.

The Lower Ogooué valley

The Lower Ogooué valley stretches from Ndjolé to Lambaréné and is home to 1,681 plant taxa from 137 families. It is the best known and most collected structural zone of the Lower Ogooué. Indeed, the majority of its area is covered by dryland habitats, which are easy to explore, and many old roads cross it, facilitating botanical exploration.

Among these families, Rubiaceae represent 16% (265) of taxa recorded, followed by Fabaceae (9.5%, 160 taxa), Annonaceae, Orchidaceae, and Apocynaceae (3% on average, between 48 and 54 taxa). Seventeen families are represented by 20 to 40 taxa, and 115 families by less than 20 taxa each, forming about a third of the flora of the area. The most diverse families are mostly found in forests: Rubiaceae, Fabaceae, and Annonaceae, as well as genera of

Orchidaceae that have been collected in the region (including *Angraecum*, *Bulbophyllum*, and *Polystachya*). This is a result of the overall homogeneity of the Lower Ogooué valley, which approximately covers the upstream half of the current Ramsar site. This structural zone displays more pronounced relief than those found downstream of Lambaréné, which results in a narrow river bed. The area is essentially thus covered with forests at different stages of maturity and "naturalness." As a result, the size of families is comparable to that generally found in the rest of Gabon, where most of the vegetation comprises dryland forest. Nevertheless, not all forests in the Lower Ogooué structural zone are identical, and local variations can be found, each with its own specific flora.

According to our data, 1,054 species (50%) of the Lower Ogooué flora are only found in the lower valley. This represents two-thirds of the taxa recorded in this structural unit. This is the result of two factors that are probably linked, the most important of which is hard to identify: this structural zone is both the most diverse and most collected of Lower Ogooué. Consequently, a large number of taxa are found within the area but not in the other structural zones. This probably reflects the ecological truth for some of these genera, but for others it might simply mean they have not yet been found elsewhere. This figure could be drastically reduced if the dryland forests of the upper deltaic plain were as well studied as the lower valley.

The upper deltaic plain

The upper deltaic plain, which encompasses the current Ramsar site downstream from Lambaréné, is the largest structural area in the Lower Ogooué and has the region's greatest habitat diversity. This area comprises a complex network of wetlands resulting from many factors: the abundance of water, turbidity, the presence and nature of the soil, etc. All these factors influence the vegetation cover and therefore the flora. This area also contains a large number of dry environments, notably dryland forests and dry savannas. The upper deltaic plain is home to 1,122 taxa in 143 families (Table 3.2). The region's flora has been better understood since the first systematic botanical inventories in the 2010s, throughout all types of environments.

Although the families Rubiaceae (15%, 168 taxa) and Fabaceae (11%, 127 taxa) are once again the two most diverse in the area, Cyperaceae come in at third place (5%, 55 taxa). Poaceae are in sixth place, with 30 taxa (2.7% of taxa found in the area). The greater representation of these two families within the upper deltaic plain compared to the Lower Ogooué valley is the result of the presence of more open landscapes and more grassland environments. Since Rubiaceae and Fabaceae are both very diverse—and since Fabaceae are present in both forests and open habitats—they remain well represented in the upper deltaic plain. Families characteristically present in forests, such as Annonaceae, Euphorbiaceae, Ochnaceae, and Ebenaceae, remain well represented due to the presence of forests in the northern and southern parts of the upper deltaic plain. This can also be partly explained by a sampling bias: indeed, even though recent inventories have specifically targeted wetlands and savannas, this focus is recent and the majority of collections from the area have been made in dryland forests. Furthermore, it should be noted that fewer collections have historically been made in the upper deltaic plain because it is difficult to access. Although the main course of the Ogooué has been extensively navigated and documented since the first European expeditions in Gabon, secondary waterways such as lakes, secondary river courses, and also non-navigable wetlands such as swamps and marshes, have been largely neglected by explorers. In fact, certain areas of the upper deltaic plain had never been explored by botanists before the 2010s. Consequently, and despite its larger surface area and habitat diversity, fewer species and families have been catalogued within this phytogeographic area than the Lower Ogooué valley. Furthermore, only a third of the taxa listed from the upper deltaic plain are absent from other structural groups.

The lower deltaic plain

This plain encompasses the "maritime" part of the delta, from the current western edge of the Ramsar site to the three mouths of the Ogooué, excluding the coastal sand formations. Its surface area is smaller than that of the upper deltaic plain but it includes numerous habitats that are not present elsewhere within Lower Ogooué. In the area where the Ogooué splits into its three main channels to form the maritime part of the delta, the wetlands along its banks progressively change according to salinity. Thus, the flood forests on the banks of the river transition from *Raphia* trees into *Pandanus* thickets and then into

Table 3.2. The ten families with the most taxa in the entire Ramsar site, including the proposed limits, in terms of the number of taxa and percentage of the total number of taxa, the number in each of the structural groups, and the number of that "unique" to each group

	Proposed Ramsar Site	Lower Ogooué valley	Upper deltaic plain	Lower deltaic plain	Coastal sands	Lagoon region
Rubiaceae	302 (14.4%)	265 (15.8%)	168 (15%)	42 (13.4%)	30 (9.3%)	13 (23.6%)
Fabaceae	203 (9.7%)	160 (9.5%)	127 (11.3%)	29 (9.2%)	22 (6.8%)	8 (14.5%)
Cyperaceae	93 (4.4%)	31 (1.8%)	55 (4.9%)	20 (6.4%)	52 (16.1%)	2 (3.6%)
Orchidaceae	66 (3.2%)	51 (3%)	23 (2%)	10 (3.2%)	9 (2.8%)	4 (7.3%)
Apocynaceae	64 (3.1%)	48 (2.9%)	38 (3.4%)	7 (2.2%)	5 (1.5%)	4 (7.3%)
Annonaceae	63 (3%)	54 (3.2%)	39 (3.5%)	6 (1.9%)	5 (1.5%)	0 (0%)
Poaceae	58 (2.8%)	15 (0.9%)	30 (2.7%)	20 (6.4%)	29 (9%)	1 (1.8%)
Euphorbiaceae	56 (2.7%)	40 (2.4%)	22 (2%)	8 (2.5%)	8 (2.5%)	1 (1.8%)
Acanthaceae	42 (2%)	41 (2.4%)	17 (1.5%)	2 (0.6%)	2 (0.6%)	0 (0%)
Malvaceae	39 (1.9%)	34 (2%)	24 (2.1%)	1 (0.3%)	4 (1.2%)	0 (0%)
Unique taxa	-	1,054	413	87	109	18
Total number of taxa	2,092	1,681	1,122	314	323	55
Total number of genera	816	656	557	213	201	48
Total number of families	167	137	143	92	84	27

mangroves. In this part of Lower Ogooué, most habitats are wet, although there are some dry savannas and dryland forests, which are easier to access than the flood forests and swamps. As a result, of the 314 taxa catalogued in the area, the two best represented families mainly belong to forested areas (Rubiaceae, 13.4%, and Fabaceae, 9% of the area's flora). Nevertheless, Cyperaceae and Poaceae together represent 12% of the lower deltaic plain's flora, pointing to the significance of open environments such as various types of savanna. Finally, only 28% of the taxa catalogued within the area are absent from other structural groups.

This area's flora has not been extensively studied, despite recent efforts concentrated in less accessible areas. Further exploration of the northern part of the lower deltaic plain, especially near the Konjo and Atadali rivers, could help increase our knowledge of the area's flora since, behind the curtain of mangroves that follows the edges of river channels, there are sometimes pockets of dryland forests and sandy savannas, which remain unexplored to this day.

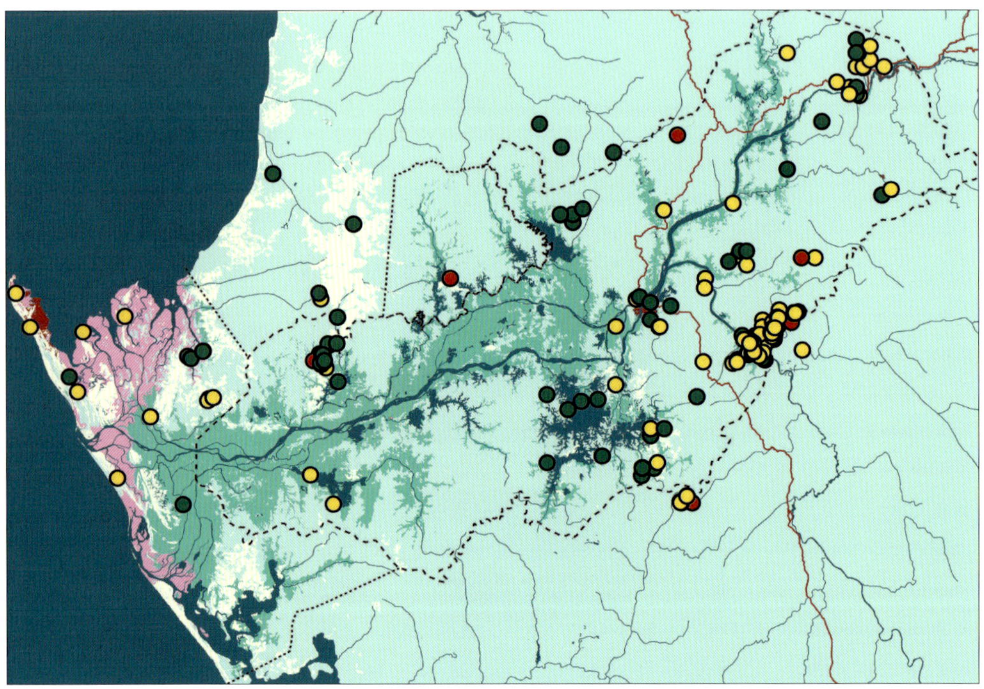

3.8. Distribution of threatened species according to the IUCN categories.

- *Critically endangered (CR)*
- *Endangered (EN)*
- *Vulnerable (VU)*

The coastal sand structures

These structures encompass a variety of the Lower Ogooué's coastal habitats and include 323 taxa recorded to date in 84 families. The four most diverse families within these sand structures are Cyperaceae (16% of the flora), Rubiaceae (9%), Poaceae (9%), and Fabaceae (7%). This suggests a strong predominance of grasslands, such as sandy savannas, especially Fabaceae, which mainly contain herbaceous genera. However, other families suggest the presence of habitats dominated by shrubs, such as Rubiaceae (9%), Euphorbiaceae, Annonaceae, Apocynaceae, and Arecaceae (each of which forms 2% of the area's known flora). These sandy coastal habitats have been explored since the start of European exploration in Gabon. However, the flora of these habitats has not been fully catalogued, and many new taxa are being found with each survey.

A third of the catalogued flora of the coastal sands is not present in other structural groups. This is due to two factors: first, there is a difference in collection efforts carried out in white sand savannas compared to the savannas of other structural groups of Lower Ogooué, which sit over substrates with more clay. Indeed, the flora of the latter form of savanna is less well known, and additional surveys would likely reveal species hitherto only known from the coastal sand structures. Second, a number of species are restricted to sandy substrates and are not found in savannas with clay substrates (as described in the chapter on the significant flora of various environments).

The lagoon region

This region encompasses the northwestern part of the Fernan Vaz Lagoon and is the least well known part of Lower Ogooué. Before 2016, only a few botanical collections had been made there: Pobéguin and Lecomte carried out some surveys in the Fernan Vaz Lagoon, but the location information provided is imprecise, and it is impossible to know whether these collections were made within the proposed limits of the Ramsar site. Collecting was done in 2016 in the forest areas within the region, so collections of Rubiaceae, Fabaceae, and Apocynaceae are numerous. However, only 55 taxa have been collected in the region, for the same number of collections. The data are therefore very incomplete, and exploration of the area has only just begun.

Significant species

With nearly 2,100 taxa belonging to over 2,000 species inventoried in the Lower Ogooué region, the flora of the Ramsar site is very rich and includes a large number of species that can be considered significant, that is to say, endemic to Gabon or with an even narrower distribution. Very often these rare species are threatened with extinction.

In fact, a preliminary assessment has identified 167 of these species as corresponding to the IUCN Red List criteria for threatened species, specifically either "Critically Endangered" (CR, 17 taxa), "Endangered" (EN, 85 taxa), or "Vulnerable" (VU, 65 taxa). With the exception of five taxa (*Entandrophragma angolense* var. *dolichocarpum*, *Sabicea sciaphilantha* subsp. *sciaphilantha*, *Cola lissachensis* subsp. nov., *Psychotria acutigemma* subsp. *acutigemma*, and *Chassalia carvalhoi* subsp. *brevifolia* subsp. nov.), all to the threatened taxa are at the rank of species. Formal assessments remain necessary for most of these taxa, and the categories defined in the list found in the appendix as well as in the text below, remain preliminary, unless otherwise indicated. These 167 taxa therefore represent the Lower Ogooué's legacy of threatened plant species, although they are not all endemic to the region. Among these taxa, 85 are endemic to Gabon and 24 are endemic to the Lower Ogooué region.

While the percentage of endemic species in the flora of Gabon is estimated at 10%, endemism is lower (7.5%) in the Lower Ogooué region. Indeed, this region of Gabon lacks the geomorphological characteristics that favor isolating evolutionary mechanisms, contrary to the Crystal Mountains and the Chaillu Mountains. The vast plain that makes up the Lower Ogooué region is not separated from the rest of the country by significant topographical barriers. Furthermore, its habitats are unique to neither Gabon nor Central Africa, though they cover a considerable area and are remarkably well conserved. It should also be noted that many of these habitats are home to species capable of long-range dispersal, which results in fairly low rates of endemism. Consequently, most of the plant species present in the Lower Ogooué region are also found elsewhere in Gabon or in Central Africa. Species strictly endemic to this region are therefore very rare, and were for the most part recently discovered and described. In fact, their known range would increase if discovered elsewhere in Gabon.

Except in the lagoon region, threatened species are present in all of the Lower Ogooué's structural areas, although not consistently. This distribution is mainly the result of the generally forested and dryland habitats of these species (125 of 167), but also reflects the areas that have been surveyed, as shown in Figure 3.6. Thus, the areas of Mabounié, the surroundings of Ndjolé, and Lakes Alombié, Evaro, and Onangué, contain numerous threatened species.

Similarly, the absence of threatened species near the lagoon region is the result of a sampling bias: it is very likely that if extensive inventories were carried out in this area, threatened species would be found, whether present elsewhere in the Lower Ogooué region or not.

Since it would be impossible to describe the entire significant flora of the Lower Ogooué region in this book, a selection of 24 species has been made in order to represent the full array of environments present. Some of these species are characteristic of environments belonging to the Ramsar site while others are threatened or are emblematic of the Lower Ogooué region, such as species from habitats rarely found elsewhere in Gabon. Among these species are some taxonomic novelties, recently described or currently being described following recent field discoveries. These are a testament to the need for ongoing efforts to develop a sufficient understanding of the region's flora: although the Lower Ogooué region has been explored for over a century, many discoveries are still being made.

Some other species from this list were discovered and described more than a century ago but had not been encountered in the field since, until the recent surveys were conducted. They are therefore illustrated here for the first time.

Finally, some of these species are neither rare nor endangered, but are noteworthy for their major ecological role in the Lower Ogooué region.

Sabicea mabouniensis (Rubiaceae)

This creeping vine, found along the roadsides (Figure 3.9), was discovered in 2013 near Mabounié, near the edge of the Ramsar site, and was described in 2015. No other corresponding specimens were found in herbaria. It is therefore considered endemic to Gabon and more specifically to the Maboumine mining concession, located in the Lower Ogooué valley. This species belongs to a taxonomically complex genus, which is currently being revised.

During its description, it was estimated to be Critically Endangered according to the Red List criteria based on its single known locality, outside of protected areas, and also because there is a risk that the only known subpopulation might decline due to possible future mining activities. However, this species could be present elsewhere in Gabon, since it was collected at the edge of a forest road, a habitat which is usually neglected during surveys.

3.9. Sabicea mabouniensis *(Photo: Ehoarn Bidault).*

3.10. Sabicea golgothae *(Photo: Ehoarn Bidault).*

3.11. Englerophytum gigantifolium *(Photo: Ehoarn Bidault).*

3.12. Plectranthus *sp. nov. (Photo: Ehoarn Bidault).*

Sabicea golgothae (Rubiaceae)

This species is very similar to *Sabicea mabouniensis* and is also a creeping vine found on the roadside (Figure 3.10). It was discovered in 2014 in the Mabounié area, with two known subpopulations, both on the same hill, and was collected on two occasions. It has never been collected elsewhere and is therefore, based on current knowledge, endemic to the Maboumine mining concession in the Lower Ogooué valley. Since its ecology is similar to that of *S. mabouniensis*, it could also be present elsewhere in Gabon. This species is distinguished from other anisophyllous *Sabicea* by its red to purple calyx.

The recent description of these two taxa shows that many species remain to be discovered in habitats often seen as degraded and uninteresting, and as such often neglected. In addition, since "historic" collections of this genus were reviewed, we know these species had never been collected before. Field surveys thus remain crucial in order to document the entire flora of Gabon.

Englerophytum gigantifolium (Sapotaceae)

This is a small dryland forest shrub, distinguished by its large silvery leaves, which measure up to 50 cm long. Its seven to eight parted flowers are unique within the genus *Englerophytum*, and are borne on the trunk or on branches at the base of leaves (Figure 3.11). This newly described

3.13. Diospyros subargentea *(Photo: Olivier Lachenaud).*

3.14. Brachystelma letestui *(Photo: Ehoarn Bidault).*

taxon is part of a taxonomic group for which collections from Mabounié helped clarify species limits. *Englerophytum gigantifolium* is endemic to Gabon and exclusive to the Mabounié area; several sterile or incomplete specimens from Nyanga (Banio Lagoon) and Ngounié (Sindara) are insufficient to confirm its presence elsewhere in the Gabon. This shrub is locally abundant in the undergrowth of these forests but it has not been found in other forested areas of the Lower Ogooué.

When first collected, this species was provisionally assessed as Critically Endangered according Red List categories and criteria. Indeed, only one location (sensu IUCN) is known for the species and the sole subpopulation is threatened by mining activities.

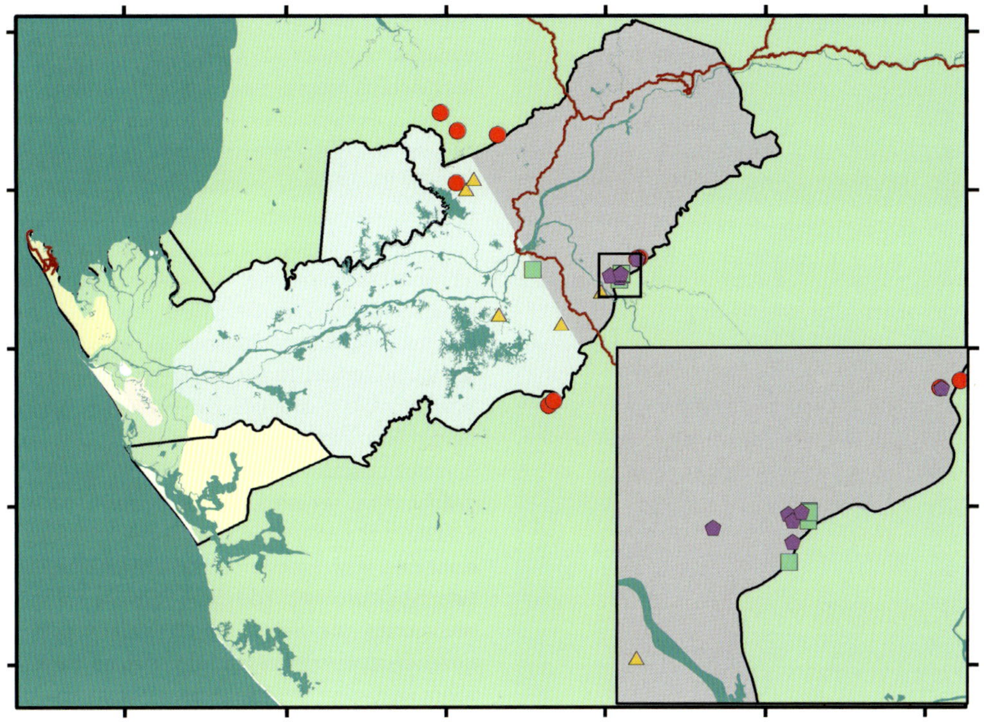

Diospyros subargentea (Ebenaceae)

This is a small tree, exceeding 8 m tall (Figure 3.13), that belongs to the same genus as the ebony, *Diospyros crassiflora*. It was identified during recent surveys in the Lower Ogooué valley, where it grows in dryland forests, always at the bottom of slopes near the edge of the floodplain. This species is distinguished from other *Diospyros* by the heavily discolorous leaves whose underside is silvery, and its narrow, spindle-shaped fruit, smaller than that of its apparent closest relative, *D. soyauxii*. It is endemic to central Gabon and is currently found from the border between the Lower Ogooué valley and the upper deltaic plain to the hills southeast of Lake Ezanga (outside of the Ramsar site). It is locally common, even dominant, in the forest undergrowth near Lake Azingo.

Diospyros subargentea is not present in any protected area and may be threatened by mining and lumbering activities, as well as agriculture. Eight subpopulations are known, occurring at eight localities according to the IUCN Red List criteria, and the species can therefore provisionally be assessed as Vulnerable according to IUCN Red List criterion B.

This taxon, which was discovered during recent expeditions in the Lower Ogooué region, was recently described. The genus *Diospyros*, which has great economic value for forestry, includes 736 species worldwide, 104 of which are found in continental Africa. Despite having been previously studied a number of times, new species of *Diospyros* are still being discovered today, including in Gabon.

Dactyladenia chevalieri (Chrysobalanaceae)

This species is a small tree, up to 15 m tall, with drooping branches, and is typical of lake shore riparian flood forests. It is distinguished by large discolorous, elliptical leaves that are beige on the underside, its erect inflorescences, and its hard, brown, ovoid fruits that have an irregularly bumpy surface (Figure 3.16). This species is found mainly in Gabon, specifically in the Lower Ogooué region, in Loango, Rabi-Kounga, and Nyanga, and is also present in Equatorial Guinea. However, most of its known subpopulations are found in the Lower Ogooué region, where it is an important element of the riparian forests of the Ramsar site Great Lakes.

Dactyladenia chevalieri is not present in any protected areas, but its habitat is not threatened by the logging, oil, or mining industries, nor by agriculture. With 10 known subpopulations (sensu IUCN), this species, although not yet assessed according to the Red List categories and criteria, should be considered as Least Concern.

Plectranthus sp. nov. (Lamiaceae)

This is a small herbaceous species with large, erect inflorescences and many small flowers ranging in color from blue to purple (Figure 3.12). It appears to be widespread in coastal regions from Cameroon to Congo (Brazzaville), and is therefore not endemic to Gabon. Well represented in herbaria, it has not yet been formally recognized as a new species. This *Plectranthus* has been

3.15. Distribution of Diospyros subargentea, Drypetes verrucosa, Scaphopetalum mannii *and* Veyretella hetaerioides:
- Diospyros subargentea
- Drypetes verrucosa
- Scaphopetalum mannii
- Veyretella hetaerioides

3.16. Fruits of Dactyladenia chevalieri *(Photo: Bonaventure Sonké).*

3.17. Pseudartabotrys letestui *(Photo: Ehoarn Bidault).*

3.18. Scaphopetalum mannii *(Photo: Ehoarn Bidault).*

3.19. Palisota alboanthera *(Photo: Ehoarn Bidault).*

collected several times during recent surveys in the Lower Ogooué region, both in coastal sand structures and in the upper deltaic plain. It is abundant at the edges of sandy forests and savannas, especially along the coast, and sporadically also further inland, around Lake Alombié for example.

This species has not been assessed based on the IUCN Red List criteria, but considering its numerous known subpopulations and its large extent of occurrence, it should be classified as of Least Concern.

This taxon, typical of savanna-forest patchworks, which are widespread in the Lower Ogooué region, had previously been confused with *Plectranthus monostachyus*, a widely distributed species found in ruderal areas that has much smaller flowers of a lighter purple color. This species was recently identified as a new following recent surveys.

Brachystelma letestui (Apocynaceae)

This is an herbaceous plant found in dry, white sand littoral savanna, with a large discoid tuber and creeping stems that produce very ornamental (but foul-smelling) flowers during the rainy season. It is unlikely to be confused for any other Gabonese species. *Brachystelma letestui* is rare and, in Gabon, only found in the littoral sand structures surrounding Port-Gentil. Though there are similar savannas in the Fernan Vaz Lagoon and near Libreville, it has not been sighted there; on the other hand, an herbarium specimen from Kouilou sand savanna, in Congo-Brazzaville, was found to belong to this taxon. It is restricted to a rare type of savanna, most of which, in Gabon, is found in the Lower Ogooué area, and it is threatened due to urban expansion around Port-Gentil and the construction of roads and pipelines.

Brachystelma letestui is not found in any protected area in Gabon or Congo and has only four known subpopulations representing for locations according to IUCN criteria, all of which are in continuing decline. It is therefore preliminary assessed as Endangered.

Only two specimens of *Brachystelma letestui* were known to exist of this species, both collected near Port-Gentil at the beginning of the 20th century, until its recent rediscovery in 2016 from three localities close to the city. This is its first published illustration (Figure 3.14).

Pseudartabotrys letestui (Annonaceae)

This species of large woody liana (over 10 m long) grows in swamps and valley-floor flood forests (Figure 3.17). The genus *Pseudartabotrys* is morphologically distinct from the rest of Annonaceae by the position of its gynoecium, and it has a single species, *P. letestui*. It is endemic to Gabon and mainly found in the southwestern quarter, although its habitat also extends north to the Mondah Forest, near Libreville. It has been collected many times, although only from six different localities; it is abundantly found around Lakes Alombié and Azingo in the Lower Ogooué upper deltaic plain.

This species has not been assessed according to the IUCN Red List criteria, but based on the number of known subpopulations and the relatively limited threats they face (swamp forests are not of interest to the logging industry), it should be considered Vulnerable.

This liana's remarkably fragrant flowers are initially yellow and then turn bright orange as they wither. It is emblematic of Gabon and especially of the Lower Ogooué region. It remains relatively unknown and probably underharvested because of its short seasonal flowering period.

Palisota alboanthera (Commelinaceae)

This herbaceous species, found in the undergrowth of both dryland and swamp forests, belongs to a complex genus that is especially diverse in Central Africa, notably in Gabon, with 30 recognized species (Figure 3.19). *Palisota alboanthera* was discovered as a result of recent expeditions in the Lower Ogooué valley and had not previously been collected. It appears to be close to *P. bogneri* but is distinguished by its pink flowers and white stamens. This species is therefore endemic to Gabon and its known range is extremely limited along the eastern edge of the Ramsar site, where it sometimes locally forms dense colonies.

Palisota alboanthera has not been assessed according to the IUCN Red List criteria, but due to its scarcity and since its known subpopulations are threatened by logging and mining activities, it should be considered as Endangered.

This new, recently published species appears to be rare and to date has only been sighted in Lower Ogooué region. Its recent discovery demonstrates the importance of the survey work undertaken in this part of Gabon.

Scaphopetalum mannii (Malvaceae)

This species is a small, humus-collecting shrub (Figure 3.18) with large, spatulate leaves and discreet flowers. It is strictly associated with swamp forests at the bottom of valleys. It is distinguished from other species in the genus by its unbranched habit and the shape of its leaves. *Scaphopetalum mannii* is abundant in the Mabounié area, in the Lower Ogooué valley, where it forms dense populations that cover river banks. It has not, however, been found in any other part of the Lower Ogooué Ramsar site. Outside the Lower Ogooué region, it has been collected on the Libreville peninsula and in Equatorial Guinea; it is only known from five localities, four of which are located in Gabon.

Scaphopetalum mannii has been provisionally assessed as Endangered according to the IUCN Red List criteria, based on five locations (sensu IUCN).

This rare shrub, restricted to swamp forests at the bottoms of valleys, is threatened by mining in the Lower Ogooué region and by urban expansion around Libreville.

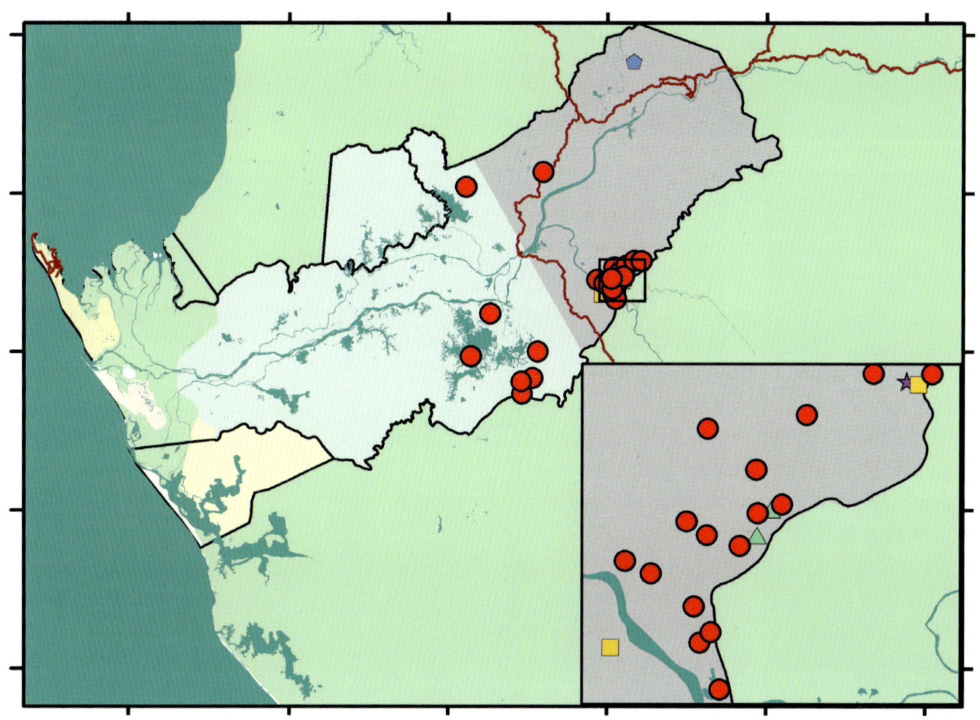

3.20. Cissus leemansii *(Photo: Ehoarn Bidault).*

3.21. Salacia diplasia *(Photo: Ehoarn Bidault).*

3.22. Distribution of Oncoba ngounyensis, Palisota alboanthera, Sabicea golgothae, Sabicea mabouniensis, *and* Salacia diplasia:

- Oncoba ngounyensis
- Palisota alboanthera
- Sabicea golgothae
- Sabicea mabouniensis
- Salacia diplasia

3.23. Strombosia fleuryana *(Photo: Ehoarn Bidault).*

3.24. Guyonia arenaria *(Photo: Ehoarn Bidault).*

3.25. Veyretella hetaerioides *(Photo: Ehoarn Bidault).*

Salacia diplasia (Celastraceae)

This species is a shrub with a very distinctive ecology; it is restricted to rocky river banks at the bottom of steep valleys. Although it belongs to a complex genus, it is easily identified by its shrub habit, winged stems, elongated fruits, and yellow flowers with reddish discs and petals that are clearly separate at the base of the flower.

Salacia diplasia is endemic to Gabon and restricted to the northwestern quarter of the country. It has been collected several times in the Mabounié area, where it forms dense populations locally; it is also found in the Crystal Mountains and the forests northwest of Ndjolé. This species was provisionally assessed as Vulnerable according the Red List criteria based on its five known locations (sensu IUCN). Since it is restricted to stream banks at the bottom of deep valleys, it is endangered by mining activities in the Lower Ogooué region.

Cissus leemansii (Vitaceae)

This species is a small climbing liana characteristically found in thickets on riverbanks of the Lower Ogooué with *Alchornea cordifolia* (Figure 3.20). It can be distinguished by its simple, slightly fleshy leaves with a cordate base, small yellow flowers, and small ovoid fruits. This species is not endemic to Gabon, since it was described from the Democratic Republic of Congo, where it is ruderal. However, it had not been found in Gabon before the inventories conducted in the Lower Ogooué region in the 2010s. This work also revealed its presence in Equatorial Guinea thanks to the study of herbarium specimens. It has only been collected twice in the Lower Ogooué region, near Lake Onangué and along the main course of the Ogooué, west of Lake Anengué. However, its apparent rarity is probably because its typical habitat is difficult to access and is thus underexplored.

This species has been assessed as of Least Concern according to the IUCN Red List criteria based on the absence of observed threats to these subpopulations.

Simirestis klaineana (Celastraceae)

This species is a large, woody vine from coastal dryland forests, sometimes also on the edge of mangroves (Figure 3.26). It is immediately distinguishable, even in a vegetative state, by the numerous woody growths present on its old stems. *Simirestis* is a small genus of vines from continental African with 15 species, most of which are rare or very localized. *Simirestis klaineana* has long been considered to be endemic to the Libreville peninsula, but was discovered in the Lower Ogooué region in 2016, north of the lower deltaic plain, in the forests bordering the Kendié River. It could be present elsewhere in the delta.

This species was assessed as Critically Endangered according to the IUCN Red List criteria (Stévart & Walters, 2016), since it is severely threatened by urban growth on the Libreville peninsula. A revision of this status is necessary due to the discovery of new subpopulations; it should be classified as Endangered given the small number of locations (sensu IUCN), the threats it faces, and the low number of known individuals (between 50 and 250).

This species, like *Psychotria wieringae* and *P. klainei,* is one of the taxa previously considered endemic to the Libreville peninsula but for which new subpopulations have been discovered in the Lower Ogooué region, illustrating the significant floristic affinity of these two areas. However, none of these species have yet been found south of the Ogooué, which could act as a barrier to dispersal.

Guyonia arenaria (Melastomataceae)

This species, a 30-40 cm tall herb with very ornate pink flowers, is restricted to the borders between forests and savannas in humid and sandy soil (Figure 3.24). It is endemic to the coastal regions of Gabon and Equatorial Guinea, and has only been found in a few localities. *Guyonia arenaria* has been described twice under two different names, and the more recently published name has been reduced to a synonym. In the Lower Ogooué region, this species is only found near the Kendié River, north of the lower deltaic plain, where two collections were made in 2016. It has not been found in similar habitats elsewhere in the region, notably around the Fernan Vaz Lagoon, but it can easily go unnoticed when not in flower.

Guyonia arenaria has been provisionally assessed as Near Threatened according to the IUCN Red List criteria on the basis of eight known subpopulations, representing five locations (sensu IUCN). Five of these subpopulations are found within protected areas (the Ndote reserve in Equatorial Guinea and Loango National Park in Gabon). However, given the small number of localities, it could become threatened in the future.

Strombosia fleuryana (Olacaceae/Strombosiaceae)

This species is a small shrub a few meters tall (Figure 3.23) and is restricted to the flood forests at the edges of lakes and rivers, where it can be found as isolated individuals. It is endemic to the Ogooué basin, where it is known from only four localities, between Ngomo (the type locality) and the lower Ivindo River. It

3.26. Simirestis klaineana *(Photo: Olivier Lachenaud).*

3.27. Drypetes verrucosa *(Photo: Ehoarn Bidault).*

3.28. *Oncoba ngounyensis (Photo: Ehoarn Bidault).*

3.29. *Chrysophyllum ogoouensis (Photo: Ehoarn Bidault).*

is frequent in parts of the flood forests that border Lake Onangué. This rare and threatened species of historical importance was only described recently, although the first specimen was collected in 1912. It appears to be rare since it flowers abundantly along the entire length of its branches and has bright red fruit, precluding any possible confusion.

Strombosia fleuryana has been provisionally assessed as Near Threatened according to the IUCN Red List criteria, since although none of its known subpopulations are found within protected areas, its habitat does not seem threatened by agriculture or logging and dry land oil exploration. However, as it is only known to exist in four localities, it may become threatened in the future.

Veyretella hetaerioides (Orchidaceae)

This species is a small creeping terrestrial orchid with small white flowers (Figure 3.25). It is restricted to sandy riverbanks in swampy forests found at the bottom of the valleys, generally dominated by *Crateranthus letestui*, and in which it forms a dense groundcover. This species is endemic to Gabon. First described in the genus *Habenaria*, it was later placed in a separate genus, *Veyretella*, with another species, *V. flabellata*, endemic to the Lower Ogooué valley. *Veyretella hetaerioides* has been collected many times in the Mabounié area in the Lower Ogooué valley, and has recently been found in the Crystal Mountains.

This species has not been assessed according to the IUCN Red List criteria. However, given the small number of known subpopulations distributed across fewer than 10 locations (sensu IUCN), including three in the Mabounié area, it should be considered Vulnerable. In addition, this species is threatened by ongoing or planned industrial projects, both in the Lower Ogooué region (mining) and the Crystal Mountains (hydroelectric dams).

Drypetes verrucosa (Putranjivaceae)

This species is a medium-sized, cauliflorous tree that grows in dryland forests, where it is generally found as isolated individuals or rarely in dense populations (Figure 3.27). It is easily distinguished by its heavily grooved bole, inflorescences located on prominent growths along the trunk, which persist after flowering, and leaves that have a distinctly asymmetrical base. *Drypetes verrucosa* is endemic to Gabon, widely found in the Lower Ogooué valley, near Mabounié and towards Lake Azingo. However, outside this area, it seems to be rare: some specimens have been collected near Sindara near the Ngounié River, and Lalara in the Ogooué-Ivindo, and there are also historic collections (made around 1900) from the Libreville peninsula, although its presence in Estuaire Province has not since been confirmed.

This species has been provisionally assessed as Vulnerable according to the IUCN Red List criteria. *Drypetes verrucosa* has been found in 11 subpopulations from nine locations (sensu IUCN), although none of them are included within a protected area. Moreover, the subpopulations documented by Klaine at the beginning of the 20th century in the Libreville region may have disappeared.

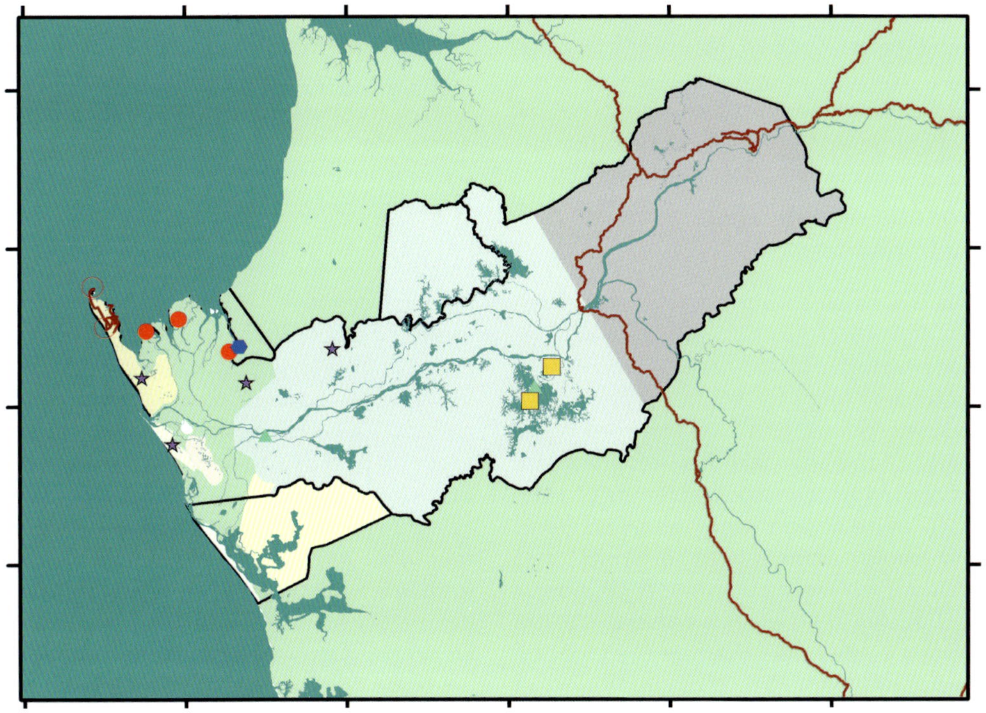

This species, endemic to the Lower Ogooué valley, is threatened by both industrial plantations in the vicinity of Lambaréné and logging activities. Although its wood is not particularly sought after, it is affected by clearing for tracks and skidding trails, and by artificial windfall.

Chrysophyllum ogoouense (Sapotaceae)

This species is a sarmentose shrub restricted to flood forests and lake and riverbanks (Figure 3.29). It is easily distinguished by its long oblanceolate leaves with dense parallel venation and acuminate, green then yellow fruits that dry on the stem, taking a reddish brown hue. Its fruits are edible although they have little taste. *Chrysophyllum ogoouense* is endemic to the Ogooué basin in Gabon (including its tributaries, the Ivindo and the Ngounié). Within the Lower Ogooué region, it is common in parts of its habitat, around Lakes Evaro, Ezanga, and Onangué. It has also been collected many times along the Ogooué in the Lower Ogooué valley, and up to the banks of the Ivindo. Curiously, it has not been observed in the riverbank flood forests near Mabounié, whereas it is present further upstream on the Ngounié, up to Fougamou.

This species has not yet been assessed according to the IUCN Red List categories and criteria. However, in view of its many subpopulations and locations (sensu IUCN), numbering more than 10, it should

3.30. *Distribution of* Brachystelma letestui, Chrysophyllum ogoouense, Cissus leemansii, Plectranthus *sp.nov., and* Simirestis klaineana.

- ● Brachystelma letestui
- ○ Brachystema letestui *(population missing)*
- ■ Chrysophyllum ogoouense
- ▲ Cissus leemansii
- ★ Plectanthrus *sp. nov.*
- ⬢ Simirestis klaineana

3.31. Cyperus papyrus *(Photo: Ehoarn Bidault).*

3.32. Rhizophora racemosa *(Photo: Ehoarn Bidault).*

be considered as Least Concern. This emblematic species of the Ogooué basin, which is endemic to Gabon, is not considered threatened because its habitat is not targeted by agriculture or logging.

Oncoba ngounyensis (Achariaceae)

This is a small shrub in the undergrowth of dryland forests (Figure 3.28). Its placement in the genus *Oncoba* is uncertain, and it could represent a separate genus. It is characterized by its large, toothed, elliptical-oblanceolate leaves, inflorescences in racemes of fairly small white flowers with yellow stamens, and ovoid, orange, smooth fruits. It is endemic to Gabon; with the exception of one locality in the vicinity of Libreville, it is limited to the lower basin of the Ogooué (including Ngounié). It is therefore an important species for conservation since almost all of its population is found in the Lower Ogooué.

Oncoba ngounyensis has not yet been assessed according to the IUCN Red List categories and criteria, but because of its numerous subpopulations found in at least 10 locations (sensu IUCN), it should be considered Near Threatened. However, the dryland forests where most of the species' known individuals have been observed are threatened by logging and mining activities, as well as industrial farming near Lambaréné. On the Libreville peninsula, it is threatened by urban growth.

This species is characteristic of Lower Ogooué dryland forests and is very abundant in the Mabounié area, where it frequently dominates the undergrowth. It is also quite common around Lake Azingo, but much less frequent near Lakes Ezanga and Onangué. It is not found in the lower part of the Ramsar site.

Cyperus papyrus (Cyperaceae)

Native to the Nile basin, but present throughout Africa, the Indian Ocean, and around the Mediterranean, the papyrus *Cyperus papyrus* is one of the distinctive plant species of the Ramsar site (Figure 3.31). It forms large, almost pure colonies along the riverbanks. It is sometimes a pioneer in the colonization of waterways but is most often found in floating meadows, mixed in with the Poaceae *Echinochloa stagnina* or the Polygonaceae *Persicaria madagascariensis*. When the current is weak, papyrus plants can colonize large areas, thus preventing navigation. Along the various channels of the Ogooué, the current can sometimes break off small sections of these "meadows," which then drift downstream and later run aground. When these small populations settle, they trap some of the river's alluvium and help form levees along the river, which separate the main river course from the surrounding wetlands. Other species then establish on these levees, including large, woody species such as *Ceiba pentandra*.

Cyperus papyrus is neither threatened nor rare, but it remains significant for its ecological role in the dynamics of the Lower Ogooué wetlands.

Chrysobalanus icaco subsp. *icaco* (Chrysobalanaceae)

This taxon is found on most tropical coasts of Africa and America; it is of great ecological importance in the Lower Ogooué region, where it takes many forms, ranging from shrubs to a woody liana. *Chrysobalanus icaco* is heliophilous and fast growing, and it actively participates in the recolonization of sandy savannas by neighboring forest. Strips comprising exclusively *C. icaco* can often be seen along the edges of savannas, behind which other species can settle, eventually forming a dense forest cover. This species is widespread in these sandy coastal forests, especially on Mandji and Ozouri islands. This taxon is not threatened or rare within Gabon, but it is significant for its important ecological role in the dynamics between the savannas and forests of Lower Ogooué.

Rhizophora racemosa (Rhizophoraceae) and *Avicennia germinans* (Acanthaceae)

In Lower Ogooué, mangroves predominantly comprise *Rhizophora racemosa* (red mangrove) and *Avicennia germinans* (white mangrove), two widespread species along the coasts of tropical Africa and also in South and Central America. While mangroves have a relatively impoverished flora and lack rare or endangered species, they are crucial from an ecological point of view, providing habitat critical for many marine and avian species. They are also essential to the delta's physical stability, since they can easily colonize areas flooded by the Ogooué, promoting alluvium build-up and protecting ecosystems further inland from the oceanic erosion. Colonization of the coast by mangroves also allows sand to accrete between the delta's channels, and islands form in places protected by a wide strip of *Rhizophora*, behind which sandy coastal forests and savannas can eventually form.

Ctenolophon englerianus (Ctenolophonaceae)

This species is a medium-sized tree, rarely reaching 30 m tall, with flutes at the base of its bole. It is found from Nigeria to Angola, and is restricted to coastal forests, sometimes on dry land, but mainly in marshes at the bottom of valleys, where it forms gregarious populations that can dominate the forest cover. *Ctenolophon englerianus* is found abundantly in the Lower Ogooué near Lake Alombié. Lower Ogooué forests with *Ctenolophon* are home to many rare or endangered species, such as *Pauridiantha gracilipes*, endemic around Lake Alombié. *Ctenolophon englerianus*, known as okip in Gabon, is a hardwood species that is not used for international trade, but which has local value for building houses, as timber for agricultural implements or for poles, and even for canoes. It is also used as fuelwood (Lemmens *et al.*, 2012). *Ctenolophon englerianus* is important for understanding the forest dynamics of coastal Gabon: it is able to colonize both dry and humid environments, including marshes and flood forests. Due to its wide ecological range, this species has probably withstood prior dry periods, which would have transformed most Gabon's coastal region into savanna; it could therefore have played a central role in forming microsanctuaries that helped preserve a forest flora and have facilitated forest recolonization of the savanna.

Palms (Arecaceae)

Few palm species are known in the Lower Ogooué region, and those present are neither rare nor threatened. However, they often play a major ecological role. The wild date palm, *Phoenix reclinata,* is a small, thorny species often found behind low mangroves–submerged at high tide–where it takes part in the transition toward dryland forests. *Eremospatha macrocarpa* is a climbing palm that is very common in the sandy coastal forests of Mandji and Ozouri islands. A few isolated populations of *Hyphaene guineensis* are spread across the coastal savannas. On the edge of the river, in places where the salinity of the water is lower, mangroves and *Pandanus* give way to large populations of *Raphia zamiana*, which participate in consolidating the river banks by forming a continuous border behind which other types of flood forests can grow.

The distinctive flora of various environments

The Lower Ogooué region has a great diversity of wetlands and of dryland habitats, which are described in the chapter on vegetation. Each habitat is distinguished by its species composition as well as specific conservation challenges. This chapter will present the principal flora of these major types of habitat.

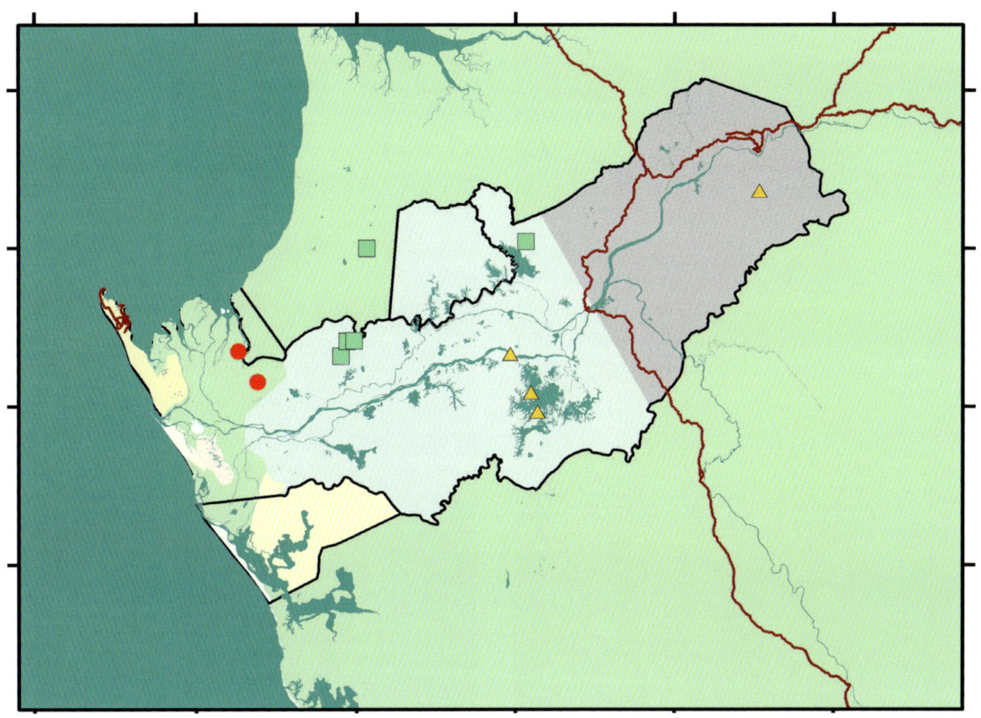

Page 158:
3.33. Raphia zamiana *(Photo: Ehoarn Bidault).*

Page 159:
3.34. Distribution of Guyonia arenaria, Pseudartabotrys letestui, *and* Strombosia fleuryana

- Guyonia arenaria
- Pseudartabotrys letestui
- Strombosia fleuryana

3.35. Vangueriella dowsettiae *(Photo: Ehoarn Bidault).*

3.36. Dry savanna on white sand (Photo: Ehoarn Bidault).

3.37. Eremospatha macrocarpa *(Photo: Ehoarn Bidault).*

Dryland forests

Coastal dryland forest

Dryland forests cover a large part of the Lower Ogooué region, and are, within Gabon, coastal in nature. This habitat is the easiest for botanists to explore and has, as a result, yielded the vast majority of botanical collections from the area. It is also probably the vegetation type with the most species. This type of forest has many local variants, which result, among other things, in varying dominance of the shrubby species most often encountered in the Lower Ogooué region, such as *Saccoglotis gabonensis, Aucoumea klaineana, Greenwayodendron suaveolens, Dichostemma glaucescens, Pycnanthus angolensis,* and *Librevillea klainei*, among others. This variation is detailed in the chapter on vegetation (pages 173–197).

Dryland forests are crucially important for conservation, since they harbor 119 of the 167 threatened taxa identified in this study (71%). Additionally, six endangered species are found in the edges of these forests.

All habitats are represented among these 125 threatened taxa. These include small and medium-sized trees, such as *Diospyros cleistantha, D. subargentea, Drypetes verrucosa, Carapa batesii, Englerophytum gigantifolium*, and three species of *Dactyladenia*; a large number of shrubs, including *Oncoba ngounyensis, Massularia stevartiana, Magnistipula devriesii, Englerodendron hallei,* and a new species of *Maesobotrya*, currently known only from the Lower Ogooué region; and herbaceous species such as *Palisota alboanthera, Tristemma rubens,* and *Lastreopsis davalliaeformis*.

The family Rubiaceae is well represented, with 20 endangered species found in the dryland forests, including three species in the genus *Eumachia* (ex-*Chazaliella*), two in *Keetia*, four in *Psychotria*, two in *Tarenna*, and two in *Pauridiantha*, among others.

These dryland forests are also important in that they contain 12 of the 17 species provisionally assessed as Critically Endangered according to the IUCN Red List categories and criteria. These species are mainly found in the Lower Ogooué valley (nine species) and specifically in the Mabounié area, where the only known specimens of *Tarenna* sp. nov. aff. *baconoides, Englerophytum gigantifolium, Uvaria* sp. nov. aff. *klainei,* and *Salacia* sp. nov. aff. *tessmannii* were collected.

Four other Critically Endangered species

are present in the dryland forests of the upper deltaic plain, specifically in the hills located southeast of Lake Ezanga, at the edge of the Ramsar site, inside the proposed limits. A submontane forest unique to these hills occurs within the Lower Ogooué region, where the only known specimens of *Cola* sp. nov. aff. *rostrata* and *Eugenia anisata* sp. nov. were collected.

Sandy coastal forests

These forests are mainly found on Mandji and Ozouri islands. They form a fairly low vegetation cover (between 10 and 20 m) and are not always continuous but sometimes punctuated with openings due to windfall and localized, more lianescent areas that resemble thickets. These forests are interrupted by savannas, forming a patchwork characteristic of the maritime parts of the Lower Ogooué.

The flora of these forests' understory is less diverse than that of dryland forests downstream in the Lower Ogooué. The dominant tree species is *Fegimanra africana* (Anacardiaceae). The shrubs *Cola gabonensis* and *Heinsia crinita* var. *scitula* are the most frequently encountered taxa. *Psychotria kimuenzae* and *Psydrax moandensis* are present along forest edges. The presence of numerous lianas makes it difficult to explore these forests: there are many individuals of *Secamone brevipes, Pararistolochia* spp., *Adenia rumicifolia, Salacia cornifolia, Landolphia* spp., *Annona glabra, Uvaria versicolor,* and *Premna lucens*. Climbing palms such as *Eremospatha macrocarpa* are also basic elements of these forests.

While the sandy coastal forests have a very disctinctive flora, with most species restricted to them, they contain few rare or endangered species. Nevertheless, characteristic species have been found there, especially at the edges of forests, including *Vangueriella dowsettiae* sp. nov., a woody vine typical of these environments, known to date from only three collections, one on Ozouri Island, one in the south of Gabon, and the last in Kouilou Province of the Republic of Congo.

Savannas

All the savannas of the deltaic complex are exclusively herbaceous. Savannas with shrubs or trees are in fact nearly absent from the coastal basin. Soil moisture content is the main criterion used to recognize the various types of savanna based on their floras throughout the Lower Ogooué region. The two main types are wet and dry savannas.

Dry savannas

Compared to other environments in the Lower Ogooué, dry savannas contain relatively few rare or endangered species. However, they have a number of significant species such as *Eragrostis sabulicola*, a rare grass exclusively known from the Lower Ogooué and Kouilou (Congo Brazzaville) regions. This species has not been collected in the Lower Ogooué since 1899. *Brachystelma letestui* (Apocynaceae) is a similarly distributed rare creeping grass, recently found near Port-Gentil. Both species are restricted to sandy to dry coastal savannas.

Rare and endangered species are found in mixed-substrate dry savannas, such as *Thesium filipes* and *Cassytha graminicola*. However, these species are probably less threatened than currently available data suggest as they have probably both been undercollected: *T. filipes* is difficult to spot below tall grasses, and *C. graminicola* is often mistaken for the pantropical species *Cassytha filiformis*. Like other mixed-substrate savannas throughout Gabon (and unlike the rarer savannas found on coastal sands or on the Kalahari sands of the Batéké

3.38. Humid savanna in a valley bottom (Photo: Ehoarn Bidault).

Plateau), the dry savannas of the Lower Ogooué contain relatively few rare species although they remain a rare and declining Gabonese habitat.

The edges of savanna gallery forests of the Lower Ogooué form a habitat with numerous rare, threatened, or significant species. These include *Guyonia arenaria* (Melastomataceae), which is found neither in the savanna nor inside the forests proper, but instead at the border between the two. The same is true for *Plectranthus* sp. nov. ined., and *Vangueriella dowsettiae* sp. nov. ined., a species which is restricted to these borders and has only been seen along the coast of Gabon up to Kouilou.

Humid and semihumid savannas

These savannas are formed when grasses grow in humid depressions at the bottom of valleys (Figure 3.38). This type of vegetation receives runoff from rainfall and is therefore periodically flooded (meadows or semihumid savannas); this environment is referred to as humid savanna. It generally escapes the fires that regularly impact dry meadows. These types of vegetation are predominantly Cyperaceae, such as *Scleria distans, Bulbostylis pusilla, Rhynchospora candida*, and *Rhynchospora holoschoenoides*, as well as Poaceae such as *Panicum brevifolium, Eragrostis squamata,* and *Acroceras zizanioides*. Other families are also present, including Lycopodiaceae (*Lycopodiella affinis*), Burmanniaceae (*Burmannia madagascariensis*), Lentibulariaceae (*Utricularia* spp.), Xyridaceae (*Xyris* spp.) and Eriocaulaceae (*Mesanthemum radicans, Eriocaulon* spp.). Some orchids are also restricted to this habitat, such as *Eulophia milnei* and *Disa* sp.

Peatlands are a variant of these humid herbaceous formations, characterized by high humidity throughout the year and with higher acidity than other humid savannas, caused by the presence of sphagnum and mosses. Some species restricted to these peat bogs—which are a rare habitat in Gabon—such as *Drosera* spp., *Lycopodiella caroliniana*, and *Genlisea africana*.

Although humid savannas are rare habitats in Gabon, and most of them are found in the Lower Ogooué region, they contain relatively few rare species. However, they have some endangered species such as *Eriocaulon nadjae* and *Burmannia liberica*.

Swamps and marshes

Swamps and marshes are very diverse throughout the Lower Ogooué region and are the main justification for classifying it as a Ramsar site. These environments can be forested or herbaceous, but are all inundated during floods. Salinity is one of the primary factors behind the floristic diversity of these environments: generally, higher salinity results in less floristic diversity.

Mangroves

Mangroves are of crucial importance to aquatic fauna and for helping to control marine erosion, but they have a relatively impoverished flora. The main woody species found in mangroves are *Avicennia germinans*, *Conocarpus erectus*, and *Rhizophora racemosa*—which often form monospecific colonies in the maritime part of the Lower Ogooué region (high or low

3.39. High mangrove with Rhizophora *(Photo: Ehoarn Bidault).*

3.40. Low mangrove with Rhizophora *(Photo: Ehoarn Bidault).*

mangrove). The large fern *Acrostichum aureum* grows under the *Rhizophora*. When going upstream on the Ogooué from the coast inland, the mangroves alongside the river transition into mixed colonies, which include varying quantities of *Pandanus candelabrum* and, farther up, *Raphia zamiana* based on the water salinity. The species that occupy these habitats are very common across mainland Africa.

In the Lower Ogooué region, no rare, threatened, or significant plant species are restricted to mangroves. However, a rare orchid, *Diaphananthe lanceolata*, was collected both in dryland forests and on one occasion in a mangrove, which demonstrates that some epiphytic

3.41. Whitfieldia letestui *(Photo: Ehoarn Bidault).*

3.42. Dense riverside papyrus stands (Photo: Ehoarn Bidault).

3.43. Swamp forest undergrowth (Photo: Ehoarn Bidault).

3.44. Coccinia racemiflora *(Photo: Ehoarn Bidault).*

3.45. Dense raffia along the river (Photo: Ehoarn Bidault).

species can occasionally establish themselves on *Rhizophora*. Furthermore, the rare shrub *Dactyladenia jongkindii*, along *Carapa mangarevensis* and the vine *Simirestis klaineana*, are sometimes found where mangroves meet dryland forests. These are endangered species endemic to the northwestern coast of Gabon, ranging from the north of the Lower Ogooué to the Libreville peninsula.

Swamp forests

Unlike the flood forests on the river banks, swamp forests are not necessarily in direct contact with water from rivers or adjoining lakes. They are also found in valley bottoms sufficiently wide and shallow for rivers to flow slowly and saturate the environment. When the milieu is open and water is abundant, the forests typically contain large trees of *Mitragyna ledermannii, Anthostema aubryanum*, and *Pandanus gabonensis*. These continuously flooded forests have a relatively homogeneous flora and contain only a few rare, endangered, or significant species, although they include the small threatened trees *Dacryodes ebatom* and *Dactyladenia eketensis*.

However, many variants exist based on the relief and therefore on the depth of the valley as well as on the water level. As such, certain swamp forests located in the bottoms of narrow valleys in dryland forests are filled with water but never flooded to more than 20 cm in depth, and their canopy is closed. These forests are comprised of *Carapa procera* s.l. and *Crateranthus le-testui* in the Lower Ogooué valley. The orchids *Veyretella hetaerioides* and *V. flabellata* grow at the foot of *C. letestui*, a small threatened tree. At the edge of these forests are also dense populations of *Scaphopetalum mannii*, a rare species, the largest known population of which is found in the Lower Ogooué region. Many other endangered, rare, and significant species are found in these swamp forests occurring within dryland forests: *Cola lissachensis* subsp.

nov., and *Craterispermum deblockianum*, among others. *Coccinia racemiflora* (Figure 3.44), *Monanthotaxis glomerulata, Thaumatococcus flavus,* and *Whitfieldia le-testui* (Figure 3.41) can also be found in the swamp forests of headwater areas.

A different floristic variant is the sandy swamp forests of valley bottoms found in the upper deltaic plain near Lake Alombié, which have soils that remain permanently swampy and become inundated during floods. These forests, rare throughout Gabon, are predominantly composed of *Ctenolophon englerianus*, sometimes with *Dactyladenia eketensis*. Their relatively sparse undergrowth is generally dominated by *Pauridiantha gracilipes*,

3.46. Flooded riparian forest (Photo: Ehoarn Bidault).

3.47. Ceiba pentandra *by the river (Photo: Ehoarn Bidault).*

3.48. Floating herbaceous mats with Echinochloa *(Photo: Ehoarn Bidault).*

3.49. *Eugenia ogoouensis (Photo: Ehoarn Bidault).*

3.50. *Begonia letestui (Photo: Ehoarn Bidault).*

3.51. *Amphiblemma setosum (Photo: Ehoarn Bidault).*

a recently discovered lianescent bush endemic to the Lower Ogooué region. The grassy undergrowth is composed of ferns like *Trichomanes crispiforme*, small suffrutex species (*Psychotria* spp., *Rauvolfia mannii*), and a large number of small myco-heterotrophic grasses (*Voyria primuloides, Sciaphila ledermannii*, and *Campylosiphon congestus*). The iconic and threatened vine *Pseudartabotrys letestui* (Figure 3.17) is also often found in these forests.

Raffia groves

Raffia groves are almost permanently flooded forests, purely or partially composed of *Raphia zamiana*, which sometimes reach 15 to 20 m in height. They form at the edges of forests along delta channels, but can also cover large flooded areas in calm water behind these channels. As a result, they are also found covering large surfaces along the edges of lakes. The flora of raffia groves is typically low in diversity. Groves of pure raffia are found in environments with water depth exceeding 50 cm. When standing water does not exceed 20 cm in depth, raffia groves form mixed swamp forest with *Raphia zamiana, Mitragyna ledermannii*, and *Anthostema aubryanum*. The permanently flooded undergrowth is not diverse, but includes vines such as *Hypselodelphys scandens* and *Cissus planchoniana*, and some shrubs including *Oubanguia africana*.

No rare, endangered, or significant species have been found in raffia groves.

Riverside flood forests

Depending on whether flood forests are located near rivers or on the edges of swamps, these riparian plant formations appear to have a high level of floristic variability. Riparian forest species are mostly exclusive to this habitat: for trees, they include *Guibourtia demeusei, Dactyladenia chevalieri, Cynometra mannii, Scytopetalum klaineanum*, and *Martretia quadricornis*. The soil in these forests is dry during the dry season and is submerged under 1 to 2 m of water during flooding. As a result, the undergrowth is almost devoid of grasses (only *Culcasia tenuifolia* survives flooding). Sometimes the banks of the Ogooué are covered in abundant flowering *Pleiocoryne fernandensis* vines.

In the upper deltaic plain and the Lower Ogooué valley, which are subject to significant variations in water level, these forests take over the majority of the Ogooué's banks and the lakes near Lambaréné, where vegetation that is not flooded year-round is able to establish. Numerous species significant for the Ogooué region are found in these forests, such as *Strombosia fleuryana, Chrysophyllum ogoouense, Eugenia ogoouensis* (Figure 3.49), and *Capparis* sp. nov., all endemic to the Ogooué basin (including the Ngounié). Other threatened species not endemic to the Lower Ogooué region are present in the area, also in these tall floodable forests, along rivers and large lakes, such as *Eugenia thollonii, Neolemonniera ogoouensis, Synsepalum fleuryanum, Psychotria acutigemma*, and *Carapa hygrophila*.

Floodable riverbanks

These are slopes, 3 m or more in height and a few meters deep, that form a clear separation between the river and the vegetation behind it, forming a barrier to water. This type of vegetation, which is clearly visible along the banks of the Ogooué, is mainly composed of *Alchornea cordifolia* groves. Some climbing liana species grow in the herbaceous and shrub cover (*Cayratia ibuensis, Commelina* spp., and *Uncaria africana* var. *domatiifera*), and the vegetation has a few isolated large trees (*Nauclea pobeguinii, Ceiba pentandra*).

There are distinctive shrubby and bushy species on the flooded edges of these slopes, facing the river, such as *Mimosa pigra, Machaerium lunatum*, and *Ficus asperifolia*, which can form dense and almost monospecific populations in places.

The twining liana *Cissus leemansii* grows in the flooded edge of these riverbank slopes; it is the only rare—but not threatened—species found in these environments within Gabon, which typically have poor species variety.

Flooded or floodable river meadows

These represent two types of strictly herbaceous vegetation frequently encountered along the Ogooué. Along the banks, along the main course of the river, floating meadows are formed of often monospecific stands predominantly composed of *Cyperus papyrus, Echinochloa stagnina* (Figure 3.48), and *Persicaria madagascariensis*, which colonize calm waters up to tens of meters back along the banks. During periods of flooding, pieces of these floating meadows can detach and settle elsewhere along the river, depending on currents. A few ancillary species are found there, such as *Ludwigia* spp., as well as some rare *Alchornea cordifolia* shrubs.

Behind the banks themselves lie large wetlands whose water is characteristically much less turbid than that of the river. These areas are flooded most of the year and contain the same species as the floating meadows situated in the river.

Between Lambaréné and Port-Gentil, there are many sandbanks in the middle of the river. A distinctive ephemeral flora grows on these sandbanks, which are inundated during floods; it contains low pioneer species such as *Leptochloa coerulescens, Mimosa pigra, Heliotropium indicum, Glinus oppositifolius, Polycarpon prostratum*, and *Hydrolea palustris*.

These flooded or floodable meadows generally have low floristic diversity, cover large areas, and are very homogeneous. They contain no rare species.

Acknowledgments

The authors thank all of the specialists who contributed to identifying plants in various groups: Henk Beentje (Royal Botanic Gardens, Kew) for Asteraceae, Pulchérie Bissiengou (Herbier National du Gabon) for Ochnaceae, Frans Breteler (Wageningen) for Anacardiaceae and Dichapetalaceae, Vincent Deblauwe (IRD) for pteridophytes, Dominique Champluvier (Botanic Garden Meise) for Acanthaceae, Thomas Couvreur (IRD) for Arecaceae, Steven Dessein (Botanic Garden Meise) for Rubiaceae, Eberhard Fischer (University of Koblenz-Landau) for Scrophulariaceae, Steven Janssens (Botanic Garden Meise) for Balsaminaceae, Carel Jongkind (NCB Naturalis, Leiden) for Lecythidaceae and Combretaceae, David Harris (Royal Botanical Garden, Edinburg) for Zingiberaceae, Alexandra Ley (University of Halle) for Marantaceae, Paul and Hiltje Maas (NVB Naturalis, Leiden) for Costaceae, Marc Sosef (Botanical Garden Meise) for Poaceae, Philippe Thery for Cyperaceae, Hans de Wilde (Herbarium Vadense and Wageningen) for Begoniaceae, and Jan Wieringa (NCB Naturalis, Leiden) for Fabaceae-Caesalpinioideae. They also thank Bonaventure Sonké for permitting the use of his photo of *Dactyladenia chevalieri*.

EB thanks Violette, Martin, Lorène, Noémie, and Sébastien for their patience.

The field work necessary for the production of this book was made possible thanks to the support of many people, institutions, NGOs, and companies, who are acknowledged in the section on vegetation.

Bibliography

AOUTIN, J., 1967. *Les pollens et les spores des sédiments du delta de l'Ogooué*. Mémoire, Ecole Pratique des Hautes Etudes, Paris, 190 pp.

AUBRÉVILLE, A., 1962. Savanisation tropicale et glaciations quaternaires. *Adansonia* 2(1): 16–84.

BACHMAN S., MOAT J., HILL A.W., DE LA TORRE J. & SCOTT B., 2011. Supporting Red List Threat assessments with GeoCAT: geospatial conservation assessment tool. *Zookeys* 150: 117–126.

BOUPOYA, A., & STÉVART, T., 2013. *Première étude sur la flore et la végétation de la région du Lac Évaro (Moyen-Ogooué-Gabon)*. Rapport, WCS, Libreville, Gabon, 86 pp.

BRETELER F.J., 2007. Novitates Gabonenses 66. Strombosia fleuryana (Olacaceae) new from Gabon. *Systematics and Geography of Plants*, 77: 119–127.

CABALLÉ, G., & FONTES, J., 1978. Inventaires forestiers du Gabon : applications à la phytogéographie. *Bois et Forêts des Tropiques*, 177: 15–33.

CABALLÉ, G., 1978. Essai phytogéographique de la forêt dense humide du Gabon. *Adansonia*, 17: 425–440.

CABALLÉ, G., 1983. Végétation. Pp. 34–37 in: *Géographie et cartographie du Gabon: atlas illustré*. Paris: EDICEF.

CHEVALIER, A., 1917. La Forêt et les Bois du Gabon. *Les Végétaux Utiles de l'Afrique Tropicale Française*, IX, 470 pp + 28 pl.

DEWIT J., 1960. Une nouvelle espèce congolaise du genre Cissus L. (Vitaceae). *Bulletin du Jardin Botanique de l'Etat à Bruxelles*, 30: 95–96.

DOUMENGE, C., NDINGA, A., FOMETE NEMBOT, T., TCHANOU, Z., MICHA ONDO, V., ONA NZE, N., BOUROBOU BOUROBOU, H. & NGOYE, A., 2003a. Conservation de la biodiversité forestière en Afrique centrale atlantique. II - Identification d'un réseau de sites critiques. *Bois et Forêts des Tropiques*, 276(2): 43–58.

DOUMENGE, C., FOMETE NEMBOT, T., TCHANOU, Z., MICHA ONDO, V., ONA NZE, N., BOUROBOU BOUROBOU, H. & NGOYE, A., 2003b. Conservation de la biodiversité forestière en Afrique centrale atlantique. III - Gestion et priorités d'investissement dans les sites critiques. *Bois et Forêts des Tropiques*, 277(3): 53–67.

DROISSART, V., SONKÉ, B., SIMO, M. & STÉVART, T., 2009. New orchid records from Atlantic Central Africa. *Edinburgh Journal of Botany*, 66(1): 1–18.

DU CHAILLU, P.B., 1863. *Voyages et aventures dans l'Afrique équatoriale*. Michel Lévy ed., Paris. 1996. Réédition Sépia-Centre culturel français, Libreville, 588 pp.

EXELL, A.W., 1944. *Catalogue of the Vascular Plants of S. Tomé (with Principe and Annobon)*. London, British Museum (Natural History).

FROMARD, F., 1998. L'écosystème mangrove. Biodiversité, fonctionnement, restauration et gestion. *Bois des Forêts des Tropiques : symposium, of the mangrove ecosystem*, 256(2): 80–86.

FROMARD, F., FONTES, J. & LOUIS, A., 1994. *Structure et dynamique des mangroves de la région de Port-Gentil, Gabon. Analyse de l'impact lié à l'activité pétrolière*. Libreville, rapport scientifique à la demande de Elf-Gabon, 52 pp + annexes.

FUH, M., 1999. *Structure et dynamique de la forêt côtière du Gabon : Implication pour une succession secondaire dérivant de la forêt monodominante à okoumé* (Aucoumea klaineana Pierre). Montpellier, Université des Sciences et Techniques du Languedoc Montpellier II, 183 pp.

GIRARDIN, N., 1978. *La mangrove*. Ministère de l'Education National, Libreville, 31 pp.

GIRARDIN, N., 1982. *Végétation de la côte sableuse*. Ministère de l'Education National, Libreville.

GOYDER, D.J., 2009. A synopsis of Asclepias (Apocynaceae : Asclepiadoideae) in tropical Africa. *Kew Bulletin*, 64: 369–399.

HALL, J.B., 1966. A new species of *Brachystelma* R. Br. (Asclepiadaceae) from Ghana. *Kew Bulletin*, 20: 251–254.

HALLÉ, N., 1986. *Flore du Gabon. Vol. 29. Celastraceae Hippocrateoideae*. Paris, Muséum National d'Histoire Naturelle.

HARRIS, D.J., ARMSTRONG, K.E., WALTERS, G.M., WILKS, C., MOUANDZA MBEMBO, J.-C., NIANGADOUMA, R., WIERINGA, J.J. & BRETELER, F.J., 2012. Phytogeographical analysis and checklist of the vascular plants of Loango National Park, Gabon. *Plant Ecology and Evolution*, 145 (1): 242–257. http://dx.doi.org/10.5091/plecevo.2012.641

HUA, H., 1900. Les explorations botaniques dans les colonies françaises de l'Afrique tropicale, d'après les collections conservées au Muséum d'Histoire naturelle de Paris. *Actes du Congrès International de Botanique de 1900*, 239–247.

IUCN, 2012. *IUCN Red List Categories and Criteria, Version 3.1*. Second edition. Gland, Switzerland and Cambridge, UK. Available from: https://portals.iucn.org/library/efiles/documents/RL-2001-001-2nd.pdf (accessed: 03 September 2014).

IUCN Standards and Petitions Subcommittee, 2014. *Guidelines for Using the IUCN Red List Categories and Criteria*. Version 11 (February 2014). Prepared by the Standards and Petitions Subcommittee. Available from: http://www.iucnredlist.org/documents/RedListGuidelines.pdf. (accessed: 03 September 2014).

JACQUES-FÉLIX, H., 1988. Un nouvel *Heterotis* (Melastomataceae) du Gabon. Bulletin du Muséum National d'Histoire Naturelle, 4e série, sect. B, *Adansonia*, 9: 255–257.

LACHENAUD, O., STÉVART, T., IKABANGA, D., NGAGNIA NDJABOUNDA, E.C. & WALTERS, G., 2013. Les forêts littorales de la région de Libreville (Gabon) et leur importance pour la conservation: description d'un nouveau Psychotria (Rubiaceae) endémique. *Plant Ecology and Evolution*, 146: 68–74.

LACHENAUD, O. & ZEMAGHO, L., 2015. Two new anisophyllous species of Sabicea Aubl. (Rubiaceae) from Gabon. *Candollea*, 70: 219–229. doi: http://dx.doi.org/10.15553/c2015v702a7

LEBIGRE, J.-M., 1990. *Les marais maritimes du Gabon et de Madagascar. Contribution à une étude d'un milieu naturel tropical*. Université De Bordeaux III, thèse de doctorat d'état, 3 tomes, 615 pp.

LEBIGRE, J.-M., 1980. Les mangroves (*Rhizophora spp.*) du Gabon. Un milieu méconnu et potentiellement riche. *Revue Gabonaise d'Etudes politiques, économiques et juridiques*, 4: 37–42.

LEBIGRE, J.-M., 1983a. Les mangroves des rias du littoral gabonais: essai de cartographie typologique. *Revue Bois et Forêts des Tropiques*, 199: 3–28.

LEBIGRE, J.-M., 1983b. *Le littoral du Gabon: Aspects Géomorphologiques et Biogéographiques*. Institut Pédagogique National, Libreville, 58 pp.

LEBRUN, J.-P. & STORK, A.L., 1984. Une nouvelle espèce de *Brachystelma* (Asclepiadaceae) du Mali. *Bulletin du Muséum National d'Histoire Naturelle de Paris*, 4e série, sect. B, *Adansonia*, 6: 491–494.

LEBRUN, J.-P. & STORK, A.L., 1989. Notes sur deux *Brachystelma* (Asclepiadaceae) d'Afrique tropicale. *Bulletin du Muséum National d'Histoire Naturelle de Paris*, 4e série, sect. B, *Adansonia*, 11: 71–77.

LEBRUN, J.-P., STORK, A.L., WÜEST, J. 1994. *Brachystelma letestui* Pellegrin, une rare Asclepiadaceae d'Afrique équatoriale. *Candollea* 49: 183–186.

LEMMENS, R.H.M.J., LOUPPE, D. & OTENG-AMOAKO, A.A. (Editeurs), 2012. *Ressources végétales de l'Afrique tropicale 7(2). Bois d'œuvre 2*. Fondation PROTA, Wageningen, Pays-Bas / CTA, Wageningen, Pays-Bas. 891 pp.

MARCHE, A., 1878. Voyage au Gabon et dans l'Ogooué. *Tour du Monde*, 36: 369–416.

MASINDE, P.S., 2007. A Revision of *Brachystelma* Sims (Apocynaceae: Asclepiadoideae-Ceropegieae) in East Africa. *Kew Bulletin*, 62: 37–84.

MENGUÉMEDOU, C., ONDAMBA OMBANDA, F., NDJOKOUNDA, C., MOUNGANGA, M.-D., BAYANI, E., & MIKALA, R., 2008. *Site Ramsar du bas Ogooué*. Fiche descriptive sur les zones humides Ramsar (FDR), version 2006–2008. 22 pp. Accessible sur http://www.ramsar.org/ris/key_ris_index.htm.

NICOLAS, P., 1977. *Contribution à l'étude phytogéographique de la forêt du Gabon*. Paris: Laboratoire de sociologie et de géographie africaine, thèse de 3e cycle, 351 pp.

NICOLL, M., & LAGRAND, O., 1986. *Conservation et utilisation rationnelle des écosystèmes forestiers du Gabon*. Rapport 3247 IUCN-WWF. Gland, Suisse, 143 pp.

NTORE, S., LACHENAUD, O., JANSSENS, S. & DESSEIN, S., 2010. Four new *Pauridiantha* species (Rubiaceae) reflect the richness of Gabon's rainforests. *Belgian Journal of Botany*, 142(2): 177–193.

ONANA, J.M., 2008. A synoptic revision of *Dacryodes* (Burseraceae) in Africa, with a new species from Central Africa. *Kew Bulletin*, 63: 385–400.

ONDO ASSOUMOU, E., 2006. *Dynamique des paysages végétaux du littoral centre-ouest du Gabon autour de Port-Gentil: approche spatiale et analyse des données de terrain*. Thèse de Doctorat, Université Paul Valéry Montpellier III, 302 pp.

PELLEGRIN, F., 1926. Plantae Letestuanae novae ou Plantes nouvelles récoltées par M. Le Testu de 1907 à 1919 dans le Mayombe congolais. XII. *Bulletin du Muséum National d'Histoire Naturelle*, 32: 393–395.

RABENKOGO, N., 1998. *L'évolution de la presqu'île Mandji, de 1957 à nos jours*. Mémoire de Maîtrise, Université Omar Bongo, Libreville, 115 pp. + annexes.

RABENKOGO, N., 2000. Les zones humides du Gabon: typologie et localisation. *Le cri du Pangolin*, 24: 9.

RABENKOGO, N., 2001. *Dynamique des paysages dans le département d'Etimboué, Ogooué-Maritime (Gabon)*. Rapport final, programme "Valorisation des écosystèmes humides en Afrique Centrale (PVEHAC)," Libreville, p 10–35.

RAPONDA-WALKER, A., & SILLANS, R., 1961. *Les plantes utiles du Gabon: essai d'inventaire et de concordance des noms vernaculaires*. Editions Paul Lechevalier, Paris, 614 pp.

RAYNAL, J., 1968. Itinéraires et lieux de récolte de Georges Le Testu. *Flore du Gabon*, 14: 17–66.

SOSEF, M.S.M, HARRIS, D.J. & ARMSTRONG, K., 2007. Novitates Gabonenses 64. A new species of *Campylospermum* (Ochnaceae) from coastal Gabon. *Blumea*, 52: 15–19.

SOSEF, M.S.M, WIERINGA, J.J., JONGKIND, C.C.H., ACHOUNDONG, G., AZIZET ISSEMBE, Y., BEDIGIAN, D., VAN DER BERG, R.G., BRETELER, F.J., CHEEK, M., DEGREEF, J., FADEN, R., GOLDBLATT, P., VAN DER MAESEN, L.J.G., NGOK BANAK, L., NIANGADOUMA, R., NZABI, T., NZIENGUI, B., ROGERS, Z.S., STÉVART, T., VAN VALKENBURG, J.L.C.H., WALTERS, G. & DE WILDE, J.J.F.E., 2006. Check-list des plantes vasculaires du Gabon. *Scripta Botanica Belgica*, 35. Meise, Jardin Botanique National de Belgique.

STÉVART, T. & WALTERS, G., 2016. *Simirestis klaineana*. The IUCN Red List of Threatened Species, 2016: e.T203777A2771328. http://dx.doi.org/10.2305/IUCN.UK.2016-1.RLTS.T203777A2771328.en. (accessed: 13 March 2017).

STÉVART, T., BARBIER, N., BIDAULT, E., BOUPOYA, A., VIENNOIS, G., LACHENAUD, O. & TEILLIER, L., 2015. *Projet de caractérisation de la végétation du Massif Forestier d'Evaro*. Rapport final de l'inventaire multi-ressources relative aux résultats botaniques. 77 pp.

UICN, 1990. *La Conservation des Ecosystèmes forestiers du Gabon*. Basé sur le travail de C. Wilks. UICN, Gland, Suisse et Cambridge, Royaume-Uni, xiv + 215 pp, illustré.

VAN DER ZON, A.P.M., 1992. *Graminées du Cameroun*. Volume II. Flore. Wageningen Agricultural University Papers, 92(1): 1–558.

VANDE WEGHE, J.P., 2003. Les milieux humides côtiers du Gabon: les rias du nord et le bas Ogooué. *Canopée*, 25: 3–8.

VANDE WEGHE, J.P., 2007. *Les Parcs Nationaux du Gabon: Loango, Mayumba et le bas Ogooué*. WCS Gabon, Libreville, Gabon, 320 pp.

VANDE WEGHE, J.P., BIDAULT, E. & STÉVART, T., 2016. *Plantes à fleurs du Gabon*. Libreville, Agence Nationale des Parcs Nationaux.

VILLIERS, J.-F., 1973. *Etude floristique et phytosociologique d'une mangrove atlantique sur substrat rocheux du littoral gabonais*. Thèse de doctorat, Université de Paris VI, 169 pp.

WHITE, L., 1995. *Etude de la végétation*. Rapport final projet "ECOFAC-Composante," Gabon, 140 pp.

WILKS, C., 1990. *La Conservation des Ecosystèmes forestiers du Gabon*. UICN, Gland.

Appendix: List of endangered plant species of the Ogooué Delta

IUCN status (provisonal): CR = Critically Endangered; EN = Endangered; VU = Vulnerable
Habitat: DLF = Dryland forest; SF = Swamp forest; FF = Floodable forest
Distribution: A = Lower Ogooué valley; B = Upper deltaic plain; C = Lower deltaic plain; D = Coastal sands; E = Nkomi Lagoon (or Fernan Vaz Lagoon)
Endemism: GCo = Guineo-Congolian; BG = Lower Guinea; G = Gabon; BOg = Lower Ogooué

	IUCN	Habitat	A	B	C	D	E	End.
ACANTHACEAE								
Justicia tigrina	EN	SF in valley bottoms, DLF	■					G
Pseudocalyx macrophyllus	EN	DLF	■					G
Whitfieldia letestui	EN	SF in valley bottoms	■					G
ANACARDIACEAE								
Trichoscypha argentea sp. nov. ined.	EN	DLF	■					G
ANNONACEAE								
Artabotrys crassipetalus	VU	SF in valley bottoms	■					G
Artabotrys letestui	VU	DLF		■				BG
Monanthotaxis glomerulata	EN	SF in valley bottoms	■		■			G
Piptostigma oyemense	VU	DLF	■					BG
Pseudartabotrys letestui	VU	SF		■				G
Uvaria aff. *klainei* sp. nov.	CR	Ruderal	■					BOg
Uvaria annickiae	VU	Edge of DLF	■	■				G
Uvaria bipindensis	VU	DLF		■				BG
Uvaria clavata	VU	DLF, edges		■				BG
APOCYNACEAE								
Alafia falcata	EN	Ruderal	■					G
Baissea longipetiolata	EN	DLF	■					BG
Brachystelma letestui	EN	Dry savanna			■	■		BG
Pleiocarpa picralimoides	EN	DLF		■				BG
Pleioceras zenkeri	EN	DLF	■					BG
ARACEAE			■					
Culcasia obliquifolia	VU	DLF	■					BG
Culcasia rotundifolia	VU	DLF, SF in valley bottoms	■					BG
ASPARAGACEAE								
Dracaena nyangensis	EN	DLF	■					BG
BEGONIACEAE								
Begonia aspleniifolia	VU	DLF	■					G
Begonia erectotricha	EN	DLF	■					G
Begonia letestui	EN	DLF	■					G
Begonia mbangaensis	VU	DLF, ravines	■					BG
Begonia vankerckhovenii	VU	DLF, ravines	■					BG
Begonia wilksii	EN	DLF	■					G
BURMANNIACEAE								
Burmannia liberica	VU	Humid savanna		■				GCo
Gymnosiphon constrictus	EN	DLF		■				BG
BURSERACEAE								
Dacryodes ebatom	EN	SF	■					BOg

	IUCN	Habitat	A	B	C	D	E	End.
CAPPARACEAE								
Capparis sp. nov.	CR	Riparian FF	■					BOg
CELASTRACEAE								
Pristimera mouilensis	EN	DLF	■					G
Salacia coronata	EN	Ruderal, DLF	■					G
Salacia diplasia	VU	DLF, ravines	■					G
Salacia sp. nov. aff. *tessmannii*	CR	DLF	■					BOg
Simirestis klaineana	EN	DLF			■			G
CHRYSOBALANACEAE								
Dactyladenia eketensis	EN	SF	■	■				BG
Dactyladenia jongkindii	EN	DLF	■					G
Dactyladenia letestui	VU	SF in valley bottoms		■				BG
Dactyladenia ndjoleensis	EN	DLF	■					G
Dactyladenia sp. nov.	EN	DLF	■					G
Magnistipula devriesii	EN	DLF	■					G
COMBRETACEAE								
Combretum clarense	VU	DLF		■				G
Combretum echirense	EN	DLF	■					BG
Combretum erosum	EN	DLF	■					G
Combretum exellii	VU	Ruderal	■					BG
Combretum ndjoleense	EN	DLF	■					BOg
Combretum oudenhovenii	VU	DLF		■				G
Combretum polyanthum	EN	DLF	■					G
Combretum sp.	CR	DLF	■					BOg
COMMELINACEAE								
Palisota alboanthera	EN	DLF	■					G
CONNARACEAE								
Cnestis uncata	VU	Ruderal	■	■				G
CUCURBITACEAE								
Coccinia racemiflora	VU	SF in valley bottoms	■	■				BG
CYPERACEAE								
Mapania raynaliana	EN	DLF	■					BG
DICHAPETALACEAE								
Dichapetalum findouense	EN	Ruderal	■					G
Dichapetalum pulchrum	VU	DLF	■					BG
Dichapetalum rabiense	EN	DLF	■					G
Tapura carinata	VU	DLF		■				BG
Lastreopsis davalliaeformis	VU	DLF	■					BG
EBENACEAE								
Diospyros cleistantha	EN	DLF		■				G

	IUCN	Habitat	A	B	C	D	E	End.
Diospyros subargentea	VU	DLF	■	■				G
ERIOCAULACEAE								
Eriocaulon nadjae	EN	Humid savanna				■		BG
EUPHORBIACEAE								
Afrotrewia kamerunica	EN	DLF	■					BG
Aubletiana leptostachys	VU	DLF, SF in valley bottoms	■					BG
Crotonogyne parvifolia	EN	Riparian FF	■					G
FABACEAE								
Anthonotha mouandzae	EN	DLF		■				G
Baphia cymosa	EN	DLF		■				G
Calpocalyx brevifolius	VU	DLF	■					G
Dalbergia crispa	VU	Ruderal		■				GC
Dalbergia stenopetala	VU	Edges of DLF	■					G
Daniellia oblonga	EN	DLF	■					BG
Entandrophragma angolense var. *dolichocarpum*	EN	DLF		■				BOg
Gilbertiodendron imenoense	VU	DLF		■				G
Gilbertiodendron quinquejugum	VU	DLF, SF in valley bottoms	■					BG
Englerodendron hallei	EN	DLF	■					G
Oddoniodendron normandii	VU	DLF	■					G
Ormocarpum klainei	EN	DLF		■				BG
HUACEAE								
Afrostyrax macranthus	VU	DLF		■				BG
LAURACEAE								
Beilschmiedia alata	EN	DLF		■				BG
Beilschmiedia minutiflora	VU	DLF		■				BG
Cassytha graminicola	VU	DRY savanna		■	■			G
LECYTHIDACEAE								
Crateranthus le-testui	EN	SF in valley bottoms	■	■				G
Napoleonaea gabonensis	VU	DLF	■					G
LORANTHACEAE								
Agelanthus glaucoviridis	VU	DLF	■					GC
Phragmanthera longiflora	VU	DLF	■					BG
MALPIGHIACEAE								
Acridocarpus caparados	EN	DLF	■					G
MALVACEAE								
Cola lissachensis subsp. nov.	EN	SF in valley bottoms, DLF	■	■				G
Cola sp. nov. aff. *rostrata*	CR	DLF		■				BOg
Scaphopetalum mannii	EN	SF in valley bottoms	■	■				BG
Scaphopetalum vanderystii	EN	DLF	■					BG
Tarrietia densiflora	VU	DLF	■	■				G
MARANTACEAE								
Marantochloa alba	EN	DLF	■					G
Thaumatococcus flavus	EN	SF in valley bottoms	■					G
MELASTOMATACEAE								
Amphiblemma hallei	VU	SF in valley bottoms	■					G
Amphiblemma setosum	VU	DLF	■					G
Amphiblemma soyauxii	VU	SF in valley bottoms		■				G
Memecylon aff. *lateriflorum* sp. nov.	VU	DLF	■					G
Memecylon diluviorum	VU	Riparian FF						G
Memecylon salicifolium	EN	DLF	■					G
Tristemma rubens	VU	DLF		■				BG
Warneckea macrantha	VU	Riparian FF	■					G
MELIACEAE								
Carapa batesii	VU	DLF	■					BG
Carapa hygrophila	VU	Riparian FF	■					BG
Carapa mangarevensis	EN	SF			■			G
MENISPERMACEAE								
Tiliacora gabonensis	VU	DLF		■				BG
Tiliacora sp. nov. aff. macrophylla	EN	DLF	■					BOg
MORACEAE								
Ficus adolfi-friderici	VU	DLF	■					BG
MYRTACEAE								
Eugenia anisata sp. nov.	CR	DLF		■				BOg
Eugenia dewevrei	VU	Riparian FF		■				BG
Eugenia ogoouensis	EN	Riparian FF	■					G
Eugenia thollonii	EN	Riparian FF	■					G
ORCHIDACEAE								
Bulbophyllum pauwelsianum	EN	DLF	■					G
Diaphananthe lanceolata	EN	DLF	■					BG
Diaphananthe sarcorhynchoides	VU	DLF, mangrove	■					GC
Hetaeria mannii	VU	DLF	■	■				BG
Polystachya testuana	VU	DLF	■					BG
Veyretella flabellata	CR	SF in valley bottoms	■					BOg
Veyretella hetaerioides	VU	SF	■					G
PANDANACEAE								
Pandanus gabonensis	EN	SF	■					G
PERACEAE								
Chaetocarpus gabonensis	VU	DLF	■	■	■			G
PERIDISCACEAE								
Soyauxia sp. nov.	EN	DLF			■			G
PHYLLANTHACEAE								
Maesobotrya sp. nov. (Boupoya)	CR	DLF		■				BOg
Phyllanthus gabonensis	VU	Dry savanna				■		BG
POACEAE								
Eragrostis sabulicola	EN	Dry savanna				■		BG
PUTRANJIVACEAE								
Drypetes verrucosa	VU	DLF	■	■				G
RHIZOPHORACEAE								
Cassipourea dinklagei	VU	DLF, SF	■					BG
Cassipourea pumila	CR	DLF	■					BOg
RUBIACEAE								
Chassalia carvalhoi subsp. *brevifolia*	EN	DLF	■	■				BOg
Craterispermum deblockianum	EN	SF in valley bottoms	■					G
Empogona bequaertii	VU	DLF		■				BG
Eumachia auriculata sp. nov.	CR	DLF	■					BOg
Eumachia eriocephala sp. nov.	VU	DLF		■				G
Eumachia mabouniensis sp. nov.	EN	DLF	■					G

	IUCN	Habitat	A	B	C	D	E	End.
Heinsia pilosa	EN	Ruderal, DLF	■					G
Hymenocoleus bracteosus sp. nov.	EN	SF in valley bottoms	■	■				G
Hymenocoleus pseudoscaphus sp. nov.	EN	SF in valley bottoms, DLF	■					G
Keetia bracteosa sp. nov.	EN	DLF		■				G
Keetia sp. nov.	EN	DLF		■				G
Massularia stevartiana	EN	DLF	■					BG
Pauridiantha smetsiana	EN	DLF		■				G
Pauridiantha gracilipes	CR	SF		■				BOg
Pauridiantha uniflora	EN	DLF	■					G
Pavetta mayumbensis	VU	DLF	■					BG
Psychotria acutigemma subsp. *acutigemma*	EN	Riparian FF	■					BOg
Psychotria klainei	EN	DLF			■			G
Psychotria koumounaboualiensis	EN	DLF		■				G
Psychotria neurosticta	EN	DLF		■				G
Psychotria wieringae	VU	DLF			■			G
Rutidea gabonensis	VU	DLF		■				G
Sabicea golgothae	CR	Ruderal	■					BOg
Sabicea mabouniensis	CR	Ruderal	■					BOg
Sabicea mapiana	EN	Ruderal	■					BG
Sabicea sciaphilantha subsp. *sciaphilantha*	EN	Ruderal, SF in valley bottoms	■	■				BOg
Tarenna ogoouensis	CR	DLF	■					BOg
Tarenna sp. nov. aff. *baconioides*	CR	DLF	■					BOg
Tricalysia sp. nov. *mabounié*	EN	DLF	■					G
Vangueriella dowsettiae sp. nov.	EN	DLF				■		BG
SAPINDACEAE								
Allophylus hallei	VU	DLF		■				G
Allophylus le-testui	VU	DLF	■					G
Allophylus oyemensis	EN	DLF	■					BG
SAPOTACEAE								
Englerophytum gigantifolium	CR	DLF	■					BOg
Neolemonniera ogouensis	EN	Riparian FF		■				G
Synsepalum fleuryanum	VU	Riparian FF	■	■				G
Tulestea koulamoutouensis	EN	DLF	■					G
VIOLACEAE								
Allexis batangae	VU	DLF	■					BG
Rinorea abbreviata	CR	DLF	■					BOg
VITACEAE								
Cissus prunifera	EN	Ruderal, DLF	■					G

4.1. Pandanus *thickets (Photo: Tariq Stévart).*

4.2. Raffia palm stand that has developed in front of a mangrove with high Rhizophora *trees (Photo: Tariq Stévart).*

4.3. Riparian forest with Cynometra mannii *(Photo: Tariq Stévart).*

4.4. Crinum natans *mats in full bloom in June 2015 in the northern delta's Kondjo River (Photo: Tariq Stévart).*

4. The vegetation of the Ogooué Delta

Archange Boupoya, Ehoarn Bidault & Tariq Stévart

A bit of history

Gabon, with an estimated forest cover of 85%, is known for its large expanses of forest and their role in maintaining Africa's biodiversity. However, despite a large number of research studies on the forest habitats of Gabon (Chevalier, 1916; Saint-Aubin, 1961, 1963; Fontès & Caballé, 1977; Caballé, 1978; Reitsma, 1988; Dauby *et al.*, 2013), many of its forest types are still poorly known. This is the case, for example, of hydromorphic or periodically flooded forests. Indeed, Gabon contains large expanses with hydromorphic environments, most often in its coastal basin. Furthermore, Gabon is also home to non-forest habitats such as savannas, humid grasslands, forest clearings, inselbergs, and rocky slabs. These relatively poorly known habitats greatly contribute to the country's flora diversity since there are species restricted to each of these environments (Boupoya, 2010).

Although knowledge of Gabonese vegetation is being gradually accumulated, information on some of the country's regions remains limited. This was the case of the Ogooué Delta, which has the largest surface of wetland habitats in the country, as well as one of the largest in tropical Africa, which explains why actors in the field of conservation had shown interest in this diversity of wetland habitats for some time, resulting in the various inventories that this book aims to synthesize.

While the flora of the Lower Ogooué Ramsar site is poorly documented, this site has a long history of botanical exploration (pages 142–171). The first published work about this flora was written by Chevalier and released in 1917 as part of a broad study that included the Lower Ogooué region. A few smaller studies were subsequently written. It should, however, be noted that they are mainly focused on the flora and only superficially mentioned vegetation.

Starting in 1980, many studies were conducted focusing primarily on the conservation in the Lower Ogooué site, which included partial vegetation data: Lebigre, 1983b; Nicoll & Langrand, 1986; Wilks, 1990; Doumenge *et al.*, 2003a, 2003b. Teams from the Missouri Botanical Garden (MBG) undertook a vast vegetation sampling and description program, first with WCS-Gabon, later as part of the initial assessment of the Maboumine mining concession (in the Mabounié area), and more recently with WWF-Gabon and the National Parks Agency (ANPN), for the production of this book. Since 2011, MBG and its partners have thus carried out numerous studies regarding both the flora and the vegetation of the Lower Ogooué site (Boupoya & Stévart, 2013; Stévart *et al.*, 2015).

The Ogooué Delta and its regional context

Any discussion of African phytogeography has to mention the work of White (1978, 1979, 1983). This botanist's major contribution was the preparation of a chorological map, which remains the most widely used tool of its kind to this day. White (1983) describes six regions characterized by high rates of endemism and separated from one another by transition zones with lower endemism. He published the first vegetation map of Africa (White, 1983), in which the Lower Ogooué is included in the Guinean-Congolese regional center of endemism, and more precisely in the Lower Guinean domain (Figure 4.5).

Several studies have also focused on the forest typology of Gabon (de Saint-Aubin, 1961; Nicolas, 1977; Caballé, 1978) by analyzing the presence/absence and abundance/scarcity of certain species, and in particular okoumé, *Aucoumea klaineana*, a species subendemic to Gabon and also the most exploited by the logging industry. A common element among Gabon in the various proposed forest typologies is the distinction between forests with okoumé and forests without okoumé, which are located in the northeastern part of Gabon. The Lower Ogooué is thus located in the area in which forests contain okoumé. Caballé described

4.5. The Lower Ogooué in the Lower Guinean domain (phytogeographic synthesis proposed by Dauby, 2012).

- *Mangroves*
- *Savannas*
- *Continental forest*
- *Forest of the Crystal-Ngovayang range*
- *Coastal forest*
- *Forest of the Du Chaillu massif*
- *Ogooué-Ivindo Forest*
- *North-Sanaga Forest*
- *Semideciduous forest*

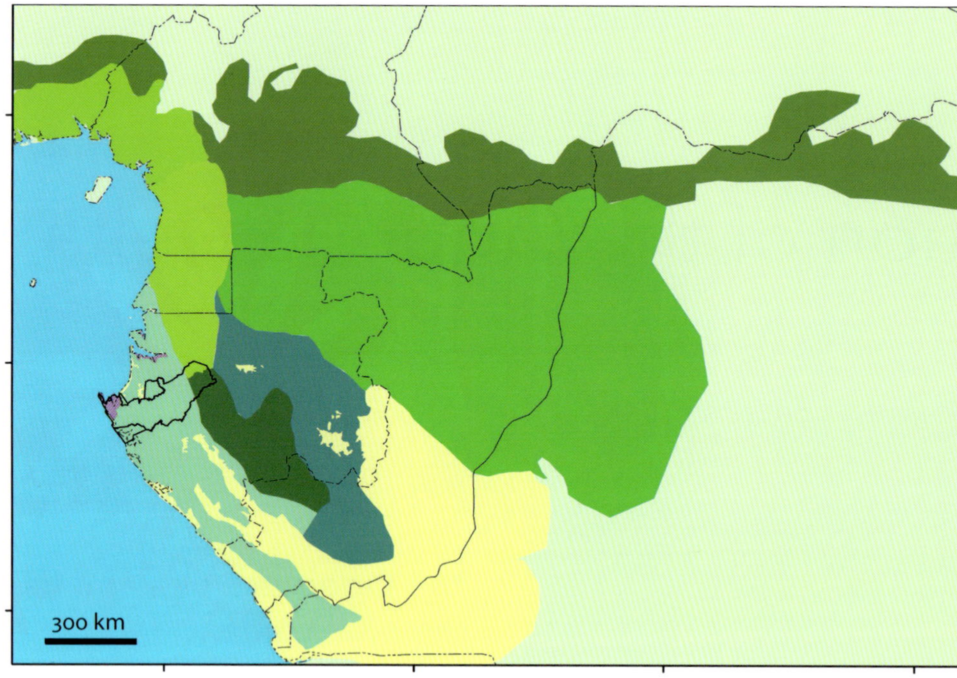

4.6. The Lower Ogooué according to the map of Caballé (1978).

- *Evergreen hill forest with* Aucoumea *and* Desbordesia
- *Evergreen hill forest with* Aucoumea *and* Paraberlinia
- *Evergreen hill forest with* Scyphocephalium *and* Pycnanthus
- *Evergreen coastal forest with* Aucoumea *and* Sacoglottis
- *Evergreen coastal forest with* Aucoumea *and* Desbordesia
- *Flooded and swamp forests*
- *Mangrove with* Rhizophora *and* Avicennia
- *Papyrus beds and swamps*
- *Rural complex*
- *Savanna*
- *Inner steppe of the Batéké Plateaux*

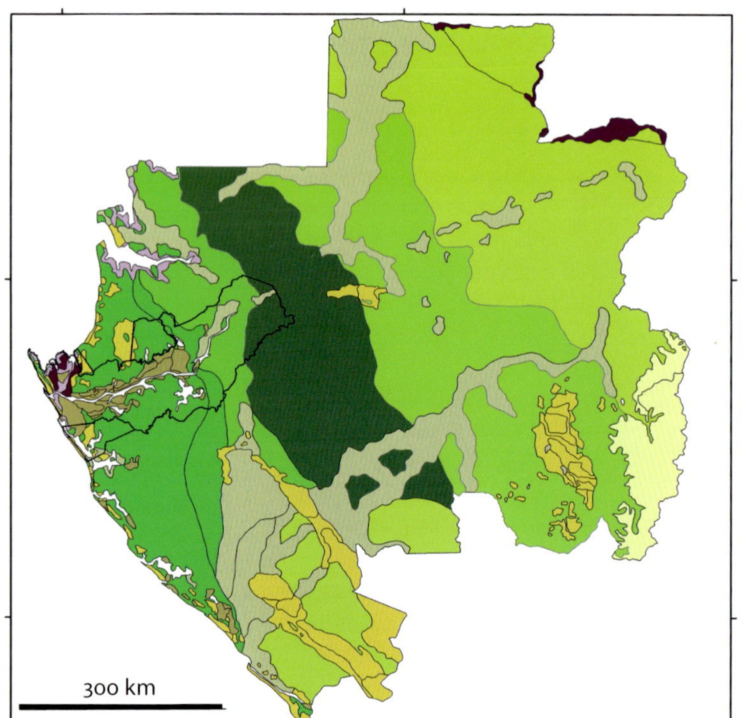

two types of vegetation in Lower Ogooué, one with ozouga, *Sacoglottis gabonensis* and the other with alep, *Desbordesia glaucescens*, and ozigo, *Dacryodes buettneri* (Figure 4.6).

A second delineation is much less obvious, distinguishing forests from the sedimentary basin from the flat and sandy area by the sea, from inland forests located on higher ground. Moreover, this delineation is handled differently in the phytogeographical systems proposed by Nicolas (1977) and Senterre (2005). Senterre (2005) includes forests from the base of mountains, which he describes as transitional forests under coastal/western influence, illustrating the gradual nature of this delineation. Nicolas (1977) subdivides the forests of the sedimentary basin into several units, mainly emphasizing each unit's edaphic properties in order to explain the boundaries he recognizes, specifically the texture of the soil (clay or sand), which varies considerably throughout the sedimentary basin.

The first steps toward a classification of vegetation

As previously mentioned, studies of the vegetation in the Lower Ogooué region started relatively recently compared to floristic studies. The first attempts to classify the vegetation of Lower Ogooué were published by Raponda-Walker & Sillans (1961). In addition to describing the coastal plant formations, they gave a first overview of the types of hydromorphic forest present in Lower Ogooué by describing certain plant formations. These included flood forests with *Oxystigma mannii* or *O. dewevrei*, which are found in the Nkomi Lagoon, as well as two types of floodable forests found in Gabon at the borders of flood forests and dryland forests. These formations occupy the low-lying parts of the terrain, on flat, humid, and poorly drained soil with sand or clay: on one hand, forests with *Sacoglottis gabonensis* and *Aucoumea klaineana*, alongside *Symphonia globulifera, Anthostema aubryanum, Copaifera coleosperma, Berlinia bracteosa, Klainedoxa gabonensis, Staudtia gabonensis, Calpocalyx heitzii, Erythrophloeum africanum, Irvingia grandifolia, Coula edulis, Nauclea diderrichii,* and *Entandrophragma utile*; and on the other hand, forest with *Sacoglottis gabonensis* and *Ctenolophon englerianus*, which are particularly present along the entire right-hand side of the upper deltaic plain (the Alombié Lake region). These authors also link the forest of the Nkomi Lagoon to these two formations, although the data they had for this area were quite limited. This first classification of hydromorphic forests was, unfortunately, not based on any statistical analysis. A statistically based classification of vegetation types was later possible thanks to the works of Ondo Assoumou (2006) and especially Boupoya & Stévart (2013) based on work carried out in the Lake Evaro area.

The eastern part of Lower Ogooué is especially affected by the lower deltaic plain, where fresh water mixes with marine water. The presence of mangroves differentiates this coastal part of the site from the remainder. Raponda-Walker & Sillans (1961) provided a first attempt at defining the littoral plant formations of Gabon and mentioned that mangroves are typical of the Ogooué Delta, where they are the most abundant within the country. According to these authors, mangroves cover entire swamps of widely varying sizes, which are covered with water at high tide. They are mainly found in the Ogooué Delta, the Gabon Estuary, and Mondah Bay, and are most often nearly monospecific populations of *Rhizophora racemosa* with *Avicennia nitida* and *Conocarpus erectus*. Brackish waters reach the base of these mangroves at high tide, and at low tide a dense network of arched aerial roots is revealed, which supports the very straight and long barrels (Figure 4.7).

Many authors have described Gabon's coastal vegetation with varying degrees of precision (Girardin, 1978, 1982; Fromard, 1998; Villiers, 1973), but Lebigre, in his numerous publications about the coast of Gabon (1980, 1983a, 1983b, 1990), provided the most precise description of the vegetation of the lower part of the Ogooué Delta, and in particular the littoral or sub-littoral habitats. He describes vegetation growing on sandy ridges, coastal thickets adjoined by coastal forests, tropical coastal steppes, mangroves, and tannes.

This classification was then largely confirmed and clarified by more recent studies based on statistical methods, including Ondo Assoumou's thesis (2006) on the dynamics of botanical landscapes on the coasts surrounding Port-Gentil. Recently, many studies have focused on natural coastal environments, from Port-Gentil to Omboué, including Fromard *et al*. (1994), Fromard (1998), and Rabenkogo (2001), among others.

Taking advantage of the inventories carried out as part of the Maboumine baseline study and the project carried out in collaboration with WWF-Gabon in the Mpaga region south of Wonga-Wongué, a more exhaustive study made it possible to define the region's dryland habitats, mainly forests and savannas, as well as wetlands (swamp forests, flood forests, grasslands, raffia groves, and *Pandanus* formations), and to highlight the fact that humidity, distance from rivers, and altitude contribute to habitat variability. Moreover, dryland forests are heavily influenced by anthropogenic activities and the effects of topography, ridges, and ravines. At the scale of the Lower Ogooué region, there are few dry savannas, which only cover small areas. However, they are abundant in the southern part of the Wonga-Wongué Reserve. Certain types of plant formations inaccessible by foot were only identified thanks to the aerial surveys carried out in 2012 and 2013. This work, coupled with the use of high definition satellite photography, enabled the establishment of the first vegetation map of the Lower Ogooué region in 2015. This map served as a precursor for those provided in the chapter on mapping the region's vegetation (pages 198-221).

Before presenting the descriptions of the vegetation, we present the main natural or anthropogenic processes that have led to the structuring and distribution of habitats in the Lower Ogooué.

Natural history of the vegetation of the Lower Ogooué

During the recurring ice ages that have affected Central Africa, the climate cooled and became drier, leading to a decrease in forest cover. However, forests have never entirely disappeared, and stands persisted in areas that remained sufficiently humid. These remnants are referred to as forest refuges because of their role in maintaining forest species. Although the theory of glacial forest refuges is widely accepted by the scientific community, their location, size, and dynamics remain controversial to this day.

In Gabon, forest refuges were located in mountain ranges, including the Chaillu Mountains and the Crystal Mountains. These large regions have significant environmental and topographic heterogeneity, allowing them to remain humid despite the decrease in precipitation during the ice ages. These mountain ranges may therefore have served as refuges for forest species, allowing them to recolonize abandoned areas once favorable conditions had returned. Similarly, smaller forest clusters may have subsisted in steep,

4.7. Pure mangrove with large Rhizophora *in the northern delta (Photo: Tariq Stévart).*

4.8. *Gallery forest in the Mpaga region (Photo: Tariq Stévart).*

humid areas or along rivers; these areas are referred to as microsanctuaries. They may have resembled gallery forests (Figure 4.8) or small clusters of forest in marshy environments. Finally, one can imagine that, during ice ages, Gabon would have been covered with large expanses of savannas broken by forest-covered mountains in humid areas, and also by small forests when the microtopography was suitable. Most depictions of the landscapes of Gabon at that point in time resemble those observed today in the Wonga-Wongué Reserve, although there is no evidence to support this.

It is difficult to imagine what the Ogooué Delta looked like during ice ages, but a strong reduction in humidity and river flow seems likely, as well as a lowering of the sea level, all of which had a major impact on the vegetation. Indeed, there is not much environmental heterogeneity in the delta, and the fact that dry savannas persist today, along with young and pioneer forests—as can be seen in the Lake Evaro region—suggests that the environment was mostly dry, with savannas, during the last ice age. We can therefore assume that forest refuges, supposedly located east of the Ramsar site, would have harbored species that recolonized the area currently occupied by forest located in the center and east of the site.

What about the large forest located south of Wonga-Wongué and the coastal forest? The species that comprise these forests on sandy soils are different from those of the forests of central Gabon, and the vegetation also has fewer species overall. These forests do not include species frequently found in the forests of Gabon, such as *Santiria* ssp. and *Lophira alata*. On the other hand, they contain some endemic species that are absent from the mountain refuges usually mentioned in publications and situated in the center of Gabon, and it remains unclear how they survived during the ice ages.

A partial answer may come from analyzing the dynamics of forest and savanna in the region of Lake Alombié (Mpaga), as well as the presence of dune ridges in the coastal forests. Dune ridges have occurred in the area for as long as the delta has existed, as indicated by their presence at various locations on the lower deltaic plain, despite the coastline having shifted. These landscapes alternate between open vegetation, such as wet savannas and marshes, and strips of dryland forest, rich in *Chrysobalanus icaco*, and which include swamp forests

4.9. *Lake Alombié in the Mpaga region. The savannas in the background are the southern extension of those of the Wonga-Wongué (Photo: Tariq Stévart).*

in localized microdepressions that allow water to accumulate. These forests are very difficult to access, but have a distinctive flora that could have been maintained by the presence of humidity in the low areas of these humid troughs or by the ability of the species to adapt to dry, sandy environments, rendering them resilient to the drier ice ages. This all remains very hypothetical, however, and the succession of vegetation in this area is less well documented than around Mpaga.

The forest-savanna dynamic in southern Wonga-Wongué

The Mpaga region was the subject of an in-depth study in 2014 as part of a series of inventories commissioned by WWF-Gabon. A team of about 15 people explored the site for two weeks and carried out a study of its vegetation and flora. The goal was to collect data in an area that had been identified in one of the high-definition satellite images the project had purchased, in order to extend the resulting interpretations and mapping to the entire site.

This expedition generated significant new knowledge and made it possible to highlight the presence of species with limited distributions, to describe the dynamics of how forests colonize the savanna, and especially to show the importance of *Ctenolophon englerianus* (Figure 4.11) in maintaining the region's dryland forests. This species is characteristic of both flood forests, where it forms monospecific colonies, and of dryland forests, where it is more sparsely distributed. However, the discovery of marshes inside large gallery forests, where the ground is so flat that it is difficult for water to drain from the area, offers potential new explanations of how these forests may have been maintained during drier periods. These areas are mainly populated with impressively sized individuals of *Ctenolophon englerianus*, which constitute the oldest known population in the coastal part of the delta, certainly more than 300 years in age. These large trees form significant mounds of dead leaves to avoid stagnant water. These forests therefore accumulate large amounts of organic matter while the trees create conditions suitable for marshes that could have remained humid throughout long dry periods of at least 300 years. Other species find refuge and survive on these mounds of dead leaves at the foot of the tree trunks, including shrubs or forest grasses. When a tree dies, the dead leaves persist and can host another species of large tree, which will then serve as a launching point for the colonization of nearby savannas. Finally, it should be noted that the presence of a forest edge rich in okoumé and ozouga, two pioneer species that are regarded as structural elements in coastal forests.

It is apparent that forests once covered a larger area, as revealed by topographical deformations in the savanna that are probably the result of humus build-up caused by forest species. Forest therefore seems to have been more extensive in

4.10. Natural succession in the savannas. When savannas are no longer burnt, forest edges slowly encroach on the savanna, and some trees are able to grow directly inside the savanna (Photo: Tariq Stévart).

the past, and following a sharp regression due to periods of drought, it is once again colonizing the savanna, at least when it is not burned by humans. In fact, most of the savannas of Mpaga are burned, which prevents forest vegetation from establishing, and specifically prevents the forest edge from slowly encroaching on the savanna. When the savanna is not burned, a strip of herbaceous plants forms quickly along the forest edge, which creates conditions favorable for the growth of okoumé, a species that dominates forest edges. Forest therefore colonizes savanna along its edge, but also by spreading ozouga seedlings, which grow from seeds dispersed in the excrement of elephants in the area. Ozouga fruits are in fact one of the few eaten by elephants in the region, which illustrates the lack of plant diversity. Burnt ozouga seeds are often found in the savanna, indicating that fire has prevented the establishment of this species. On the other hand, in areas untouched by fire, isolated ozouga individuals can facilitate the establishment of connections between forested areas. Monospecific stands of old ozouga trees can also be found within these forests. This type of vegetation has been reported previously in publications, in which two explanations are offered for their presence. The first suggests that this species releases chemicals into the soil, which prevents other species from growing, an idea that is reinforced by the fact that these monospecific populations are clear of undergrowth. The second explanation suggests that pockets of ozouga, initially established from elephant excrement, are used by these amimals for feeding, reinforcing the pioneer populations. The two explanations are not mutually exclusive. It is therefore reasonable to assume that small pockets of forest remained in the wetlands, sheltering mainly *Ctenolophon*, okoumés, and ozougas, all of which have a wide ecological range. Okoumé and ozouga trees are also found in floodplains.

Beyond these ancient dynamics dating from the ice ages, a much more recent colonization of the savanna has taken place in Mpaga, which points to rapid changes in vegetation, on the order of a few dozen years at most. This brings us to the subject of swamp vegetation recently colonizing lakes.

4.11. Flood forest with Ctenolophon englerianus *in the Mpaga region (Photo: Nicolas Barbier).*

4.12. Dryland forest north of Lake Alombié (Photo: Ehoarn Bidault).

Dynamics of the expansion of riparian forests

One of the most significant species along the banks of the Lower Ogooué is *Ceiba pentandra*. This tree has a massive stature and a large trunk, and flowers once leafless, which makes it easily identified by the uninitiated. It is regularly present in riparian forests along riverbanks. This pioneer species contributes significantly to the colonization of lakes.

Sediment build-up can be observed in most of the lakes of the Upper Ogooué deltaic basin, where they form around the channels that carry the brown waters of the Ogooué and flow into the black waters of lakes during the rainy season. These rivers slowly fill up the lakes with their sediment, facilitating rapid colonization from their edges by floating meadows with *Echinochloa stagnina*. With the help of sediments, these meadows are replaced by shrubs, including *Alchornea cordifolia*, which are in turn colonized by monospecific populations of *Ceiba*. Dry banks form rapidly, allowing other species to colonize populations of shrubs as well as dense and diversified riparian forest. The *Ceiba* slowly die, but natural regeneration occurs. This colonization process divides the lakes into smaller units. It is easy to observe this dynamic along the Rivière des Pélicans and at the point where it flows into Lake Onangué. It is, however, linked to sedimentation influenced by the Ogooué. Some rivers, fed by rainwater runoff from nearby forests and savannas, do not carry sediment and therefore have black water. Along these rivers, the banks are lower and are usually permanently flooded. Furthermore, these riparian forests in black water have low species diversity since fewer diaspores are carried by the river. Few rivers in the area resemble the Ogooué in this regard since most of them originate in the Wonga-Wongé Plateau and the northern delta near Port-Gentil, including the Kendjé tidal river.

4.13. *Alchornea cordifolia* shrub cover with *Ceiba pentandra* along one of the large channels of the Ogooué (Photo: Ehoarn Bidault).

Temporal and geographic succession in the lower deltaic plain

As previously described, the vegetation of Lower Ogooué region exhibits both fast and slow succession; it is generally both complex and poorly documented, resulting in an array of habitats unique within Gabon. However, the most spectacular dynamic found in the Lower Ogooué region, and which is unique within Central Africa, is the slow and gradual geographic succession that begins at the mudflats found at river mouths and along the coasts, and ends with dense swamp forests in the nonsaline environments at the center of the coastal delta.

A gradual environmental transition can be observed from the marine edge of the estuary to the center of the lower deltaic plain. It starts with pure, high mangroves by the sea, which include areas of dryland forest and savanna in places where the altitude exceeds a few meters, and ends with swamp forests. As previously noted, the Ogooué Delta has evolved significantly over time as a result of fluctuations in sea level, and this succession corresponds to a gradient formed by pioneer vegetation colonizing the maritime part of the delta, i.e. the mangrove. Pure mangroves, which colonize the salt flats along the sea, are slowly being replaced by large shrub populations, including *Pandanus candelabrum* and *Raphia* along various channels of the Ogooué. Although these areas start as a few isolated individuals, *Raphia* and *Pandanus* colonies slowly form dense populations that very often become pure stands. *Raphia* are much more abundant than *Pandanus*, which especially prefers the extremities of islands. Mangroves are heavily flooded, and the gradual establishment of *Raphia* and *Pandanus* colonies, which stabilize the soil, increases ground levels and makes it possible for other species to come in. The resulting environments are still wetlands since they remain inundated during floods. This is only possible due to the presence of seeds carried by the Ogooué. Indeed, as previously indicated, tidal rivers, which mostly originate from savannas and do not carry diaspores, are bordered by pure

4.14. Cross section through the vegetation bordering one of the main rivers of the northern delta, with a strip of Pandanus *(A), a strip of* Raphia *sp. (B), and the band of tall mangroves with* Rhizophora *(C) (Drawing: Jean P. Vande weghe).*

4.15. Cross section through the vegetation bordering a tidal channel, with a ridge of high mangroves (A) and low mangroves behind it (B) (Drawing; Jean P. Vande weghe).

mangrove along most of their course. Therefore, the *Raphia* cover slowly gives way to swamp forest or riparian forest, depending on the level of the ground. On the other hand, most of this spatial and temporal succession takes place along the rivers, and the resulting dynamics behind the high mangroves are very different. While the river's edge is being rapidly colonized by *Raphia*, the low mangroves behind it, with less sediment and no contact with diaspores, remain unchanged for much longer. It will also be slowly colonized by riverbank species, including *Raphia* and others, and will typically be replaced by a swamp forest (although not exclusively, with a lot of *Raphia*) when the *Rhizophora* die.

The succession that is taking place slightly inland from the river's edge along the various channels of the Ogooué is therefore a process that largely depends on the survival of *Rhizophora*. The mangroves of the northern delta also appear to have higher swamp species diversity, while the swamp forests located in the southern part of the delta, at the mouth of the Fernan Vaz Lagoon, contain numerous *Raphia*, which often form monospecific populations. There are also large stands of high mangrove, the "undergrowth" of which is composed of *Raphia*.

Finally, behind these high mangrove strands that surround the rivers, where nutrients it provides are less abundant, low mangrove is present, which sometimes forms tannes, low-lying, ultrabasic formations. There are also several strips of savanna along a humidity gradient that ends in dry savanna.

Approach used to characterize vegetation

The characterization of vegetation in the Lower Ogooué region has been based on a classic approach, which consists of collecting data in all environments, identifying species, and statistically analyzing the records. Between 2011 and 2017, 11 field expeditions were carried out in the Lower Ogooué region during which 275 floristic surveys were undertaken to characterize the grassland or wetland environments, such as riverbanks, *Raphia* groves, thickets, and shrubby ridges. As for dryland forests, a separate method was used based on linear transects, defined according to topographic, edaphic, and structural criteria.

For grasslands and nonforest environments, the first step consisted in targeting an area defined using maps and satellite images, then once in the field, participants looked for homogeneous portions of vegetation—that is, areas without structural or physiognomic difference. Once this step was completed, in the case of grasslands, the floristic survey consisted of listing all species present in the selected section of homogeneous vegetation. While establishing this floristic list, a pre-established coefficient was assigned to each species in order to assess its frequency in the chosen section of vegetation. This value indicates whether a given species occupies 50%, 60%, or even 100% of the area. Given the lack of absolute certainty regarding the precise identity of the species observed during these surveys, material was systematically collected of unknown taxa in order to make subsequent identifications in key herbaria, mainly those in Libreville, Brussels and Meise (Belgium), and Paris (France), as well as in Wageningen and Leiden (Netherlands). Figure 4.16 shows the location of all the surveys carried out in the Lower Ogooué.

Regarding the characterization of forests, each transect comprised four sections measuring 50 m in length and 5 m in width, for a total area of 0.1 ha. Trees and shrubs with a diameter greater than 5 cm at breast height were measured and identified within the 50 × 5 m area. For trees whose identity was not known, a specimen, comprising a branch or a leaf, was collected. All of the transects and floristic surveys were georeferenced using a GPS.

In order to develop a typology of the vegetation of the Lower Ogooué Ramsar site, several studies were carried out, including an ascending hierarchical classification in order to distinguish the main groups. Then, the surveys focusing on each large group were conducted to understand why various sites were assigned to a given group. This was made possible by calculating average collection frequency for each species. Three large groups were thus obtained. A first set of 35 surveys corresponds to floating marshy meadows or floating reed beds (Vande weghe, 2007). As described in the chapter on cartography, these meadows cover 37.2 km², or 0.6%, of the Lower Ogooué valley, whereas they cover 907.9 km², or 11.5%, of the upper deltaic plain. In total, they cover 968.6 km², or 6.8%, of the Ramsar site. These formations, as described in the chapter on flora, are characterized by a low floristic diversity and significant homogeneity; they do not contain any rare species. The main distinctive species are *Echinochloa stagnina* (Poaceae), *Persicaria madagascariensis* (Polygonaceae), and *Cyperus papyrus* (Cyperaceae), which form vast floating meadows along the banks of the Ogooué. Sometimes they can commingle, but this is quite rare and most populations tend to be quasi-monospecific. *Echinochloa stagnina* and *Persicaria madagascariensis* formations, which are directly in contact with water, are followed by a formation of *Cyperus papyrus* and, just behind it, there are usually shrub thickets with *Alchornea cordifolia* (Euphorbiaceae) or *Anthocleista liebrechtsiana* (Gentianaceae). These formations are found almost throughout the Lower Ogooué Ramsar site, from upstream all the way down to Ndjolé and Abanga, although they are especially present in the Great Lakes region. Indeed, the banks of the Evaro, Ezanga, and Oguémoué rivers are colonized by this type of plant formation which becomes less frequent toward the sea or the lower deltaic plain and is replaced by mangroves with *Rhizophora racemosa* (Rhizophoraceae).

The second set, comprising 240 survey areas, is subdivided into two subsets, riparian and flood-prone marshes on the one hand, and dry and semihumid formations on the other hand. These are dense, closed swamp forests. As described in the chapter on cartography, they cover 1,043.9 km², or 7.4%, of the entire Ramsar site. As shown in the chapter on flora, these dense swamp forests contain many rare or endangered species that are important for conservation. Vande weghe (2007) compared this type of forest to the Igapó forests of the Amazon, which do not receive alluvium, unlike the várzea forests, where alluvium is abundant. The species most found in this type of environment is *Anthostema aubryanum* (Euphorbiaceae), a large tree with a whitish cylindrical trunk that reaches 25 m tall and 35 cm in diameter. One of the defining criteria of this species is its toxic white sap, which flows abundantly as soon as the bark is cut.

Riparian formations are forests growing alongside rivers, referred to as flood forests in the chapter on cartography. As described in the section on flora, these forests show significant variability. They cover 470.6 km², or 3.3%, of the Ramsar site. The dominant species in this type of environment are *Baphia laurifolia* (Fabaceae), *Spondianthus preussii* (Phyllanthaceae), *Synsepalum brevipes* (Sapotaceae), and *Duguetia confinis* (Annonaceae); sometimes they also include *Treculia africana* (Moraceae) and *Pentaclethra macrophylla* (Fabaceae). These floodable plant formations correspond to the dense swamps described in the chapter on cartography. They also exhibit great variability. They can be found along the rivers just beyond the riparian forests or at low points between valleys. The main dominant species are *Raphia hookeri* (Arecaceae), but variants are found with predominantly *Pandanus candelabrum* (Pandanaceae) or *Nauclea pobeguinii* (Rubiaceae).

The dry and semihumid formations are dry savannas that include smaller, more humid areas. Dry savannas cover 158.9 km², or 1.1%, of the total area of the Ramsar site. Although they lack species diversity, as indicated in the chapter on flora, as a whole these dry savannas contain many rare and endangered species that are important for conservation. The main dominant

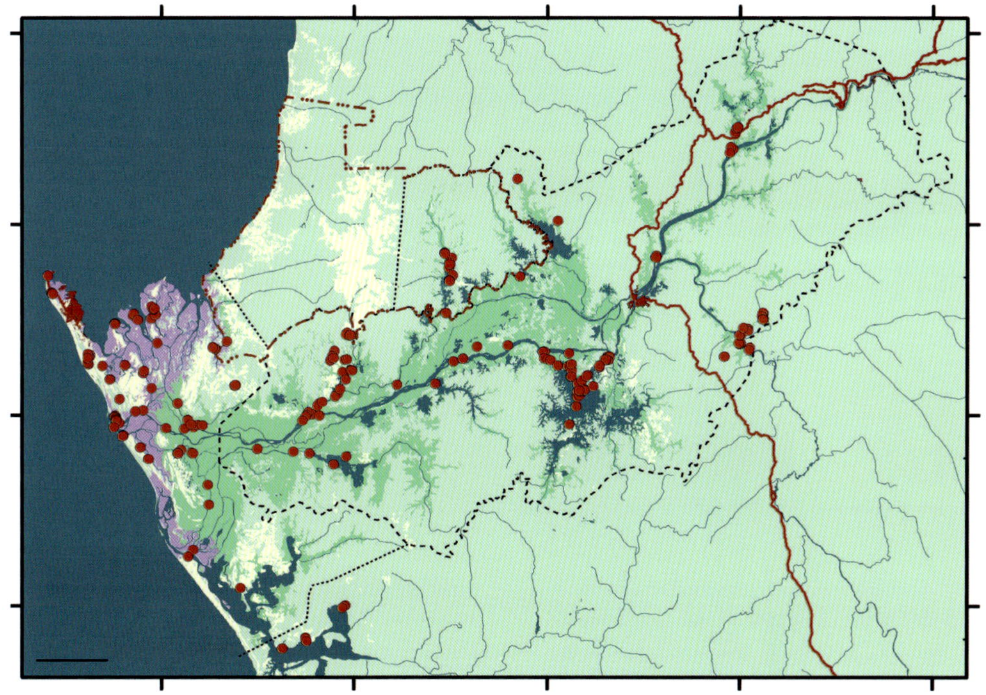

4.16. Locations of surveys carried out in the Lower Ogooué region.

4.17. An alluvial bed at the edge of a main channel of the river, with its dense thickets of Alchornea cordifolia *from which large* Ceiba pentandra *emerge; floating herbaceous mats with* Echinochloa stagnina *extend behind the alluvial bed (Drawing: Jean P. Vande weghe).*

species in these environments are three Poaceae, *Melinis nerviglumis, Anadelphia leptocoma*, and *Ctenium newtonii*, as well as *Mesanthemum radicans* (Eriocaulaceae), *Eriosema glomeratum* (Fabaceae), *Stylosanthes erecta* (Fabaceae), and *Schwenckia americana* (Solanaceae).

These three large groups are then subdivided into numerous vegetation subtypes according to the presence or absence of certain species and their densities. These subtypes are described below.

Current state of the vegetation

Based on the analysis described above, 14 major types of vegetation have been identified in the Lower Ogooué. These are present in six environment types (Table 4.1) and are described in the rest of the chapter based on the environment in which they are found. They are divided into three main groups: wetlands, savannas, and dryland forests.

Wetlands

These humid formations cover 6,093.1 km², or 30.7%, of the surface of the Lower Ogooué, and include vegetation types in restricted areas in which water is present. The 20 families with the most species are listed in Table 4.2. The great diversity of Fabaceae and Rubiaceae should be noted, as they are also very abundant, along with Melastomataceae and Orchidaceae, which are much less common but still very diverse.

Floating meadows

The analyses separated these formations into two groups: those dominated by *Echinochloa stagnina* (Poaceae) and those dominated by *Cyperus papyrus* (Cyperaceae). Nearly monospecific herbaceous floating meadows of *E. stagnina* occur almost throughout the delta, both along large rivers and in calm lakes. However, these plant formations form a gradient between floating meadows, shrub mats, and flooded forests. This type of vegetation is thus found throughout the Lower Ogooué, but most often along the banks where the waters

4.18. Permanently flooded formations with Echinochloa stagnina *(Poaceae) (Photo: Archange Boupoya).*

4.19. Permanently flooded formations with Persicaria madagascariensis *(Polygonaceae) (Photo: Archange Boupoya).*

are calm. The dominant species is *E. stagnina*, but *Persicaria madagascariensis* (Polygonaceae) can also be dominant, forming large mats that develop along riverbanks (Figure 4.19).

These formations also include climbing species such as *Aniseia martinicensis* (Convolvulaceae) and above all, they contain the small aquatic ferns *Azolla africana* (Azollaceae) and *Salvinia nymphellula* (Salviniaceae), which colonize small pools of standing water found between floating stems of *P. madagascariensis*.

Behind these floating meadows, there are shrubby stands of mixed floristic composition. Their most numerous species are *Alchornea cordifolia* (Euphorbiaceae), *Melastomastrum segregatum* (Melastomataceae), *Persicaria senegalensis* (Polygonaceae), and sometimes also *Anthocleista liebrechtsiana* (Gentianaceae). This is the same arrangement as that encountered along the banks that are in direct contact with water, and they are frequent within the delta, especially in its central part.

This type of plant formation acts as a transition between floating grassy meadows and marshy tree formations, which also usually contain climbing species belonging to the family Vitaceae, such as *Cayratia ibuensis* and *Cissus leemansii*.

Finally, there are swamp forests (sometimes completely flooded) with *Nauclea pobeguinii* (Rubiaceae), *Morelia senegalensis* (Rubiaceae), *Phyllanthus reticulatus* (Phyllanthaceae), and *Oubanguia laurifolia* (Lecythidaceae). These formations colonize the banks of rivers and have been observed in the Mabounié region, east of the Lower Ogooué, but have also occasionally been observed elsewhere. These formations sometimes shelter an herbaceous layer typically formed by floating mats of *Echinochloa stagnina* (Poaceae), which include *Persicaria madagascariensis*, *Ludwigia adscendens* (Onagraceae), and some aquatic ferns such as *Ceratopteris thalictroides* (Pteridaceae). The shrubs in these areas support a wide variety of epiphytic plants ranging from Orchidaceae (*Calyptrochilum emarginatum, Polystachya odorata,* and *Bulbophyllum saltatorium*) to Cactaceae (*Rhipsalis baccifera*). When the herbaceous vegetation becomes dense, the resulting formations are referred to as open swamp forests, but in fact they form a gradient between the white waters and closed swamp forests.

The plant formations predominantly comprised of *Cyperus papyrus* are also divided into three subtypes. Nearly monospecific floating meadow formations of *C. papyrus* form along the banks of streams in the Lower Ogooué, just behind the first strip of vegetation formed by the Poaceae *Echinochloa stagnina* (Figure 4.20).

This is the most common type of vegetation in permanently flooded formations. *Echinochloa stagnina* forms a first strip, directly in contact with the water, behind which are found large tufts of *Cyperus papyrus*, which can reach 3 m high in places. It should be noted that this species is terrestrial and therefore does not form true floating meadows. Some other species are present but occur sporadically and at very low density. The second subtype is riparian formations of *Ficus trichopoda* (Moraceae), which also occur along the riverbanks. These are bushy

Table 4.1. Main vegetation types of the Ogooué Delta

Habitats	Characteristic species of the environment	Vegetation type	Dominant species
Permanently flooded	*Echinochloa stagnina, Cyperus papyrus*	Floating meadow	*Echinochloa stagnina*
		Floating maedow	*Persicaria madagascariensis*
		Papyrus	*Cyperus papyrus*
Swamps	*Anthostema aubryanum, Raphia hookeri, Lasimorpha senegalensis*	Swamp forest	*Anthostema aubryanum* and *Alstonia boonei*
		Raphia groves	*Raphia hookeri* and *Pandanus candelabrum*
		Swampy meadows	*Lasimorpha senegalensis* and *Nephrolepis undulata*
Riparian habitats	*Cynometra mannii, Guibourtia demeusei, Baphia laurifolia* and *Alchornea cordifolia*	Riparian forest	*Cynometra mannii* and *Guibourtia demeusei*
		Riparian shrub	*Alchornea cordifolia, Ceiba pentandra*
		Alluvial levee	*Pandanus candelabrum* and/or *Raphia hookeri*
		Mangrove	*Rhizophora racemosa*
Floodable habitats	*Uapaca heudelotii*	Foodable forest	*Uapaca heudelotii, Duguetia confinis*
		Floodable thickets	*Dactyladenia chevalieri, Chrysobalanus icaco*
Dry savannas	*Melinis nerviglumis, Ctenium newtonii, Elionurus muticus*		
Dryland forest	*Sacoglottis gabonensis* and *Aucoumea klaineana*		

4.20. Cyperus papyrus *(Cyperaceae) (Photo: Tariq Stévart).*

Table 4.2: The 20 families with most species and highest number of occurrences in the Lower Ogooué region				
Families	Number of species	%	Number of occurrences	%
Fabaceae	25	9.1	64	7.8
Rubiaceae	25	9.1	92	11.2
Melastomataceae	17	6.2	34	4.2
Orchidaceae	15	5.5	21	2.6
Cyperaceae	11	4	42	5.1
Euphorbiaceae	11	4	44	5.4
Poaceae	8	2.9	24	2.9
Vitaceae	7	2.6	19	2.3
Araceae	6	2.2	19	2.3
Commelinaceae	6	2.2	13	1.6
Marantaceae	6	2.2	23	2.8
Onagraceae	6	2.2	22	2.7
Arecaceae	5	1.8	24	2.9
Gentianaceae	5	1.8	23	2.8
Annonaceae	4	1.5	10	1.2
Apocynaceae	4	1.5	16	2
Asteraceae	4	1.5	6	0.7
Convolvulaceae	4	1.5	10	1.2
Polygonaceae	4	1.5	13	1.6
Acanthaceae	3	1.1	6	0.7

4.21. Open swamp forest with Nauclea pobeguinii *(Rubiaceae) (Photo: Tariq Stévart).*

thickets that can be seen in the middle of *Cyperus papyrus* meadows (Figure 4.20) or directly along the water's edge. Some ferns are often present in these environments, and in some areas there are also trees such as *Nauclea pobeguinii* (Rubiaceae) and *Alstonia boonei* (Apocynaceae). Finally, the third subtype is swamp meadows with *Cyclosorus striatus* (Thelypteridaceae), *Lasimorpha senegalensis* (Araceae), and *Machaerium lunatum* (Fabaceae). They grow in marshes or along the river in calm and stagnant waters or even in the middle of the lagoon. In addition to *C. papyrus*, these contain very few other species, which are usually present in juvenile form.

Submerged formations

Certain herbaceous formations of the Lower Ogooué grow under water; this is notably the case of immersed meadows with *Crinum natans* (Amaryllidaceae) and partially immersed meadows found on a sandbanks. Submerged *C. natans* meadows are monospecific aquatic formations. They grow in calm waters along rivers and have been spotted in the northern delta (Kondjo River). The immersed meadows found on sandbanks, on the other hand, are aquatic herbaceous formations subject to either river flooding/recession or tides. They develop on some sandbanks and are completely submerged during part of the year. Those that have been studied in the coastal part of the delta characteristically contain *Sphenoclea zeylanica* (Sphenocleaceae) and *Hydrolea palustris* (Hydroleaceae), as well as various Cyperaceae and Onagraceae. In fact, these are ephemeral environments and are therefore very difficult to access. They colonize certain sandbanks along various Ogooué River channels where this type of vegetation can grow during the dry season.

Swamp forests

The marsh formations most encountered in the Lower Ogooué region are *Raphia* groves, forests with *Anthostema aubryanum* (Euphorbiaceae), and marshy meadows. These formations are difficult to access and exploration has very often been done via aerial overflights and drone photography. The *Raphia* groves (Figure 4.22) are large swamps found throughout the Ogooué Delta. They are situated along rivers, in swamp forests behind riparian forests, or in waterlogged shallows. *Raphia* usually grows alongside other tree species, forming mixed populations, and pure *Raphia* only covers a small area, about 81 km², or 0.4%, of the deltaic complex, as mentioned in the chapter on cartography. Floristic analyses have shown that these environments do not contain rare or endangered species. *Raphia hookeri* is the main species colonizing these environments, a species that can reach 10 m in height, although they also include *Anthostema aubryanum, Pandanus candelabrum* (Pandanaceae), *Spondianthus preussii* (Phyllanthaceae), and *Mitragyna ledermannii* (Rubiaceae) forming the canopy, while the undergrowth predominantly comprises *Marantochloa cordifolia* (Maranthaceae), *Hypselodelphys scandens* (Marantaceae), *Lasimorpha senegalensis* (Araceae), *Diospyros preussii* (Ebenaceae), and *Crateranthus talbotii* (Lecythidaceae).

Marsh forests with *Anthostema aubryanum* are very similar to *Raphia* groves (Figure 4.23) but do not contain *R. hookeri*, or if present, it occurs at very low density. These forests also grow in the bottoms of valleys and along rivers, and are found throughout the Lower Ogooué region. Apart from *A. aubryanum*, which dominates the upper stratum, other species are found in large numbers, such as *Alstonia boonei* (Apocynaceae), *Spondianthus preussii* (Phyllanthaceae), and *Anthocleista vogelii* (Gentianaceae), as well as *Spathandra* sp. nov. (Melastomataceae), which has not yet been described. Species such as *Vitex doniana* (Lamiaceae) or *Pandanus candelabrum* (Pandanaceae) are also found in these marshy environments. *Pandanus candelabrum* sometimes forms circular shrub mats around ponds. In some forests with *A. aubryanum*, dryland species are also present, such as *Dacryodes klaineana* (Burseraceae) and *Cola* sp. (Malvaceae), a species currently being studied that colonizes part of the undergrowth. These swamp forests have a fairly heterogeneous undergrowth with many shrub species belonging to a variety of families. However, some species are more numerous, such as *Lasianthera africana* (Stemonuraceae) and *Xylopia quintasii* (Annonaceae).

Besides these two main types of swamp formation, other much smaller formations occur, namely swamp meadows, which

4.22. In a raffia grove viewed from above, with Raphia *palm trees mixed in with other species (Photo: Tariq Stévart).*

4.23. An Anthostema aubryanum *forest viewed from above (Photo: Tariq Stévart).*

4.24. Swamp forest with Anthostema aubryanum *(Euphorbiaceae). This tree can be recognized by its slightly fluted trunk, which is bottle-shaped at the base (Photo: Jean P. Vande weghe).*

4.25. The riparian forest is flooded part of the year (here in the Mabounié region). In the dry season, the undergrowth is open, allowing for easy access (Photo: Vincent Droissart).

were observed numerous times in the dunes of Mandji Island. They are grasslands that vary significantly, both in their physiognomy and floristic composition. Although only about 30 cm tall, they can in some cases be colonized by Cyperaceae over 1 m tall. As a result, many different variants of these formations have been observed, including, for example, *Persicaria senegalensis, Eleocharis variegata* (Cyperaceae), or *Lasimorpha senegalensis,* which occur along small, permanent pools of standing water (Figure 4.26).

Another variant of these marshy meadows forms a low vegetation, rarely more than 30 cm tall, with *Laurembergia tetrandra* (Haloragaceae), *Spermacoce verticillata* (Rubiaceae), and *Rhamphicarpa fistulosa* (Orobanchaceae), and is found around small lakes (Figure 4.27) on Mandji Island.

Finally, a facies with *Melochia corchorifolia* (Malvaceae), *Mesanthemum radicans* (Eriocaulaceae), and *Cyperus erectus* (Cyperaceae) also covers large, flat areas along the dune ridges. These three species coexist, but depending on the site, one usually becomes nearly dominant (Figures 4.28–4.30).

4.26. Swampy meadow with Persicaria senegalensis *(Polygonaceae),* Eleocharis variegata *(Cyperaceae), and* Lasimorpha senegalensis *(Araceae) (Photo: Archange Boupoya).*

4.27. Marshy meadow with Laurembergia tetrandra *(Haloragaceae),* Spermacoce verticillata *(Rubiaceae), and* Rhamphicarpa fistulosa *(Orobanchaceae) (Photo: Archange Boupoya).*

4.28. Marshy meadow predominantly with Melochia corchorifolia *(Malvaceae) (Photo: Archange Boupoya).*

4.29. Swampy meadow predominantly with Mesanthemum radicans *(Eriocaulaceae) (Photo: Archange Boupoya).*

4.30. Wet meadow predominantly with Cyperus erectus *(Cyperaceae) (Photo: Archange Boupoya).*

Riparian formations

Riparian structures are, by definition, found alongside rivers. Their vegetation is flooded during part of the year, and they are therefore accessible when the water level is low during the long dry season. As described in the chapter on flora, these forests contain many species emblematic of, and endemic to, the Ogooué basin. The main riparian formations in the Lower Ogooué region are forests, thickets, and shrubs patches. Riparian forests are predominantly composed of *Cynometra mannii* (Fabaceae) and form along the Ogooué and the Ngounié. The other species of the upper stratum are *Synsepalum brevipes* (Sapotaceae), *Baphia laurifolia* (Fabaceae), *Guibourtia demeusei* (Fabaceae), *Uapaca heudelotii* (Phyllanthaceae), and *Duguetia staudtii* (Annonaceae). The understory often has poor diversity of species, and apart from a few seedlings of the previously mentioned taxa, there are almost no other species.

Another type of riparian forest is found at the edges of lakes and small bodies of calm water. In this case, *Dactyladenia chevalieri* (Chrysobalanaceae), *Chrysobalanus icaco* (Chrysobalanaceae), and *Spathandra blakeoides* (Melastomataceae) are predominant, forming compact shrubby ridges. One of the peculiarities of these forests is that they constitute natural shade houses for orchids. Indeed, the shrub populations serve as supports for the growth of a wide variety of orchids.

Riparian shrub formations are in direct contact with the water of large rivers, colonizing their banks, and they often have

4.31. The large trees dominating riparian forests are home to a great diversity of epiphytic orchids (here along the Ngounié) (Photo: Vincent Droissart).
4.32. Anthropogenic flood formations. The embankment causes accumulations of water, which floods the forest and transforms it into a marsh (Photo: Tariq Stévart).

the appearance of bushy thickets. Several types can occur but the most frequent in the Lower Ogooué is comprised of *Alchornea cordifolia*. They are similar to those described for floating meadows, but shrub thickets with *Alchornea cordifolia* are directly in contact with water and do not include floating meadows, although the latter can, however, take shelter behind this plant barrier, which hides the horizon. The few other species encountered within these tufts are *Cayratia ibuensis* and *Phyllanthus reticulatus*. Individuals of *Ceiba pentandra* (Malvaceae) often occur in these formations. This is especially visible along the banks of rivers that flow into large lakes, such as the one that feeds Lake Evaro.

Another variant of these shrub formations is made up of *Ficus trichopoda* (Moraceae), in particular along the channels of the Ogooué and in small lakes channels. They also form tufts that can be completely submerged. Mats of

4.33. Cross section of an alluvial bed with Uapaca heudelotii *and* Cynometra mannii, *as well as a swamp forest with* Anthostema *and* Raphia *(Drawing: Jean P. Vande weghe).*

Anthocleista liebrechtsiana (Gentianaceae) form another facies, which occurs along the channels of the Ogooué and especially in the small lakes north of Lambaréné. The other species that are frequently found in this environment are *Ficus asperifolia* (Moraceae) and *Machaerium lunatum* (Fabaceae).

The final type of riparian formation found in the Lower Ogooué region comprises mats of *Pandanus candelabrum* and *Raphia zamiana*. These formations also colonize the banks of the Ogooué in the lower deltaic plain. They can be nearly monospecific, exclusively comprising *Pandanus candelabrum* (Figure 4.1) or *Raphia zamiana* (Figure 4.22), although sometimes can contain both species simultaneously.

Anthropogenic flood formations

These are the result of human activity (Figure 4.32). Along roads, near bridges and landings, earthworks favor water retention and consequently specific types of vegetation colonize. In this case the main species are *Marantochloa congensis* (Marantaceae), *Rhynchospora corymbosa* (Cyperaceae), *Nephrolepis biserrata* (Davalliaceae), *Ochthocharis dicellandroides* (Melastomataceae), and *Lasimorpha senegalensis* (Araceae).

Mangroves

These are specific plant formations that form at the interface between land and sea. A very complete summary of the Gabonese mangroves is provided by Vande weghe (2005, 2007). In the Lower Ogooué region, mangroves are found in the lower delta and cover 784.1 km², or 26.3%, of the lower deltaic plain. As described by Vande weghe regarding the Akanda and Pongara National Parks, there are two forms of mangroves in the Ogooué Delta. The first, referred to as high mangrove, is found on the edge of open water, both on the seafront and along rivers. It is a dense and tall plant formation in which the vegetation can reach 20 m in height. Areas of high mangrove are not very wide, forming a strip of large trees behind which extensive areas of low mangroves form, with the vegetation rarely exceeding 5 m.

These mangroves generally have poor species diversity, and very often they are nearly monospecific, with *Rhizophora racemosa* (Rhizophoraceae) as the predominant species. In high mangrove, the few other encountered species are *Acrostichum aureum* (Pteridaceae), a large fern that can reach 2 m tall, and *Phoenix reclinata* (Arecaceae), the date palm, which is often found behind the mangrove. Low mangroves, which are likewise not very diverse, have higher species diversity than high mangroves. Indeed, they include species such as *Avicennia nitida* (Acanthaceae), *Conocarpus erectus,* and *Laguncularia racemosa* (Combretaceae).

Between these low mangroves and dry land are hypersaline areas, which are almost devoid of vegetation; these are known as tannes. They represent the last transition zone between marine and dryland environments. There are multiple species that colonize this environment, in particular Cyperaceae such as *Rhynchospora holoschoenoides* and *Eleocharis variegata*; Xyridaceae, in particular *Xyris filiformis;* and Lentibulariaceae (*Utricularia subulata* and *U. andongensis*).

In contrast with the plant formations linked to the presence of water and which fully designate the Lower Ogooué as a Ramsar site, the region is in fact mostly covered with dryland plant formations, covering approximately 12,580.2 km², or 63.4%, of the surface. These dryland formations are found throughout the Ramsar site and encompass both dryland forests and dry savannas inserted between pockets of forest.

4.34. Low mangrove with a mature forest at the background (Photo: Tariq Stévart).

4.35. Back of the low mangrove with small Rhizophora, *a clump of* Phoenix reclinata, *and tufts of* Pandanus *(Photo: Tariq Stévart).*

Dryland forests

The dryland forests of the Lower Ogooué region can be classified into two main types depending on the nature of the soil—sandy or clay. Forests on clay soils are mostly found east of the Lower Ogooué region, while those on sandy soils occur along the coast and in the interior of the delta. Another forest type is found within the clay forests, at higher altitudes, forming the lower part of the Crystal Mountains, between Lambaréné and Ndjolé. It should be noted that these forests connect to Waka National Park to the south, the gateway to the Chaillu Mountains, taking advantage of the heterogeneous topography, which ensures sufficient humidity for plants that prefer higher altitudes.

Within the Lower Ogooué region, the families of trees most represented in our inventories are Fabaceae, and in particular Caesalpinioideae (52 species and morphospecies)—a subfamily that is also very abundant in terms of individuals—along with Rubiaceae (45 species and morphospecies) and other groups of Fabaceae (37 species and morphospecies); few individuals of these last two families were recorded in the inventories. Other families such as Annonaceae and Ebenaceae

Table 4.3: Number of species and morphospecies in the 20 most diverse families in the Lower Ogooué region, across the entire site and for the four forest areas identified in this study

Family	Evaro-Mpaga	Mabounié-Azingo	Maurel-Gomé-Odimba	Nkomi Lag.	Total
Caesalpinioideae	5	45	25		52
Rubiaceae	11	36	16	10	45
Fabaceae	1	16	7	23	37
Annonaceae	9	26	14	11	30
Ebenaceae	5	26	17	8	28
Phyllanthaceae	5	17	9		27
Sapindaceae	1	22	3	5	26
Euphorbiaceae	6	21	9	8	25
Malvaceae		19	9		21
Meliaceae	3	12	5	5	20
Chrysobalanaceae	4	13	6	4	19
Sapotaceae	4	12	7	4	18
Anacardiaceae	1	9	9		14
Olacaceae	2	10	10	5	14
Burseraceae	3	9	8	4	12
Mimosaceae		12	6		12
Salicaceae	4	9	7	1	12
Clusiaceae	1	9	5	4	10
Putranjivaceae		6	5		10
Apocynaceae	2	7	2	2	8
Total	92	438	232	146	590

4.36. Cross section of a low mangrove transitioning to dry savanna: (A) low mangrove, (B) shrub edge with Phoenix reclinata *beds, (C) wet savanna, and (D) dry savanna. The fine line indicates the maximum water level (Drawing: Jean P. Vande weghe).*

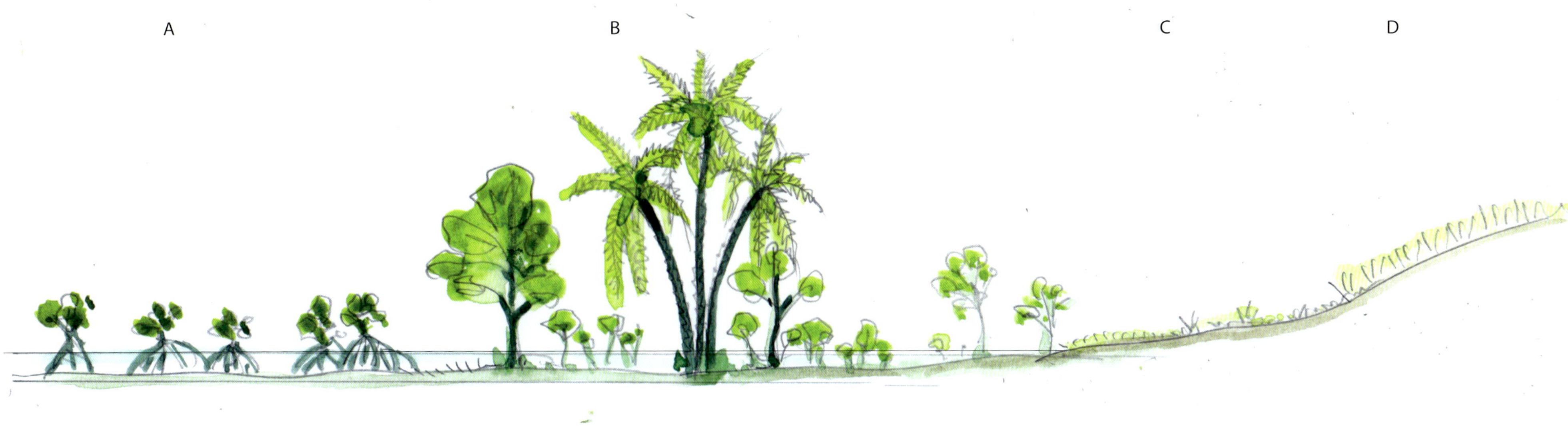

Table 4.4: Abundance of species and morphospecies in the 20 most collected families of the Lower Ogooué region, within the entire site and for the four forest areas identified in this study. Abundance is calculated by dividing the total number of individuals in a family by the total number of individuals in a given geographic area

Family	Evaro-Mpaga	Mabounié-Azingo	Maurel-Gomé-Odimba	POG-Fernan Vaz	Total
Annonaceae	16.9	12.1	10.2	5.9	3,262
Caesalpiniaceae	9.9	13.4	13.0	0.0	3,022
Ebenaceae	1.9	14.0	10.6	3.6	2,725
Euphorbiaceae	9.9	7.6	6.0	10.6	2,193
Olacaceae	1.3	10.3	10.1	2.3	2,069
Burseraceae	4.5	5.1	8.9	6.8	1,501
Ctenolophonaceae	24.0	0.0	0.0	4.1	1,363
Rubiaceae	1.9	4.9	3.2	1.4	1,003
Fabaceae	0.3	2.2	0.8	17.2	867
Humiriaceae	6.3	0.6	0.7	10.0	729
Myristicaceae	0.3	2.8	3.4	4.8	704
Irvingiaceae	0.0	3.1	2.5	1.8	612
Clusiaceae	0.1	1.0	2.9	7.8	479
Ochnaceae	4.0	1.2	0.8	0.6	434
Lecythidaceae	2.0	1.4	2.6	0.0	408
Mimosaceae	0.0	1.7	3.6	0.0	370
Malvaceae	0.0	2.1	0.9	0.0	359
Sapotaceae	2.6	0.6	2.4	1.2	332
Phyllanthaceae	0.8	1.4	1.7	0.0	315
Chrysobalanaceae	2.4	0.6	2.1	0.8	303
Total	5,187	15,847	2,918	2,828	26,780

also have high diversity (Table 4.3) and are among the most abundant in terms of inventoried individuals (Table 4.4). The most abundant species in the forest inventory data is *Ctenolophon englerianus* (Ctenolophonaceae) (Table 4.5), which is restricted to humid environments but is capable of colonizing dryland forests, and which is predominant in the southwestern forests of Wonga-Wongué.

The hierarchical analysis of flora data from the transects studied revealed four large forest groups. The first are coastal sand forests, which are abundant along the coast and which have been inventoried in the lower deltaic plain and in the Fernan Vaz Lagoon. The second group comprises forests of *Ctenolophon englerianus* and pioneer forests on sand, which occur in the upper deltaic plain, south of Wonga-Wongué and around Lake Evaro. These are related to the third group, littoral forests, but prefer humid basins from which they begin colonizing old savannas. Finally, the fourth type, mature forests, occur in certain parts of the upper deltaic plain and are similar to the coastal forests with *Desbordesia insignis* found in Caballé. This last group is separated into two subgroups, one of which transitions into littoral forest and has been inventoried south of Lake Onangué (Maurel and Prom concession) and Lake Gomé (Maurel-Gomé-Odimba group), and the other is found at Lake Azingo and at Mabounié.

Coastal forests

These are mature forests that have often been logged several times. They contain a rich diversity of species and habitats, and display substantial variation in habitats over short distances. Their tree layer, comprising the species that reach the canopy, consists of large individuals that can reach 30 m tall and is predominantly composed of Caesalpinioideae and Rubiaceae (Table 4.3). The most abundant species in this stratum are mainly *Dialium angolense*, *Coula edulis*, *Santiria trimera*, and *Staudtia kamerunensis*. Other species, although less frequent, are clearly distinguished by

Table 4.5: Abundance of the 20 most collected species and morphospecies of the Lower Ogooué region, within the entire site and for the four forest areas identified in this study. Abundance is calculated by dividing the total number of individuals of a species by the total number of individuals in a given geographic area

Species	Evaro-Mpaga	Mabounié-Azingo	Maurel-Gomé-Odimba	Nkomi Lag.	Total
Ctenolophon englerianus	24.0	0.0	0.0	4.1	1,363
Dialium angolense	0.0	5.0	7.4	0.0	1,014
Coula edulis	1.3	4.7	5.7	1.0	996
Anthostema aubryanum	4.7	2.2	0.0	8.2	820
Meiocarpidium lepidotum	0.0	4.6	0.4	0.0	746
Sacoglottis gabonensis	6.3	0.6	0.7	10.0	729
Greenwayodendron suaveolens	0.0	3.6	4.5	0.2	711
Dichostemma glaucescens	4.1	1.6	4.2	2.0	639
Diospyros	0.0	3.6	0.6	0.2	592
Aucoumea klaineana	3.5	0.5	2.5	3.9	446
Santiria trimera	0.0	2.0	3.3	0.0	408
Diogoa zenkeri	0.0	2.4	0.6	0.0	402
Dialium bipindense	7.4	0.1	0.0	0.0	399
Scytopetalum klaineanum	0.5	1.4	2.4	1.8	365
Strombosia pustulata	0.0	1.9	2.0	0.0	365
Staudtia kamerunensis	0.3	1.5	2.3	1.4	353
Diospyros iturensis	0.0	2.1	0.5	0.0	344
Xylopia quintasii	3.4	0.5	1.1	1.7	332
Duguetia confinis	5.0	0.2	0.0	1.4	331
Klaineanthus gabonii	0.9	1.5	0.8	0.0	300
Total	5,187	15,847	2,918	2,828	26,780

the size of their individuals. These include *Klainedoxa gabonensis* (Irvingiaceae) and *Uapaca guineensis* (Phyllanthaceae).

The undergrowth is dominated by plants from the family Ebenaceae, counting at least 11 species (*Diospyros hoyleana, D. dendo, D. rabiensis, D. piscatoria, D. gabunensis, D. simulans, D. zenkeri, D. conocarpa, D. melocarpa,* and D. *suaveolens*). Euphorbiaceae are also well represented in the understory, including *Dichostemma glaucescens*, which is part of the undergrowth of nearly all dryland forests in the Lower Ogooué region. Other families such as Olacaceae s.l. (*Strombosia pustulata, Strombosia scheffleri,* and *Heisteria parvifolia*) and Annonaceae (*Greenwayodendron suaveolens* and *G. suaveolens* var. *gabonica)* add to the floristic composition of these forests' undergrowth.

One of the differences between the forests of Mabounié and Azingo, which form a unit, and those located further inland in the delta, along certain lakes, is the greater abundance of okoumé found inland. Indeed, many forests of the deltaic plain are formed by colonizing savannas, and okoumé is one of the main pioneer species. The presence of *Garcinia smeathmannii* (Clusiaceae) is also notable, which is very abundant in certain mature forests located on the higher areas to the south of Lake Onangué.

Beach forests

The second main group of forests—coastal forests and forests with *Ctenolophon englerianus*—is characterized not only by the presence of *Ctenolophon*, but mainly by the abundance of *Sacoglottis gabonensis* (ozouga, Humiriaceae) and to a lesser extent *Aucoumea klaineana*. These are certainly the most abundant forest formations on the sandy substrates of Lower Ogooué. They often surround small islets of savanna, which they begin colonizing as soon as conditions are favorable.

4.37. A cross section of a dune ridge with a dry forest interspersed with swamp forest, as well as a dry meadow interspersed with floodable prairies (Drawing: Jean P. Vande weghe).

4.38. A forest with Desbordesia insignis *near Lake Azingo (Photo: Tariq Stévart).*

4.39. Dry savanna with Imperata cylindrica *(Poaceae) and* Ctenium newtonii *(Poaceae) north of Lake Evaro (Photo: Archange Boupoya).*

Table 4.6: Families by the number of species and the number of occurrences in the Lower Ogooué savanna.

Families	Number of species	%	Number of occurrences	%
Poaceae	44	17.8	171	20.8
Cyperaceae	41	16.6	146	17.8
Fabaceae	21	8.5	79	9.6
Rubiaceae	18	7.3	43	5.2
Melastomataceae	6	2.4	23	2.8
Orchidaceae	6	2.4	15	1.8
Apocynaceae	5	2	12	1.5
Asteraceae	5	2	10	1.2
Commelinaceae	5	2	14	1.7
Convolvulaceae	5	2	14	1.7
Polygonaceae	5	2	20	2.4
Xyridaceae	5	2	23	2.8
Lentibulariaceae	4	1.6	23	2.8
Araceae	3	1.2	8	1
Cucurbitaceae	3	1.2	8	1
Droseraceae	3	1.2	9	1.1
Eriocaulaceae	3	1.2	27	3.3
Euphorbiaceae	3	1.2	3	0.4
Gentianaceae	3	1.2	13	1.6
Scrophulariaceae	3	1.2	8	1
Total	241	100	822	100

Forests with *Ctenolophon* have mostly been inventoried south of Wonga-Wongué, near the village of Mpaga. In addition to *Ctenolophon englerianus* and *Sacoglottis gabonensis*, they are characterized by an abundance of *Dialium bipindense*. The understory of these forests predominantly comprises *Greenwayodendron suaveolens* var. *gabonica* (Annonaceae). Humid forests are wedged in at the bottom of valleys within these dryland forests, which grow both on hills or flat terrain; these humid forests receive runoff, and they are most often waterlogged and typically contain small streams that wind between the trees. They generally have peat soils. A single species largely dominates the upper stratum of this forest type, *Ctenolophon englerianus*. This large tree can reach 90 cm in diameter and forms dense, almost monospecific stands. Built-up mounds of dead herbaceous species are found at the base of these trees and a dense carpet of moss grows on a thick layer of plant litter that can exceed 1 m in depth. Few species are found in this environment except for *Rhabdophyllum arnoldianum* (Ochnaceae), which is very common.

Finally, the formation referred to as coastal forests are found almost everywhere along the ocean. Often, they are only separated from the sea by a small strip of savanna. The canopy is dominated by Fabaceae, including *Librevillea klainei*, which can form almost monospecific stands. Another species exhibits similar behavior in these forests, *Pinacopodium congolense* (Erythroxylaceae). The diameter of this tree rarely reaches 30-40 cm, but they can reach heights exceeding 20 m. It is easily recognizable by its light green leaves, which stand out in the canopy foliage. The main other species that make up the flora of the coastal forest tree strata are *Vitex doniana* (Lamiaceae), *Didelotia unifoliolata*, and *Guibourtia pellegriniana* (both Fabaceae).

A variant of these coastal forests is found mainly on Mandji Island, on coastal sand structures; these are sclerophyllous thickets that develop between two savannas atop dune ridges. Very often, their vegetation is degraded and does not exceed 10 m in height. The most encountered species are *Fegimanra africana*, *Chrysobalanus icaco, Vitex doniana*, and *Syzygium littorale* (Myrtaceae), as well as, to a lesser degree, *Barteria nigritana* (Passifloraceae). These forests are scattered along the coast of Gabon, but only cover small areas, except far south in the country. They are related to the coastal sclerophyllous forests cited by Vande weghe (2005).

Savannas

Besides the forest formations described above, the Lower Ogooué region is also distinguished by a great diversity of dry savannas, which incorporate some semi-humid and periodically flooded formations. Unlike other Gabonese savannas, the dry savannas of Lower Ogooué are vast and generally low grassy areas with almost no trees or shrubs. They form along the edges of the Ogooué Delta's channels. The best represented families are, of course, Poaceae and Cyperaceae (Table 4.6). While the Poaceae *Melinis nerviglumis*, *Elionurus muticus,* and *Ctenium newtonii* are predominant (Figure 4.42), other species such as *Eriosema glomeratum* (Fabaceae), *Schwenckia americana* (Solanaceae), *Cyperus tenax* (Cyperaceae), *Eulophia cucullata* (Orchidaceae), and Rubiaceae such as *Oldenlandia affinis* and *Spermacoce pusilla* are also present, forming the flora of the Lower Ogooué's dry savannas. Although they may seem to be identical, the dry savannas of Lower Ogooué have several facies.

The dry savanna south of Wonga-Wongué forms a vast, continuous grass cover that stretches as far as the eye can see and is predominantly composed of *Melinis nerviglumis, Elionurus muticus, Schizachyrium thollonii* and *Panicum* sp., all Poaceae, which are mixed in with various small herbaceous plants (Figure 4.42). Vande weghe (2005) also described these formations as the typical coastal savanna that forms on white sand along the entire Gabonese coast.

On the other hand, the dry savannas found north of Lake Evaro are characterized by the presence of two other Poaceae, *Imperata cylindrica* and *Ctenium newtonii* (Figure 4.39). They form a vast, continuous herbaceous cover dotted with termite mounds, developing between thickets and forest galleries. Even though their vegetation seems homogeneous, other species can be found in these formations, such as *Vernonia smithiana* and *Vernonia guineensis* (Asteraceae). The floristic composition of these savannas includes species commonly found in continental savannas. Large Poaceae measuring more than 1 m in height are worth mentioning, such as *Anadelphia leptocoma*. The presence of the genus *Hyparrhenia* is also noteworthy, with at least two known species represented.

4.40. Anthropogenic flooded savanna in the Lake Evaro region (center) (Photo: Tariq Stévart).

4.41. Flooded savanna with Clappertonia ficifolia *(Malvaceae) (Photo: Ehoarn Bidault).*

4.42. Dry savanna with Melinis nerviglumis *in the southern part of Wonga-Wongué (Photo: Ehoarn Bidault).*

In the northern Ogooué Delta region and west of Wonga-Wongué, dry savannas on poor, sandy soil are home to a species rarely collected in Gabon, *Brachystelma letestui* (Apocynaceae). The importance of this species for conservation is detailed in the chapter on flora. These are hardy grass formations that exceed 15 cm in height. Apart from rare tufts of *Melinis nerviglumis, Brachystelma letestui* is abundant in these formations, although they also include *Schwenckia americana* (Solanaceae), *Cyperus erectus*, and *Cyperus tenax*, as well as *Merremia tridentata* (Convolvulaceae) and sometimes *Neurotheca corymbosa* (Gentianaceae). This vegetation seems to grow following fires. In fact, it is only found in recently burnt areas, where the ground has almost been exposed. The presence of ephemeral flowering species such as the small orchid *Nervilia petraea* supports this interpretation.

Dry coastal savannas are found south of Mandji Island, with *Mesanthemum radicans* (Eriocaulaceae), *Curculigo pilosa* (Hypoxidaceae), and *Platycoryne guingangae* (Orchidaceae). These savannas are sometimes mostly shrubs, but tend to be mainly composed of herbs. The most frequent shrubs are *Heinsia crinita* (Rubiaceae), *Cola flavovelutina* (Malvaceae), and *Manilkara lacera* (Sapotaceae). In this case, the herbaceous stratum is discontinuous and consists of disjointed tufts of *Melinis nerviglumis* and *Stylosanthes erecta*, between which grow creeping species, such as *Cassytha filiformis* (Lauraceae) and *Merremia tridentata*.

Another savanna type is present around Nkomi Lagoon, with *Anadelphia leptocoma* (Poaceae). The plant cover of these savannas can in places exceed 1 m in height. Large Poaceae such as *Phacelurus gabonensis* and *Loudetia phragmitoides* are also present. *Murdannia simplex* (Commelinaceae) and *Gladiolus unguiculatus* (Iridaceae) are abundant as well. Shrubs such as *Psidium guineense* (Myrtaceae) and sometimes even young *Sacoglottis gabonensis* indicate the beginning of forests encroaching on the savanna.

Dry savannas are home to small depressions and valley bottoms that tend to retain runoff water. Flood formations develop in these environments and also on Mandji Island. These formations are characterized by *Rhynchospora candida* (Cyperaceae) and *Melastomastrum segregatum* (Melastomataceae). Apart from these two species, which are largely dominant, *Scleria aterrima* (Cyperaceae), *Polygala arenaria* (Polygalaceae), *Burmannia madagascariensis* (Burmanniaceae), *Acroceras zizanioides* (Poaceae), and *Sabicea africana* (Rubiaceae) are also present. Other formations subject to periodic flooding occur in the Kendje region. These grow on granular soils and the main dominant species are *Setaria sphacelata* (Poaceae), *Dissotis congolensis* (Melastomataceae), and *Mesanthemum radicans* (Eriocaulaceae). Almost no shrubs are found in these savannas, where vegetation sometimes exceeds 1 m high. There are savannas prone to flooding in the bottom valleys; they are dominated by Cyperaceae, *Platycoryne guingangae* (Orchidaceae), *Xyris congensis* (Xyridaceae), and *Clappertonia ficifolia* (Malvaceae). This type of savanna also contains *Thesium filipes* (Santalaceae).

Some flooded savannas are anthropogenic in nature and are associated with the presence of agricultural ridges. Indeed, amidst the dry savannas, flooded savannas have formed in areas that had previously been dug out by humans for reasons that are described in the chapter about human populations. The floristic composition of these flooded savannas is completely different from that of the dry savannas surrounding them. Indeed, the dominant species in these environments are *Rhynchospora candida* (Cyperaceae), *Melastomastrum segregatum* (Melastomataceae), and *Mesanthemum radicans* (Eriocaulaceae). Seen from above using drones, these formations are easily distinguished from the rest of the vegetation by their darker green color.

In conclusion, these savannas also include small peat bogs located in places where water stagnates. The plant cover is very low and not dominated by any species in particular. However, some taxa are significant, although quite common in these environments, including *Xyris filiformis* (Xyridaceae), *Lycopodiella affinis* (Lycopodiaceae), *Burmannia mada-*

gascariensis, Utricularia spiralis, U. andongensis, U. subulata, and *Genlisea africana* (all Lentibulariaceae) and *Drosera madagascariensis* (Droseraceae). The soil is covered with *Sphagnum* (Sphagnaceae).

Acknowledgments

The results presented in this chapter are the fruit of eight years of work on the vegetation of the Lower Ogooué region. The authors would like to thank the many people involved in this work, starting with those who accompanied them in the field: Eric Akouangou, Dimitri Bikissa, Jean-Noël Boussiengui, Gilles Dauby, Sheelah Griffith, Davy Ikabanga, Yves Issembé, John Kaparidi, Ian Lafferty, Moses Libalah, Wilfried Mbading Mbading, Jean-Claude Mouandza, Diosdado Nguema, Raoul Niangadouma, Thomas Nzabi, Christelle Nyangala, Pierre Ploton, Jean-Yves Serein, Bonaventure Sonké, Kaeli Stark, Nicolas Texier, and Donatien Zébazé.

They also thank the photographers Vincent Droissart and Nicolas Barbier for allowing the use of their images, as well as for many unforgettable moments spent in the field.

The field expeditions needed to produce this book were made possible thanks to the support of the National Herbarium of Gabon (IPHAMETRA, CENAREST). The authors would like to thank all the botanists connected to this institution who helped them, and in particular Ludovic Ngok Banak and Nestor Engone Obiang, successive curators of the National Herbarium of Gabon, as well as Henri Paul Bourobou Bourobou, former director of IPHAMETRA.

The Wildlife Conservation Society, as well as Jean-Philippe Biteau (Jardi-Gab), are warmly thanked for their unfailing support during the preparation of each field expedition, and Jean P. Vande weghe for his vegetation drawings.

Bibliography

BIDAULT, E., O. LACHENAUD & T. STÉVART, 2017. La flore. Pp. 143-171 in J.P. VANDE WEGHE & T. STÉVART. *Le delta de l'Ogooué*. ANPN, Libreville.

BOUPOYA, A., 2010. *Flore et végétation des clairières intraforestières sur sol hydromorphe dans le Parc National de l'Ivindo.* Thèse de doctorat. 246 p.

BOUPOYA, A., & Stévart, T., 2013. Première étude sur la flore et la végétation de la région du Lac Évaro (Moyen-Ogooué-Gabon). Rapport, WCS, Libreville, Gabon, 86 p.

CABALLÉ, G., 1978. Essai sur la géographie forestière du Gabon. *Adansonia. sér.* 2, 17 (4): 425-440.

CHEVALIER, A., 1916. *La forêt et les bois du Gabon.* Paris A. Challamel. 468 p.

CHEVALIER, A., 1917. *La Forêt et les Bois du Gabon. Les Végétaux Utiles de l'Afrique Tropicale Française*, IX, 470 pp + 28 pl.

DAUBY, G., 2012. *Structuration spatiale de la diversité intra- et interspécifique en Afrique centrale - le cas des forêts gabonaises*. PhD thesis, ULB.

DAUBY, G., HARDY, O.J., LEAL, M., BRETELER, F. & STÉVART, T., 2013. Drivers of tree diversity in tropical rain forests: new insights from a comparison between littoral and hilly landscapes of Central Africa. *Journal of Biogeography,* 41 (3): 574-586.

DOUMENGE, C., NDINGA, A., FOMETE NEMBOT, T., TCHANOU, Z., MICHA ONDO, V., ONA NZE, N., BOUROBOU BOUROBOU, H. & NGOYE, A., 2003a. Conservation de la biodiversité forestière en Afrique centrale atlantique. II - Identification d'un réseau de sites critiques. *Bois et Forêts des Tropiques,* 276: 43-58.

DOUMENGE, C., FOMETE NEMBOT, T., TCHANOU, Z., MICHA ONDO, V., ONA NZE, N., BOUROBOU BOUROBOU, H. & NGOYE, A., 2003b. Conservation de la biodiversité forestière en Afrique centrale atlantique. III - Gestion et priorités d'investissement dans les sites critiques. *Bois et Forêts des Tropiques*, 277: 53-67.

SAINT-AUBIN, G., de, 1961. Aperçu sur la forêt du Gabon. *Bois et Forêts des Tropiques*, 78: 3-17

FONTÈS, A. & CABALLÉ, G., 1977. Carte des formations végétales du Gabon au 1/200000, in *Atlas National du Gabon*. Berger Levrault, Nancy.

FROMARD, F., FONTES, J. & LOUIS, A., 1994. *Structure et dynamique des mangroves de la région de Port-Gentil Gabon. Analyse de l'impact lié à l'activité pétrolière*. Libreville, rapport scientifique à la demande de Elf-Gabon, 52 p + annexes.

FROMARD, F., 1998. L'écosystème mangrove. Biodiversité, fonctionnement, restauration et gestion. *Bois des Forêts des Tropiques : symposium, of the mangrove ecosystem*, 256(2): 80-86.

GIRARDIN, N., 1978. *La mangrove*. Ministère de l'Education National, Libreville, 31 p.

GIRARDIN, N., 1982. *Végétation de la côte sableuse*. Ministère de l'Education National, Libreville.

LEBIGRE, J.-M., 1980. Les mangroves (*Rhizophora* spp) du Gabon. Un milieu méconnu et potentiellement riche. *Revue Gabonaise d'Etudes politiques, économiques et juridiques*, 4: 37-42.

LEBIGRE, J.-M., 1983a. Les mangroves des rias du littoral gabonais: essai de cartographie typologique. *Revue Bois et Forêts des Tropiques,* 199: 3-28.

LEBIGRE, J.-M., 1983b. *Le littoral du Gabon : Aspects Géomorphologiques et Biogéographiques.* Institut Pédagogique National, Libreville, 58 p.

LEBIGRE, J.-M., 1990. *Les marais maritimes du Gabon et de Madagascar. Contribution à une étude d'un milieu naturel tropical*. Université De Bordeaux III, thèse de doctorat d'état, 3 tomes, 615 p.,

NICOL, M. & O. LANGRAND, 1986. *Conservation et utilisation rationnelle des écosystèmes forestiers du Gabon*. Rapport 3247 IUCN-WWF. Gland, Suisse. 143 p.

ONDO ASSOUMOU, E., 2006. *Dynamique des paysages végétaux du littoral centre-ouest du Gabon autour de Port-Gentil: approche spatiale et analyse des données de terrain*. Thèse de Doctorat, Université Paul Valéry Montpellier III, 302 p.

RABENKOGO, N., 2001. *Dynamique des paysages dans le département d'Etimboué, Ogooué-Maritime (Gabon).* Rapport final, programme "Valorisation des écosystèmes humides en Afrique Centrale (PVEHAC)," Libreville, pp. 10-35.

RAPONDA-WALKER, A., & SILLANS, R., 1961. Les plantes utiles du Gabon: essai d'inventaire et de concordance des noms vernaculaires. Série *Encyclopédie Biologique*, LVI. Editions Paul Lechevalier, Paris, 614 p.

REITSMA, J.M., 1988. Végétation forestière du Gabon/ Forest vegetation of Gabon. *Tropenbos technique series,* 1. The Netherlands. 142 p.

STÉVART, T., BARBIER, N., BIDAULT, E., BOUPOYA, A., VIENNOIS, G., LACHENAUD, O., & TEILLIER, L., 2015. *Projet de caractérisation de la végétation du Massif Forestier d'Evaro*. Rapport final de l'inventaire multi-ressources relative aux résultats botaniques. 77p.

VANDE WEGHE, J.P., 2005. *Les Parcs Nationaux du Gabon: Akanda et Pongara: plages et mangroves*. WCS Gabon, Libreville, Gabon, 208 p.

VANDE WEGHE, J.P., 2007. *Les Parcs Nationaux du Gabon: Loango, Mayumba et le bas Ogooué.* WCS Gabon, Libreville, Gabon, 320 p.

VIENNOIS, G., T. STÉVART, J.P. VANDE WEGHE, S. SAATCHI, S. SCHILL, A. ALDOUS, M.-C. PAIZ, A. BOUPOYA & N. BARBIER, 2017. 6. Cartographie de la végétation. Pp. 199-221 in: J.P. VANDE WEGHE & T. STÉVART. *Le delta de l'Ogooué*. ANPN, Libreville.

VILLIERS, J.-F., 1973. *Etude floristique et phytosociologique d'une mangrove atlantique sur substrat rocheux du littoral gabonais*. Thèse de doctorat, Université de Paris VI, 169 p.

WILKS, C., 1990. *La conservation des écosystèmes forestiers du Gabon*. Tropical Forest Program, IUCN.

Methods for water surface and aerial sampling of field sites.

5.1. Flight of a quadcopter drone (Photo: Tariq Stévart).

5.2. Survey team aboard an ANPN speedboat (Photo: Nicolas Barbier).

5.3. Botanists in a canoe (Photo: Nicolas Barbier).

5. Vegetation mapping

Gaëlle Viennois, Tariq Stévart, Jean P. Vande weghe, Sassan Saatchi, Steven Schill, Alison Aldous, Marie-Claire Paiz, Archange Boupoya & Nicolas Barbier

Mapping the Lower Ogooué, a complex issue

Mapping the vegetation of the Lower Ogooué Ramsar site and its potential extensions constitutes a major challenge, if only because of its size: 20,000 km². Among the many obstacles to overcome, mapping out such a large area would be insurmountable without the help of remote sensing, such as observations from satellites, airplanes, or drones. In addition to this relatively standard difficulty, several others are specific to the Lower Ogooué region.

Wetland inaccessibility

The area is not only vast, it is also difficult to access. Travel over land is impossible most of the time, and boats remain the best way to explore this huge site. Navigation over great distances is necessary, often through narrow channels where passage is made difficult by the presence of sandbanks and vegetation that is very dense at times. It is therefore necessary to use a combination of speedboats and lighter canoes. Finally, to cover the whole area properly, it is necessary to fly over the zone in a light aircraft or to use drones.

In recent years, the use of drones has expanded beyond the military and professional spheres and is now accessible to the general public. A wide range of small devices, including foldable drones, can be found on the market at very reasonable prices. They are easy to control with a simple mobile phone, even in windy weather. They include sensors that make it possible to take professional-level photos and videos using stabilized digital cameras. Automatic image capture and processing allow anyone to produce precise topographic maps over areas of a few dozen hectares. Indeed, by programming a flight path, it is possible to take a large number of shots of an area following a systematic grid. These images can then automatically be assembled in order to deduce the area's vertical structure (with stereophotogrammetry) and produce a centimeter-wide mosaic of data (Figure 5.4). Where it was once necessary for human operators to deduce vertical coordinates from pairs of images with polarized glasses (those used in 3D cinemas), computer programs are now able to recognize millions of corresponding points between each pair of images and deduce the structure of captured objects, also using the exact position of the camera and its focal length. This produces a detailed 3D rendering of the target surface.

A perpetually changing complex system

A third difficulty in mapping the area is that the lower section of the Ogooué River is extremely dynamic, both in terms of time and space, first of all because of seasonal variations in water level. The water regime, the duration and height of flooding, strongly influences which types of vegetation can establish themselves. This regime can be inferred through remote sensing in two distinct ways. Either directly, if a signal can help determine the presence of a layer of water with sufficient consistency over time to define the regime, or indirectly, through topographical information. Both of these approaches have been used here.

Over the years, erosion and sedimentary deposits have induced progressive transformations in plant populations, notably through the process of primary succession, resulting in the formation of a patchwork of successive mud or sediment deposits. While experienced naturalists can easily distinguish numerous plant formations from the ground thanks to the dominance of certain species, it is not always possible to obtain this level of detail when mapping a large area. Doing so requires a compromise between the level of detail of the observations used for mapping purposes, the availability of images, and the size of the area to be mapped. To illustrate this, consider the changes in perception between sets of images (Figure 5.5) acquired by a drone with centimeter resolution, multispectral images from the Pléiades satellite sensor (resolution: 2.8 m), and multispectral images from the SPOT 6-7 sensor (resolution: 6 m). Drone images make it possible to distinguish fine details such as papyrus inflorescences or *Raphia* leaflets. They nearly correspond to the information visible through ground-level exploration, with the advantage of being easier to use when drawing outlines of various areas. They make it possible to distinguish many nuances regarding the abundance of the main species within a given plant formation. For example, for marshy meadows, variations in the dominance of Cyperaceae (*Cyperus papyrus*), grasses (e.g., *Echinochloa*), Polygonaceae (e.g., *Polygonum*), and ferns (e.g., *Pteridium*) are easily observed. Likewise, in swamp forests, the areas dominated by *Raphia*, *Anthostema, Ficus,* and *Nauclea* can be properly outlined. Some of the variations between plant formations are much more subtle to observe on even the highest-resolution satellite images (Pleiades). Table 5.2 summarizes the formations that were mapped out and their different possible variants in the field.

5.4. Top: Automatic assembly of photos and colorized point cloud generation (Pix4D software).
Bottom left: Image (orthomosaic) created using this process.
Bottom right: Surface elevation of the plant cover (above sea level).

Image processing automation

The last problem to be overcome is that, in order to obtain detailed and objective mapping at large scales, it is essential for image processing to be automated, but this implies having homogeneous remote sensing data across the entire area. Assembling images acquired at different times and with various camera angles generally requires considerable technical and human investment. Indeed, atmospheric and angular effects confound the signal and must be eliminated. These effects depend on the wavelengths of the electromagnetic spectrum being used, yet no remote sensing method—neither optical, radar, active, nor passive—is exempt. This constraint takes on a new dimension along the Gabonese coast, where the level of cloudiness is among the highest in the world! Optical satellite data are the most affected by this problem. For this study, Airbus DS had to take nearly 3,500 SPOT 6-7 image captures in order to obtain approximately 30 images of sufficient quality. That said, the eastern part of the studied area, which is even cloudier, was never properly captured with this sensor, and the use of data with a lower spatial resolution was necessary (SPOT 4 and Landsat 7) (Figure 5.6).

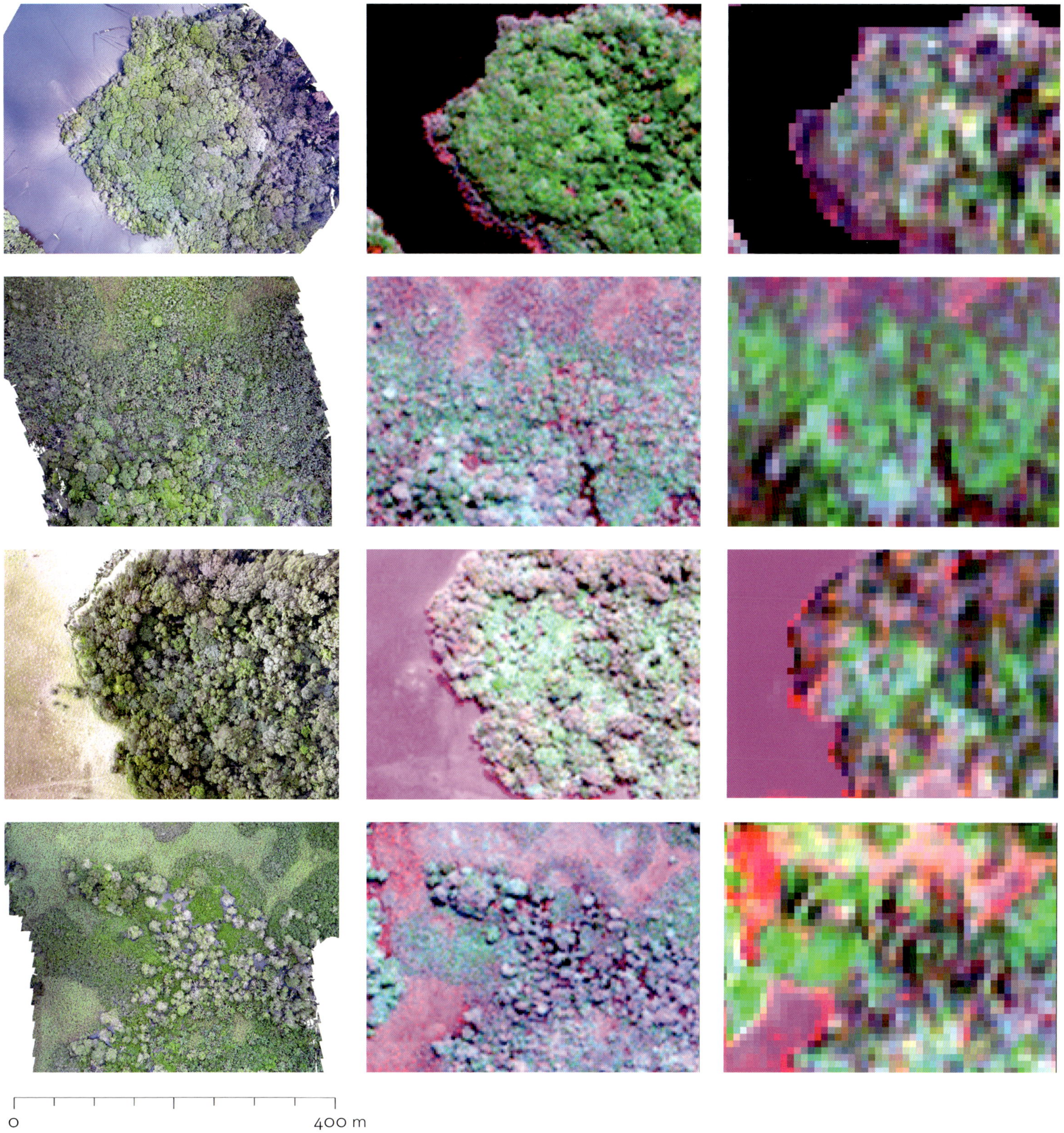

5.5. Effect of the change in spatial resolution on the perception of plant associations.
Left column: Drone image tiles (Red-Green-Blue) with centimeter resolution. Central column: Composite of Pleiades images (2.8 m resolution) in false colors (Red-NIR-G). Right column: Composite of SPOT 6-7 images (6 m resolution) in false colors (Red-Near Infrared-Green).
Line 1: Okoumé monodominated dryland forest.

Line 2: Patchwork of floodable forests on alluvial ridges, swamp forests with variable predominance of Raphia, *and various types of herbaceous or shrubby swamps.*

Line 3: Patchwork of savannas, degraded forests, and okoumé monodominated dryland forest.

Line 4: Patchwork of swamp forests with variable predominance of Nauclea *and various types of herbaceous or shrubby swamps.*

5.6. Footprints of the different optical images used to map the various plant formations. Yellow: seven SPOT 6-7 images (6 m); Red: five Pleiades images (2.8 m); Pink: four SPOT 4 images (20 m); Blue: one Landsat image (30 m).

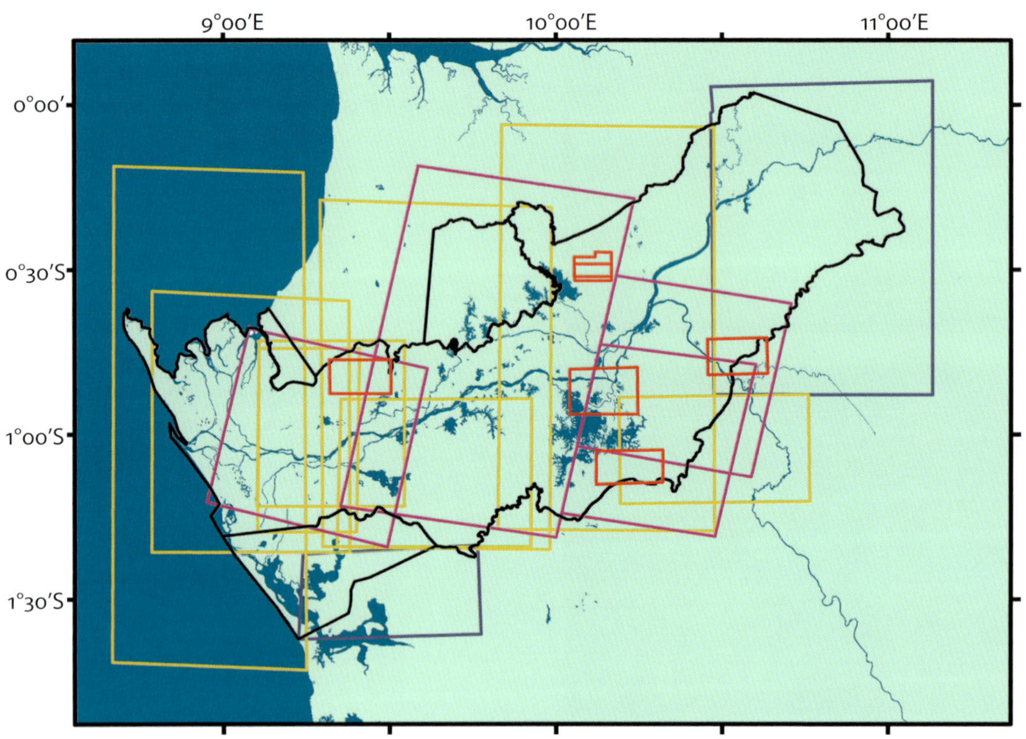

Previous efforts

The difficulties mentioned above make mapping efforts particularly difficult. This is made clear by certain inconsistencies between the different plant formation maps published in the area. The first two maps were produced at a scale of 1/200,000 by IGN ("Carte de l'Afrique centrale au 1:200.000," 1962) and then by the Department of Inventories, Development and Forest Regeneration of the Ministry of Waters and Forests of Gabon ("Carte des formations végétales au 1:200.000," 1998). Simard *et al.* (2002) mapped the coastal region of Gabon using an automated classification method based on radar images. Figure 5.7 illustrates two extracts from these different maps. With regard to the extracts from the coastal region (on the left), the three maps show the presence of mangroves. On the 1962 map (top, Figure 5.7), only low mangrove and marshy meadows are mapped. The 1998 map (middle, Figure 5.7) also indicates the presence of high mangroves along the coast, as well as of humid savannas and islets of dry land. The most recent map (bottom, Figure 5.7) also distinguishes between low and high mangrove formations, and identifies areas of wooded and herbaceous savanna, but does not show islets of dryland forest. The second set of extracts (right, Figure 5.7) corresponds to freshwater areas. Marshes, referred to as "bare wetlands," and the *Raphia* swamp forests mapped in 1998 (middle, Figure 5.7) are too extensive compared to the other two maps. Conversely, the zones of flooded vegetation mapped in 2002 (bottom, Figure 5.7) appear to be much smaller than the wetlands shown on the two other maps. These examples illustrate the difficulties encountered in mapping this wetland complex.

Combining the best optical and radar satellite images

The method that was used to define the main plant formations is based primarily on the textural properties of very high spatial resolution (VHSR) optical images. Multispectral optical images from the constellations of SPOT 6 and 7 and Pleiades 1a and b satellites were obtained for this purpose. In addition to these optical images, three types of remote sensing images from radar sensors (Table 5.1) were used. SRTM data were used as the topographic (height) information layer. A temporal series of Sentinel-1 images (band C) covering a year were assembled to analyze the seasonality of floods in open areas (savannas in particular). And finally, images from the Palsar satellite (L-band), sensitive to humidity levels in areas with dense vegetation, were used to map the areas not covered by the VHSR and SPOT 4 images as well as to distinguish between dryland or wetlands areas.

How to interpret high-resolution images?

Since the 2000s, there has been a significant increase in the availability of very high spatial resolution (VHSR) optical satellite images. These images, whose pixel size is less than 2 m, offer a high level of detail and make it possible to distinguish relatively small objects (buildings, plots, trees, etc.). The trade-off is an increased spectral variability (colors) within the objects, for example between exposed areas and shaded areas. In order to group pixels together to assign them to a thematic class, this local variability compromises conventional classification methods that simply sort pixels on the basis of their color, regardless of the surrounding context.

To counter this problem, actions naturally performed by the human brain are reproduced by including information on what is happening in the vicinity of each pixel. This can be used for both texture analysis methods, which seek to characterize the local organization of pixel values, and for object-oriented classification, which identifies relatively homogeneous objects before attempting to classify them.

FOTO Texture Analysis (Barbier & Couteron, 2015; Couteron *et al.*, 2005) was developed to summarize the local texture variations visible in an image in two or three dimensions. This method makes it possible to sort areas of an image according to the average size and regularity of observable structures (details and homogeneity). Figure 5.9 shows how, in the cases of the Lower Ogooué, this method makes it possible to distinguish types of vegetation based on the sizes of predominant plants, but also other structures (fields, dunes, populated areas). A major advantage offered by this approach is that it is relatively immune to mist or haze pollution, which heavily impacts color-based methods. It is also possible to eliminate effects linked to changes in image characteristics.

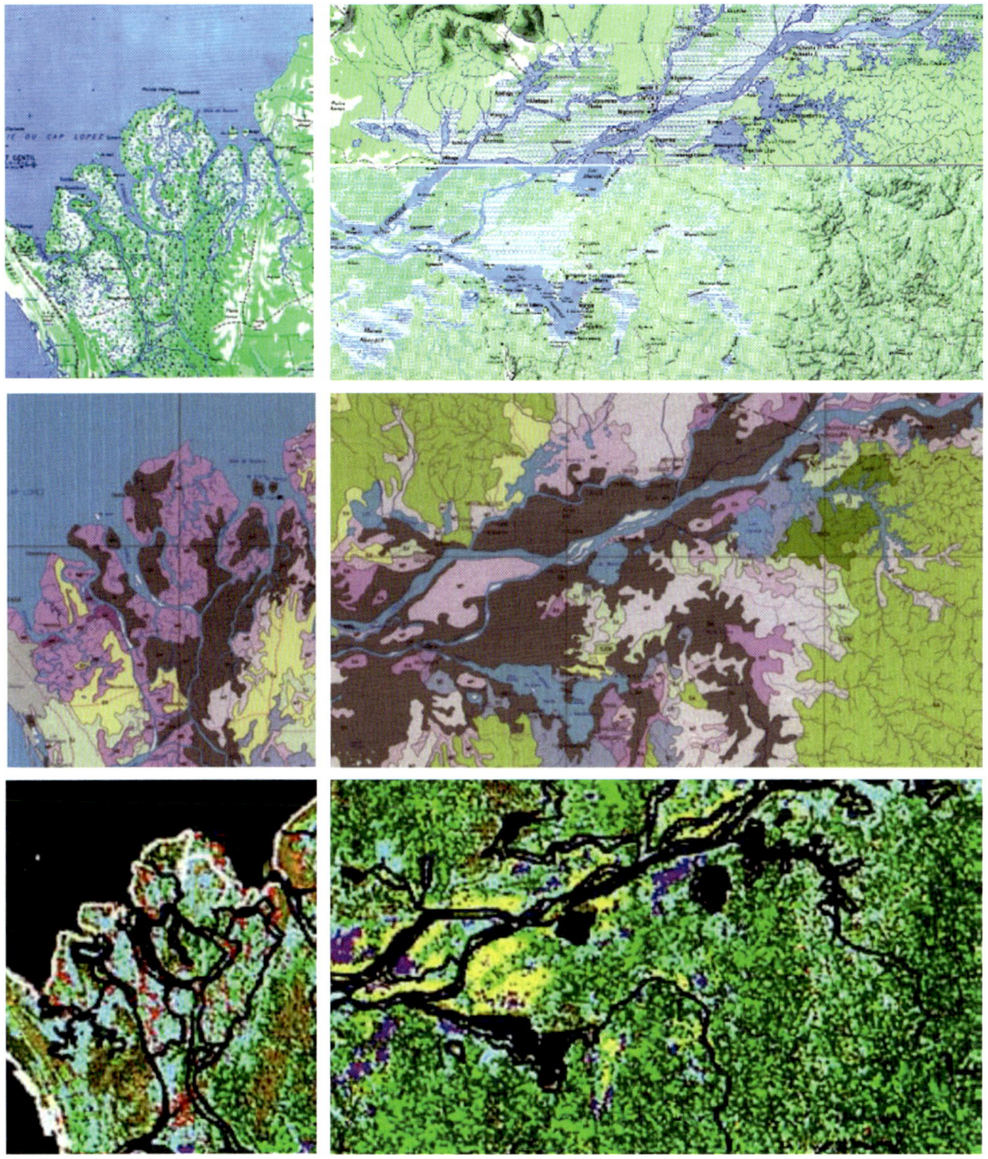

5.7. Excerpts from three different maps over two areas: the coastal region (left) and the upper deltaic plain (right). The first map (top) was produced by IGN in 1962; the second (middle) by the "Direction des inventaires, des aménagements et de la régénération des forêts" of the Gabonese Ministry for Water and Forests in 1998; the third by Simard et al. *in 2002*

It is therefore possible to analyze textural variations in a coherent manner across a large set of images, provided that the spatial resolution of the images is sufficient (less than 2 m), thereby revealing the structure of the vegetation. The SPOT 6-7 and Pleiades images used here provide a good example of this.

The object-oriented method (Blaschke, 2010) was also designed to process these new VHSR images by including other types of information besides the individual spectral information of the pixel. This approach involves two main steps. The first consists of segmenting the image into objects according to homogeneity criteria based on the luminosity, texture, and the shape of identifiable areas within the image. The second step involves classifying these objects based on a wide range of possible attributes tied to the object itself, but also to nearby objects.

The method used

Figure 5.8 summarizes the method developed to define and map out plant formations using different data sources through an object-oriented approach. The first step consists of dividing the area into segments based on a set of source data (SRTM mosaic, FOTO texture, the Infrared Band of SPOT 6-7 images and the temporal Sentinel-1 radar series) in order to delineate objects that might correspond to different

Table 5.1. Available cartographic data			
Data	Characteristics	Approach	Utilization
Pleiades Optical data	1 panchromatic channel at a resolution of 0.5 m	Photo-interpretation	Recognizing mangroves, swamp meadows, open and closed swamp forests, floodable forests, degraded forests, crops, monodominant dryland forests, and monodominant flood forests
	4 multispectral channels at a resolution of 2 m: Blue; Green; Red; Near-infrared		
SPOT 6-7 Optical data	1 panchromatic channel at a resolution of 1.5 m	Texture analysis	Distinguishing between mangroves, swamp meadows, open and closed swamp forests and flood forests, coastal forests, and monodominant dryland forests and monodominant flood forests
	4 multispectral channels at a resolution of 6 m: Blue; Green; Red; Near-infrared	Near-infrared strip utilization	Differentiate between degraded forests, monodominant crops, and dryland forests
SPOT 4 Optical data	4 multispectral channels at a resolution of 20 m: Green; Red; Near infrared; Mid-wavelength infrared	Spectral analysis	Mapping areas not covered by THR images
Landsat mosaic Optical data	4 multispectral channels at a resolution of 30 m: Mid-wavelength infrared; Near-infrared; Red; Green	Spectral analysis	Mapping areas not covered by THR or SPOT 4 images
SRTM mosaic Radar data	Evaluation data at a resolution of 30 m	Altitude thresholding	Differentiate between wetlands and dryland areas based on topography
PALSAR mosaic Radar data	Band L at a resolution of 12.5 m, HH polaritmetry	Variations in signal intensity caused by flooding	Differentiate between wetlands and dryland areas based on the radar signal
Sentinel 1 Radar data	Band C (20 m resolution) composed of two channels of VV and VH polarimetry	Signal variations over time using 12 monthly images over one year	Differentiate between open wetlands and open dryland areas based on the radar signal

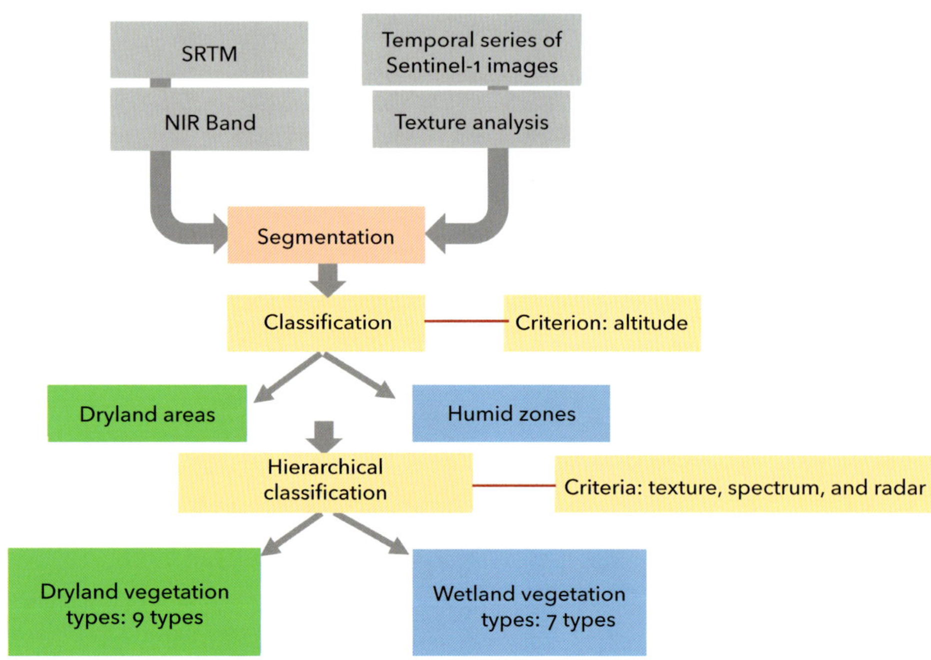

5.8. Method used to map the Ramsar site and its extensions.

plant formations (Figure 5.8). Then, a first classification based on the altitude criterion provided by the SRTM mosaic will allow these segmented objects to be sorted into two groups constituting a first hierarchical level: wetlands and dryland areas. Within each group, a second hierarchical level is created in which different plant formations have been defined based on their spectral, textural, and radar attributes.

The 16 plant formations that have been mapped are displayed in Figure 5.11 and listed in Table 5.2. Five types are identified within dryland areas, including dry savannas, dryland forests, monodominant populations such as okoumé forests, and also formations resulting from anthropogenic disturbances such as degraded forests and crops. A subgroup

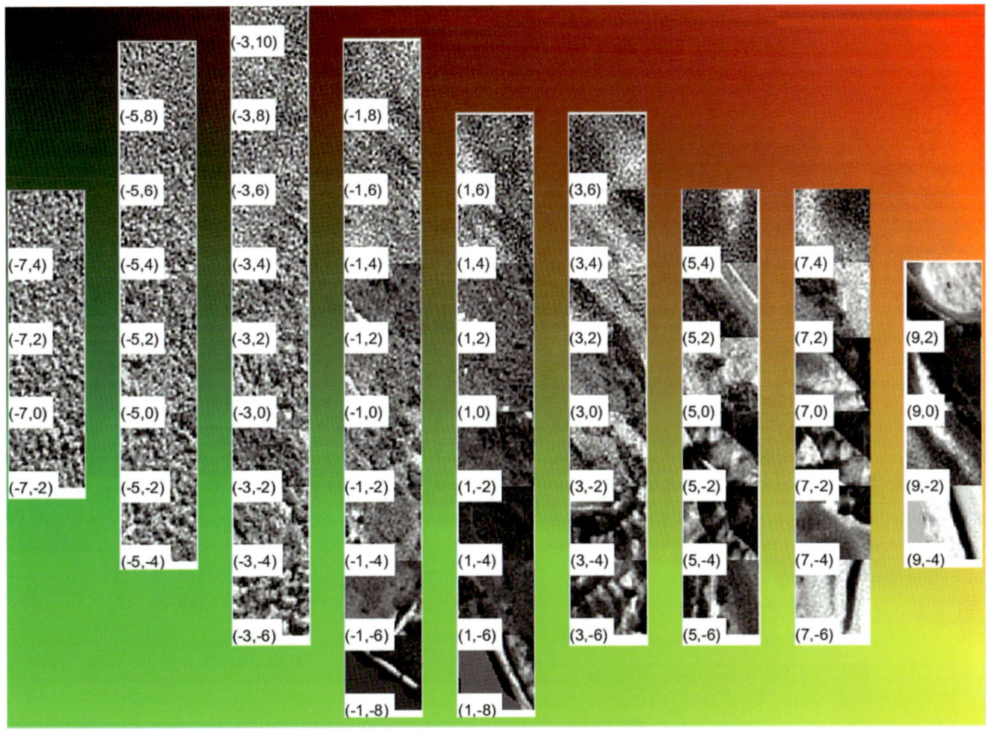

also corresponds to plant formations that are temporarily flooded, such as the humid savannas, floodable forests, littoral forests, and monodominant floodable forests such as *Ctenolophon* forests. Within wetlands, a distinction is made between low and high mangroves, and wetlands of maritime environments. In freshwater environments, there is a gradient of plant formations stretching from swamp meadows through different stages of open and closed swamp forests. In areas where SPOT 6-7 images could not be obtained (Figure 5.6), only radiometric (SPOT 4, Landsat), altitudinal (SRTM), and humidity (PALSAR) information was used, and certain types of plant formations cannot be correctly identified.

5.9. Variation of the local FOTO texture in a set of SPOT 6-7 images. At the top, sample thumbnails are arranged to show how local textures are ordered based on their smoothness and regularity. Background colors translate this variation into shades of red and green, which are then used to map these textures (shown below).
Bottom left: Overview map of texture variations over the western third of the study area.
Bottom right: More detailed map of the texture and the corresponding SPOT 6 image, showing the relationship between the identified textures and the different types of vegetation. This shows, for example, that savannas and marshy meadows have a relatively fine texture (mapped in yellow or green), which contrasts with that of forests (red-brown). The borders between forests and savannas appear in orange.

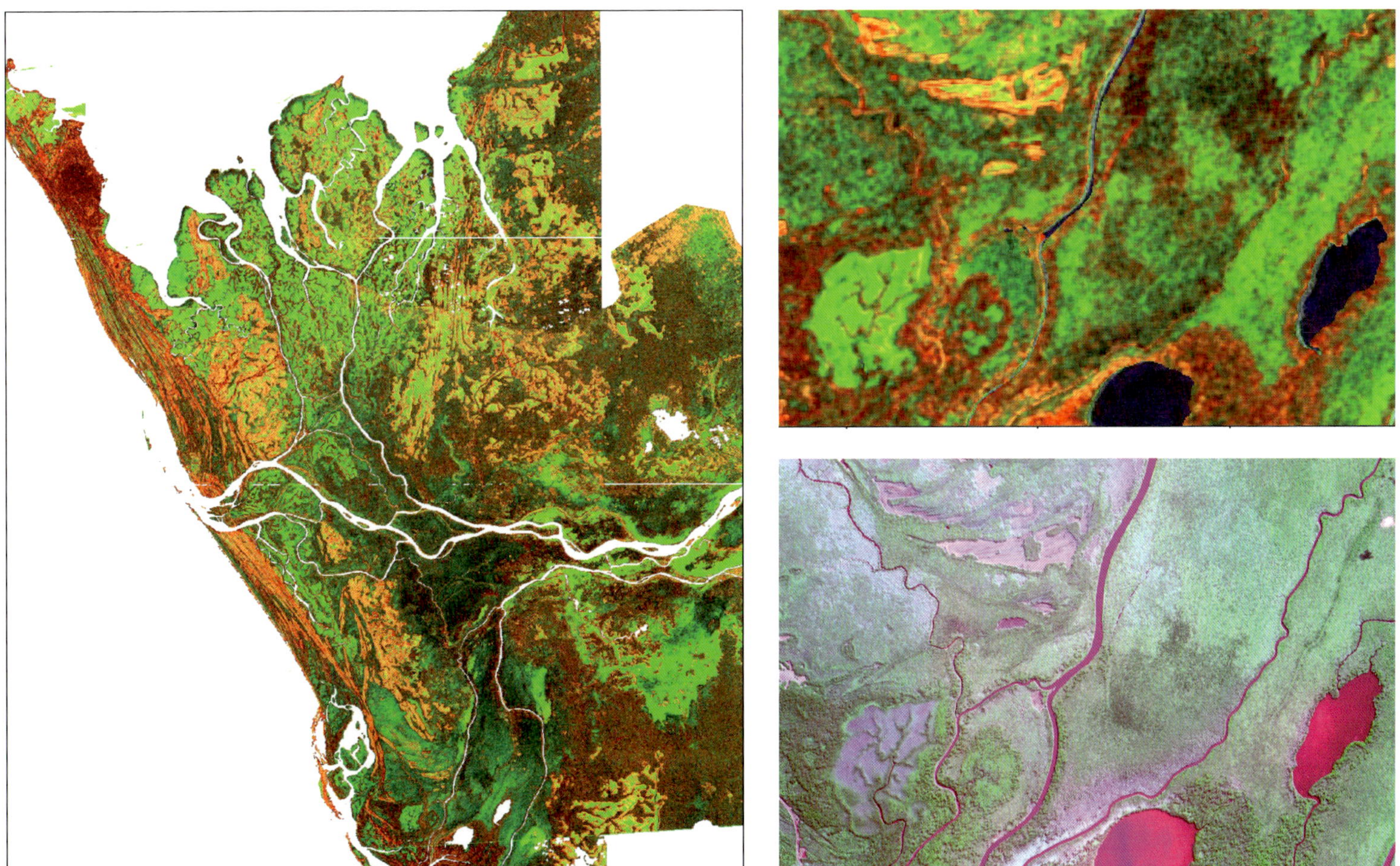

5.10. Vegetation map obtained by TNC and NASA covering almost the entire coastal basin of Gabon.

The Ogooué wetlands compared to the rest of the Gabonese coastline

A map of vegetation types at the scale of the entire coastal zone of Gabon has been produced by The Nature Conservancy (TNC) in collaboration with NASA, with the aim of covering all Ramsar sites in Gabon. This map also uses an object-oriented classification, but is based solely on mosaics of Landsat optical images (7 and 8) and ALOS PALSAR radar resampled at 30 m resolution. Classification training is based on drone acquisitions. This map allows us to put into perspective the vegetation in the Lower Ogooué area in a wider regional context (Table attached). We note that for a certain number of wetland classes, the proportions included in the Lower Ogooué are significant, despite the existence of other important sites such as the Komo Estuary (for mangroves).

TNC typology	Approximate equivalent	Area Lower Ogooué (km²)	Total area (km²)	Proportion (%)
Agriculture	Plantations + disturbed forest	56	252	22
Closed forest	Dryland forest	9,321	63,161	15
Open forest	Dryland forest	2,013	2,048	98
Degraded forest	Disturbed forest	839	2,078	40
Flooded closed forest	Closed swamp forest	978	7,450	13
Flooded open forest	Open swamp forest	3,049	9,437	32
High mangrove	High mangrove	225	772	29
Low mangrove	Low mangrove	211	546	39
Mud/barren	Herbaceous swamp	166	552	30
Sand	Sandbank	15	74	21
Dry savanna	Dry savanna	395	2,639	15
Seasonally inundated savanna	Humid savanna	199	980	20
Urban	Infrastructure	43	218	20
Water	Water	1,340	7,330	18
Emergent wetland	Wetland	170	401	42
Herbaceous wetland	Wetland	647	1,638	39
	Total	19,668	99,578	39

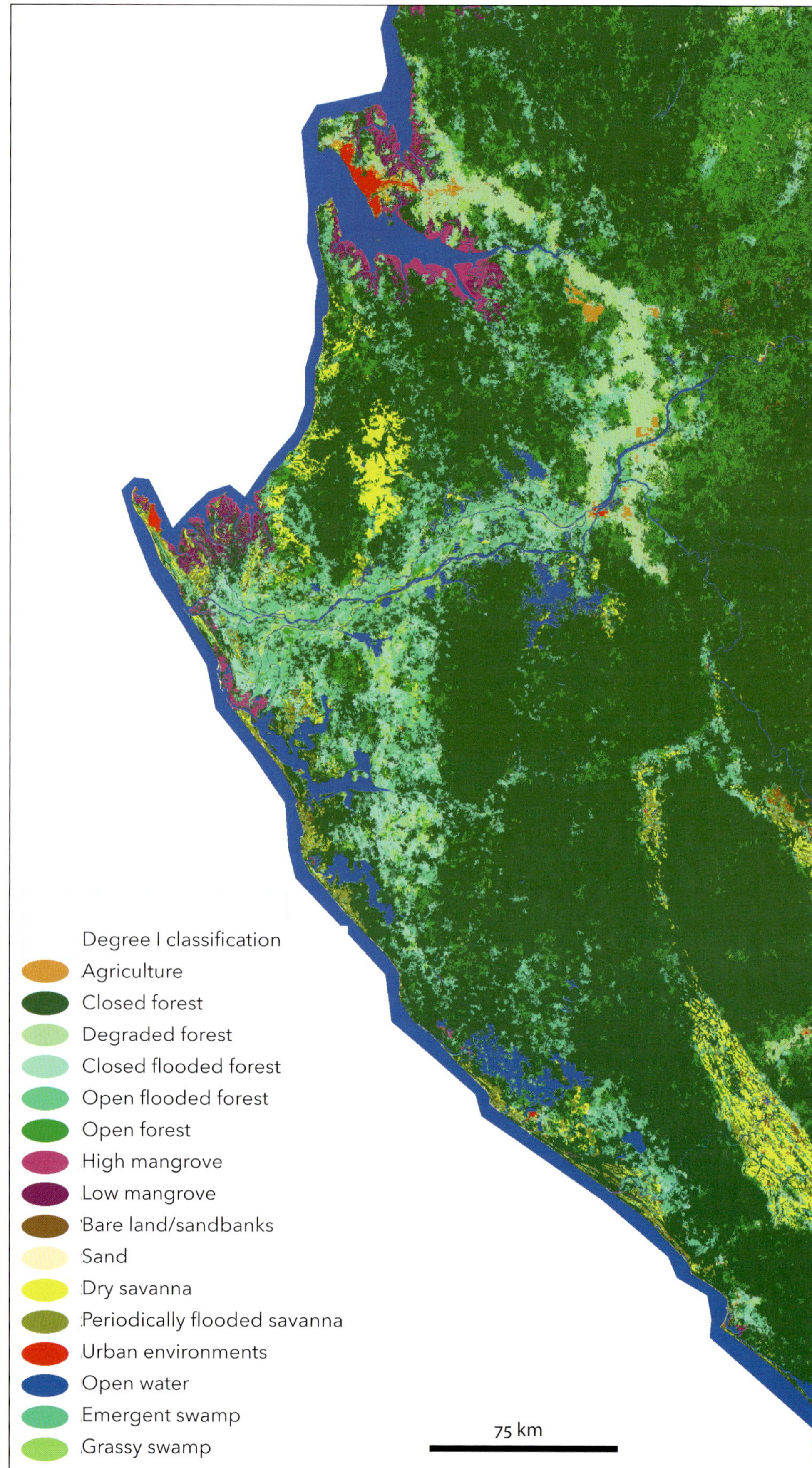

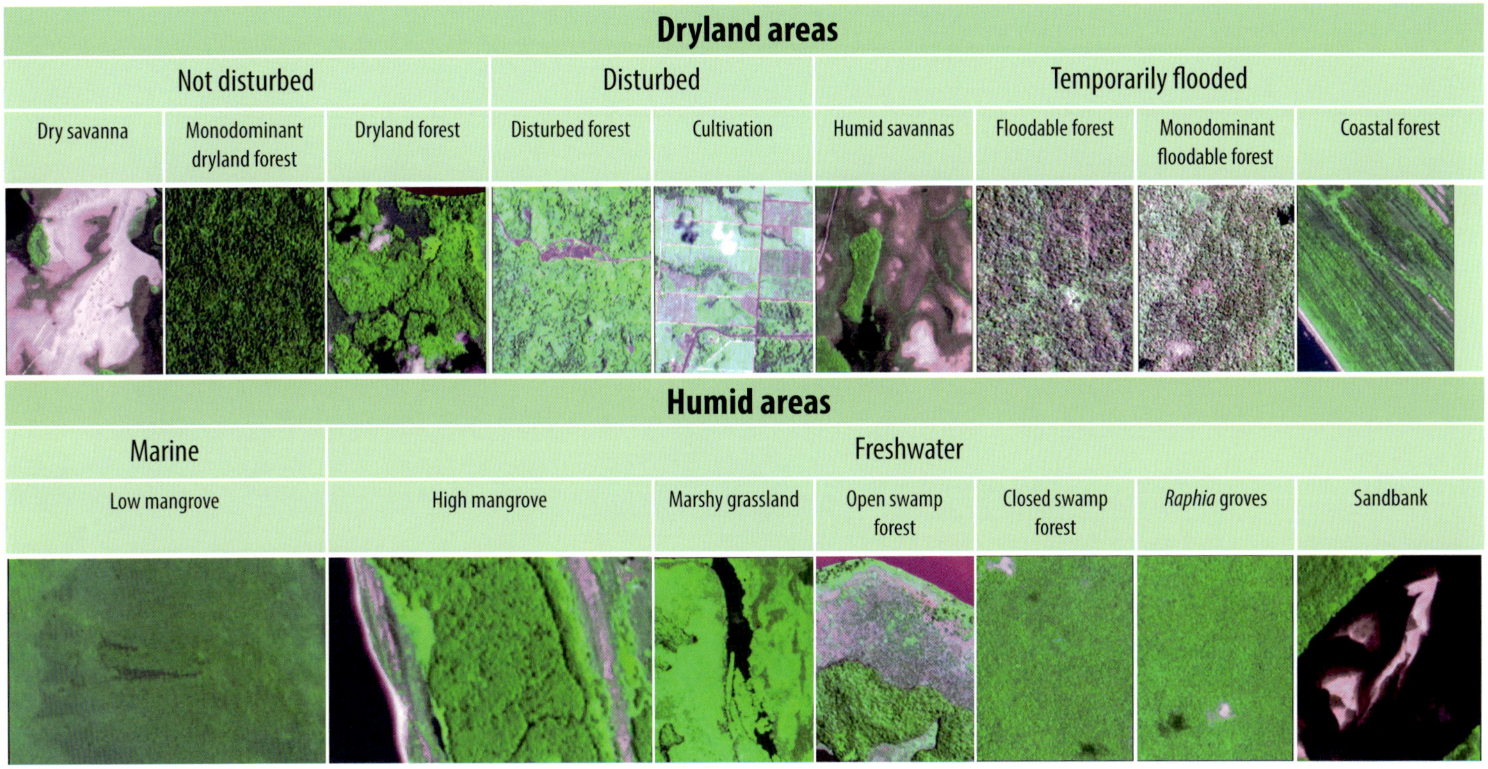

5.11. The 16 plant formations that can be distinguished in the VHSR images. All images are at the same scale.

Table 5.2. The 16 mapped vegetation types

Mapped vegetation types	Water regime	Subtypes
Dryland forest	Dry land	Diverse coastal forest
Monodominant dryland forest	Dry land	Forest with dominant okoumé or *Sacoglottis*, either juvenile or mature
Disturbed forests	Dry land	
Plantations	Dry land	Oil palm or bamboo
Dry savanna	Dry land	
Humid savanna	Temporarily flooded	Floodable savanna with Cyperaceae or *Drosera*
Coastal forest	Temporarily flooded	Forest with *Fegimanra africana, Chrysobalanus*
Floodable forest	Temporarily flooded	Forest with *Ceiba* or *Cynometra*
Monodominant floodable forest	Temporarily flooded	Most often with *Ctenolophon*
Low mangrove	Marine wetland	Low mangrove, monospecific mangrove of average height with *Rhizophora*
High mangrove	Marine wetland	Mangroves of pure *Rhizophora* or mixed in with *Raphia*
Marsh	Freshwater wetland	Floating meadow with *Echinochloa*. Mixed floating meadow with *Echinochloa* or papyrus. Meadow with *Panicum*. Marsh meadow with Marantaceae.
Open swamp forest	Freshwater wetland	Intermediate class between marshes and closed swamp forests, with an open canopy.
Closed swamp forest	Freshwater wetland	Varying shrub cover with *Alchornea, Masserium, Dugetia, Raphia, Anthocleista, Nauclea, Anthostema, Ficus*, or *Ceiba*. Riparian formations with *Pandanus* or *Raphia*. Shrub thickets with *Morelia*.
Raphia grove	Freshwater wetland	Swamp forest predominantly with *Raphia*
Sandbanks	Freshwater wetland	Periodically flooded islets

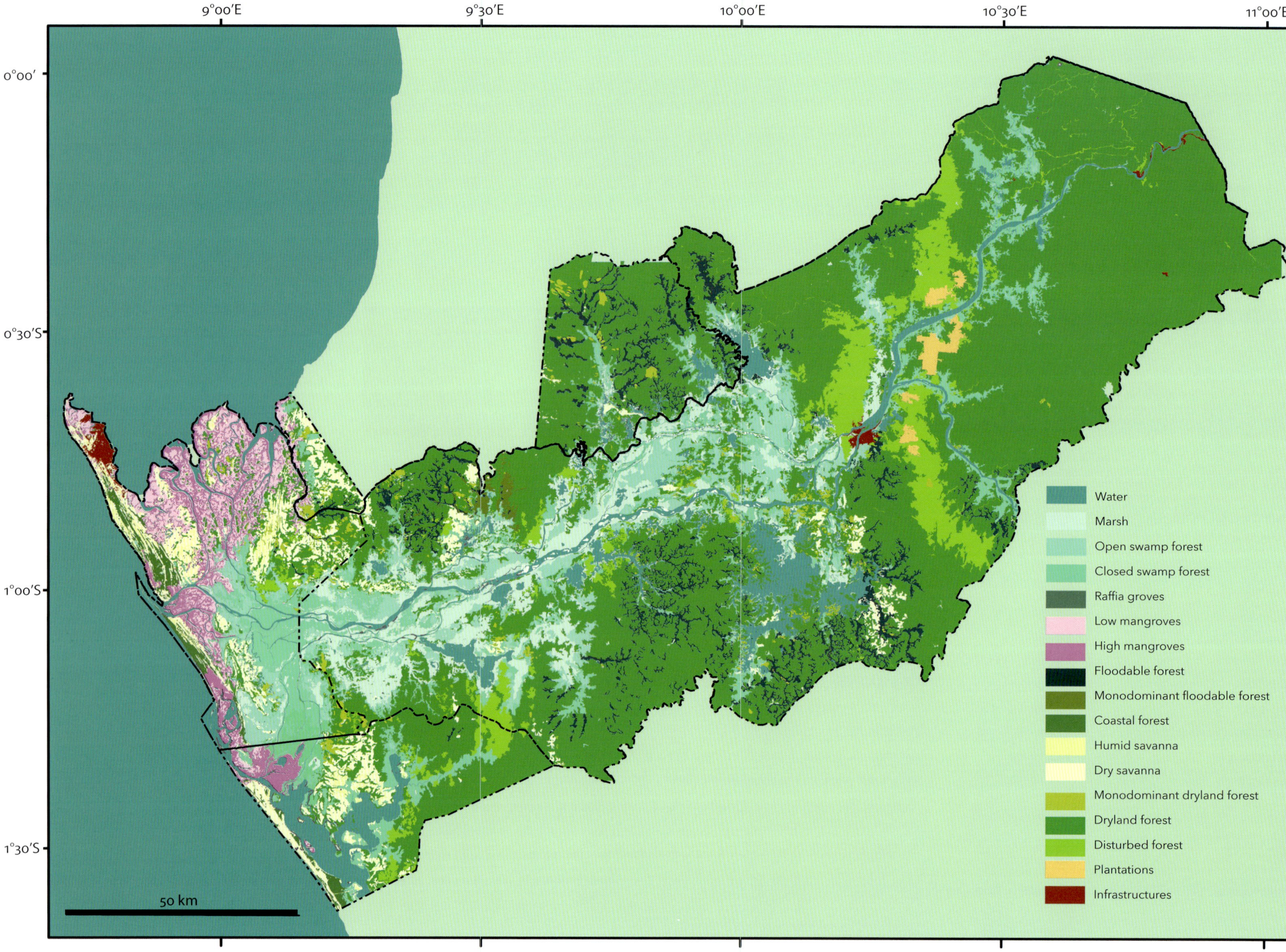
9°00′E
9°30′E
10°00′E
10°30′E
11°00′E
0°00′
0°30′S
1°00′S
1°30′S
50 km
Water
Marsh
Open swamp forest
Closed swamp forest
Raffia groves
Low mangroves
High mangroves
Floodable forest
Monodominant floodable forest
Coastal forest
Humid savanna
Dry savanna
Monodominant dryland forest
Dryland forest
Disturbed forest
Plantations
Infrastructures

Mapping the vegetation of the Lower Ogooué

The resulting map of plant formations across the entire area of study is presented in Figure 5.12. A distinction can be seen between freshwater inland wetlands and marine wetlands (mangroves) in the lower deltaic plain of the Ogooué Delta. Table 5.3 summarizes the surface area of each plant formation and its proportions throughout the entire region, the current Ramsar zone, and the proposed extensions. Wetlands and floodplains cover approximately 30% of the total surface of the current Ramsar site. Taking into account all the proposed extensions for the Ramsar site, these environments represent 36% of the total combined area. This proportion reaches 69.5% within the western proposed extension (toward Port-Gentil), which furthermore includes the highest proportion of maritime wetlands; mangroves cover 28% of the area. A series of extensions will now be proposed in multiple areas of interest.

Page 208:
5.12. Map of plant formations throughout the current Ramsar site and proposed additions.

Table 5.3. Area and overall proportion of vegetation types

	Entire zone		Ramsar site		Port-Gentil extension		Northern extension		Northern extension 2		Southern extension	
Class	km²	%	km²	%	km²	%	km²	%	km²	%	km²	%
Permanently flooded areas												
Sandbanks	26.7	0.1	21.2	0.1	0.3	0.0	5.8	0.4	0.0	0.0	0.0	0.0
Open water	1,438.5	7.3	894.2	6.3	206.2	8.2	54.8	4.1	0.5	0.2	280.7	17.5
Raphia groves	81.0	0.4	0.0	0.0	36.3	1.4	0.0	0.0	0.0	0.0	44.6	2.8
Closed swamp forest	1,671.4	8.4	1,043.9	7.4	310.6	12.3	92.8	7.0	27.8	12.5	195.7	12.2
Open swamp forest	1,022.1	5.2	823.8	5.8	184.1	7.3	9.8	0.7	0.0	0.0	5.3	0.3
Marsh	1,034.4	5.2	968.6	6.8	35.9	1.4	26.0	2.0	0.0	0.0	6.8	0.4
Low mangrove	362.4	1.8	0.0	0.0	333.2	13.2	0.0	0.0	18.0	8.1	11.2	0.7
High mangrove	456.7	2.3	0.0	0.0	375.3	14.9	0.0	0.0	12.5	5.6	68.6	4.3
Subtotal	6,093.1	30.7	3,751.8	26.5	1,481.9	58.8	189.2	14.2	58.7	26.5	612.9	38.3
Temporarily flooded areas												
Floodable forest	626.0	3.2	470.6	3.3	8.0	0.3	140.4	10.5	5.6	2.5	1.4	0.1
Monodominant floodable forest	52.4	0.3	52.4	0.4	0.0	0.0	0.0	0.0	0.0	0.0	0.0	0.0
Coastal forest	143.1	0.7	0.1	0.0	104.6	4.2	0.0	0.0	0.0	0.0	38.4	2.4
Humid savanna	253.3	1.3	6.0	0.0	157.2	6.2	2.3	0.2	0.0	0.0	87.8	5.5
Subtotal	1,074.8	5.4	529.1	3.7	269.7	10.7	142.7	10.7	5.7	2.6	127.6	8.0
Dryland areas												
Dryland forest	10,507.7	53.0	8,583.0	60.6	313.6	12.4	959.3	72.1	76.2	34.4	575.6	36.0
Disturbed forest	1,052.5	5.3	913.1	6.4	37.5	1.5	0.0	0.0	0.0	0.0	102.2	6.4
Plantations	91.8	0.5	91.8	0.6	0.0	0.0	0.0	0.0	0.0	0.0	0.0	0.0
Dry savannas	716.3	3.6	158.9	1.1	310.6	12.3	11.2	0.8	65.3	29.5	170.2	10.6
Monodominant dryland forest	211.9	1.1	101.4	0.7	52.5	2.1	28.4	2.1	15.6	7.1	12.4	0.8
Subtotal	12,580.2	63.4	9,848.1	69.5	714.2	28.4	998.9	75.0	157.1	70.9	860.3	53.7
Others												
Infrastructure	72.7	0.4	23.0	0.2	49.5	2.0	0.0	0.0	0.0	0.0	0.0	0.0
Nonclassified	14.4	0.1	7.9	0.1	4.0	0.2	0.4	0.0	0.0	0.0	2.3	0.1
Subtotal	87.1	0.4	30.8	0.2	53.4	2.1	0.4	0.0	0.0	0.0	2.3	0.1
TOTAL	**19,835.2**	**100.0**	**14,159.8**	**100.0**	**2,519.3**	**100.0**	**1,331.2**	**100.0**	**221.6**	**100.0**	**1,600.8**	**100.0**

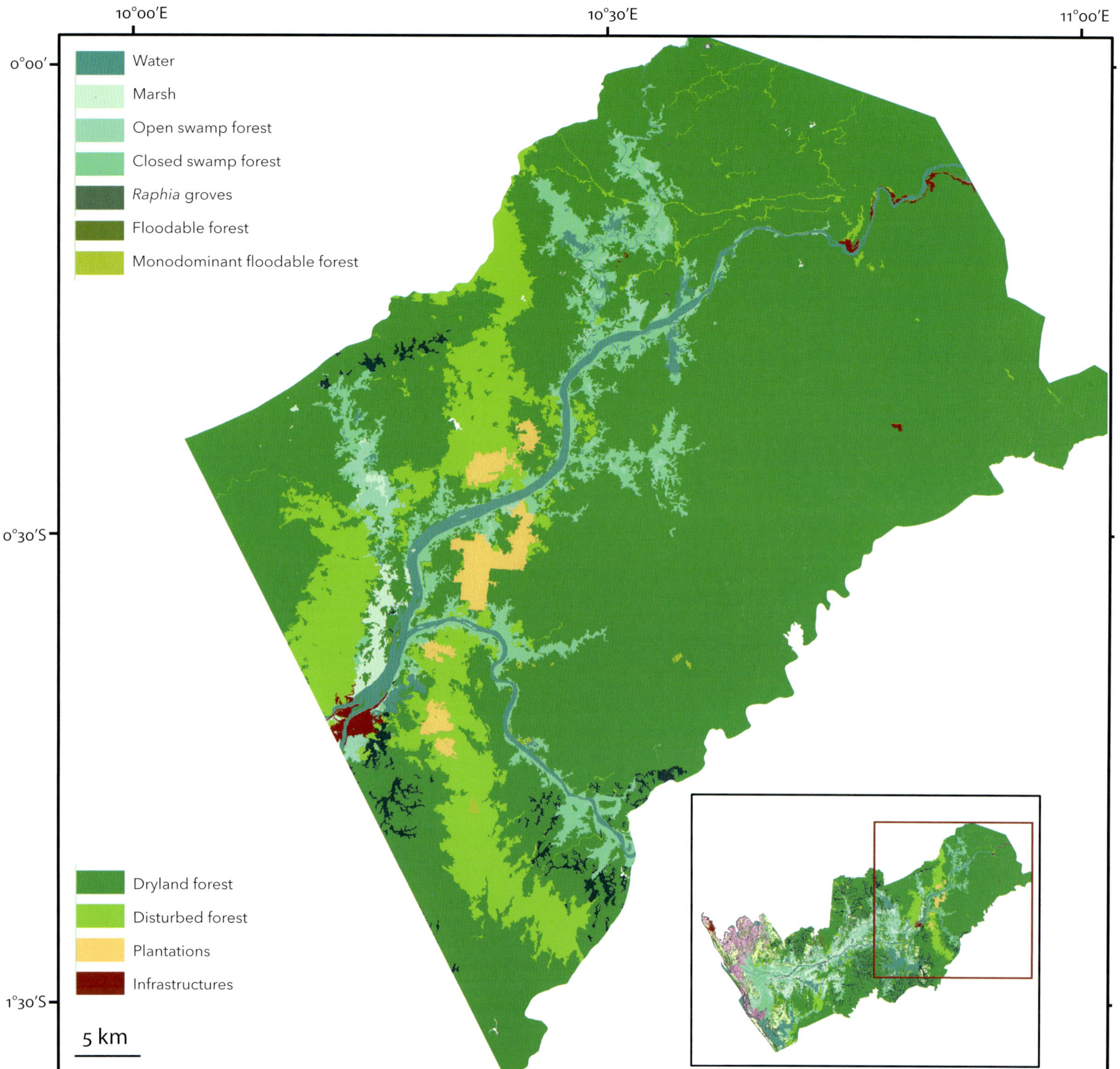

The Lower Ogooué valley

Dryland and degraded forests are predominant in the Lower Ogooué valley, between Lambaréné and Ndjolé (Figure 5.13). Indeed, this part of the Ramsar site extends into the Crystal Mountains to the northeast and into the hilly region to the southeast. It therefore makes sense that dryland formations are predominant (88%). In addition, the influence of a city like Lambaréné, along with large industrial plantations as well as National Roads 1 and 2, explains the large proportion of degraded forests (11%). There are, however, humid habitats along the Ogooué (riparian forest) and adjacent lakes, as well as a substantial proportion of dense and open swamp forests, including large *Raphia* groves, and also swamp meadows, which penetrate the dryland forests along the three major rivers that flow into the Ogooué (the Ngounié, Abanga, and Mbiné) (Figures 5.13–5.16). Technically, this area was the most difficult to capture using VHSR imagery because its proximity to the Crystal Mountains results in nearly permanent cloud cover. It is therefore likely that its typology is less accurate.

Table 5.4. Area covered and proportion of each plant formation in the Lower Ogooué valley

Class	km²	%
Water	164.6	2.6
Disturbed forest	695.7	11.1
Floodable forest	54.5	0.9
Closed swamp forest	354.9	5.6
Open swamp forest	74.8	1.2
Dryland forest	4,792.1	76.1
Infrastructures	22.4	0.4
Marsh	37.2	0.6
Monodominant dryland forest	4.9	0.1
Plantations	91.8	1.5
Dry savanna	0.2	0.0
Total	**6,293.1**	**100**

Page 210:
5.13. Map of plant formations in the Lower Ogooué valley.

Page 211:
5.14. Very open swamp forest with Nauclea pobeguinii *on the banks of the Abanga River.*

5.15. Patchwork of dryland forests, Raphia *groves, and herbaceous marshes on the Mbiné River.*

5.16. Flooded Nauclea pobeguinii *population near the mouth of the Mbiné River.*

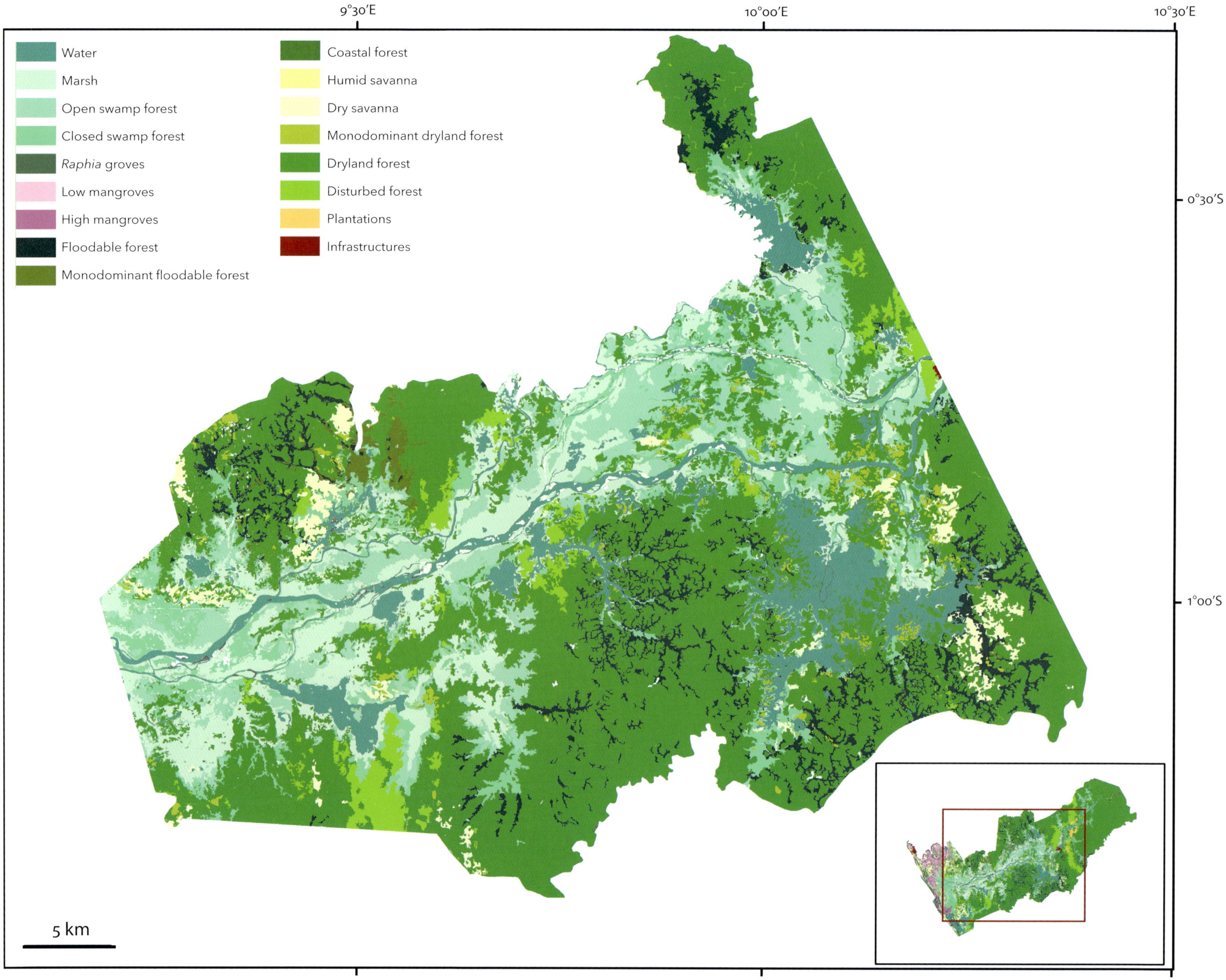

9°30'E
10°00'E
10°30'E
0°30'S
1°00'S
Water
Marsh
Open swamp forest
Closed swamp forest
Raphia groves
Low mangroves
High mangroves
Floodable forest
Monodominant floodable forest
Coastal forest
Humid savanna
Dry savanna
Monodominant dryland forest
Dryland forest
Disturbed forest
Plantations
Infrastructures
5 km

The upper deltaic plain

This part is the most complex of the region, with regard to the patchwork of freshwater wetlands, the floodable forests, and also the patchwork of savannas, forests, and monodominant forests (Figure 5.17). It includes most of the Ramsar site's large lakes, which explains the high percentages of water (9%) and wetlands (35%) (Table 5.5). A good portion of the marshy habitats are well preserved because they are difficult to access. The large lakes are connected by interlacing channels that supply them with water and in doing so deposit sediments carried by the Ogooué, slowly filling them in. This forms alluvial build-up, which is quickly colonized by pioneer vegetation (Figures 5.19 and 5.20). Very often, old river channels are filled in by marshy meadows, which can gradually evolve into denser formations (Figures 5.20 and 5.22). Sandbanks are common along the Ogooué in the dry season, particularly in the central channel and along the Ouango.

Half of the surface area is nevertheless covered by dryland forest, only very partially degraded (3%). These forests are located mainly in the south, a slightly hilly region that shelters floodable forests along small streams. The savannas in the center of the upper deltaic plain, around Lake Evaro, have been heavily modified by humans. Indeed, there are ridges covered with dry savanna (appearing pale yellow on the photo, Figure 5.21) in many areas of the flooded savannas (appearing greener on the photo). In the absence of fires, the savanna is quickly colonized by monospecific stands of okoumé or ozouga.

Table 5.5. Area covered and proportion of each plant formation in the upper deltaic plain

Class	km^2	%
Sandbank	20.9	0.3
Monodominant floodable forest	52.2	0.7
Open water	722.9	9.2
Disturbed forest	240.0	3.0
Floodable forest	418.0	5.3
Coastal forest	0.1	0.0
Closed swamp forest	650.3	8.3
Open swamp forest	731.5	9.3
Dryland forest	3,874.3	49.2
Infrastructures	0.9	0.0
Marsh	907.9	11.5
Monodominant dryland forest	91.6	1.2
Humid savanna	5.8	0.1
Dry savanna	159.2	2.0
Total	**7,875.6**	**100**

Page 212:
5.17. Map of the vegetation of the upper deltaic plain.

Page 213:
5.18. Lake Evaro (Photo: Nicolas Barbier).

5.19. Rivière des Pélicans delta in Lake Onangué (Photo: Nicolas Barbier).

5.20. Channel carrying water from the Rivière des Pélicans into Lake Evaro (Photo: Nicolas Barbier).

5.21. Patchwork of savannas and dryland forests with monodominant populations of okoumé, Aucoumea klaineana*. Old ridges are visible in the foreground (page 128) (Photo: Nicolas Barbier).*

5.22. Closed swamp forest and swamp meadow behind the riparian forest that runs along the main channel of the Ogooué (Photo: Nicolas Barbier).

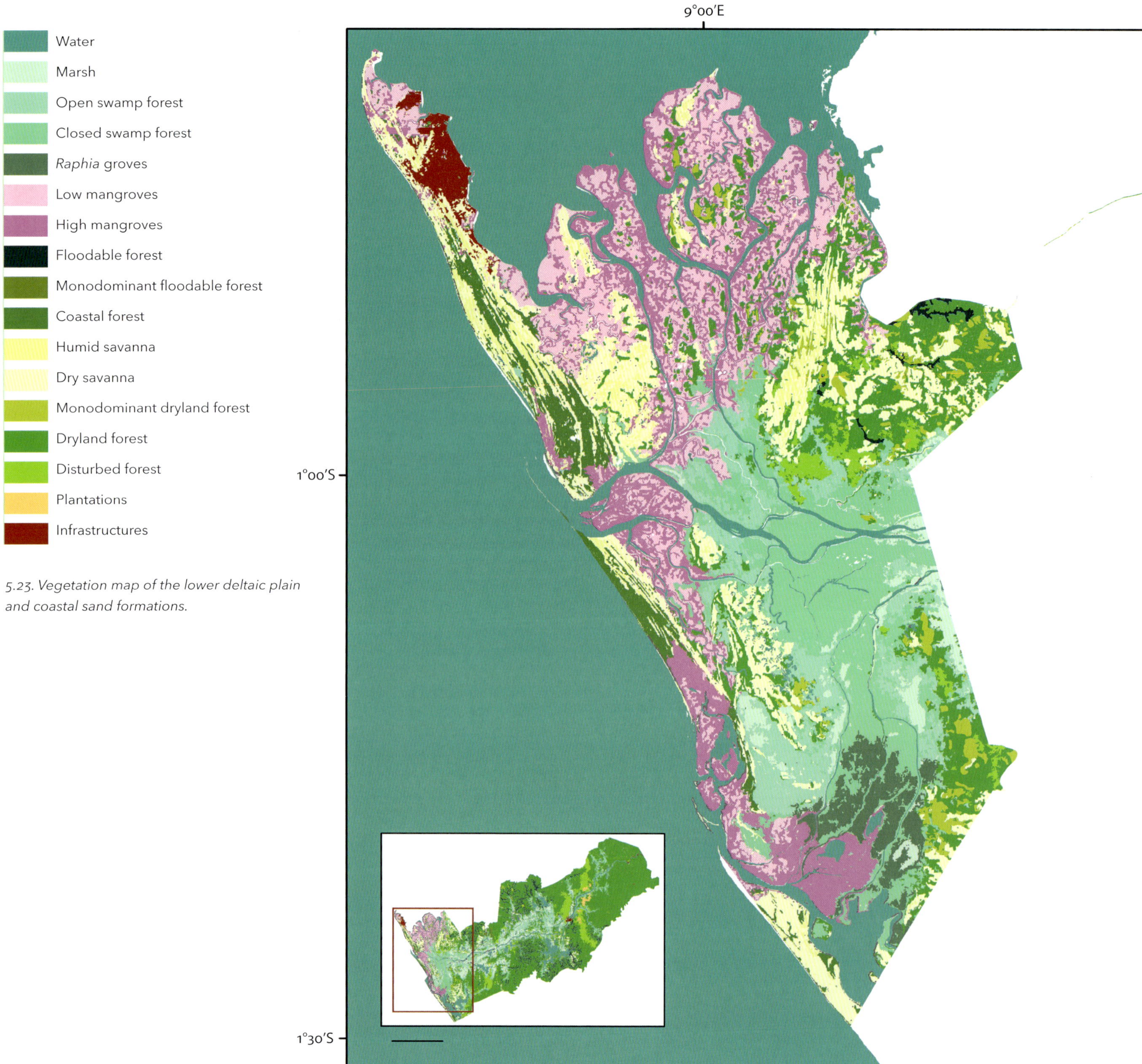

5.23. Vegetation map of the lower deltaic plain and coastal sand formations.

The lower deltaic plain and coastal sand formations

The most remarkable aspect of the lower deltaic plain is the transition between freshwater wetlands and marine wetlands, with a progressively more pronounced dominance of mangroves (26%) in the direction of the various mouths of the Ogooué (Figure 5.23). Another characteristic is the presence of large *Raphia* groves, not just along the different channels of the Ogooué, but also inside groves of swampy forest (Figure 5.26). Indeed, *Raphia* covers almost 3% of the total area. Notably, the dominant species, which has not yet been identified, is different from that found within the Ramsar site. This area is also characterized by the presence of a unique type of forest within the Ramsar site, littoral forests (Figure 5.23), which are a mixture of dry and swampy forests interspersed with dry and flooded savanna. Overall, wetland habitats represent around 60% of the area and water (lakes, lagoons, and rivers) about 10%. The presence of Port-Gentil explains why parts of these habitats are degraded (just over 1%).

Table 5.6. Area covered and proportion of each plant formation in the lower deltaic plain

Type	km²	%
Sandbank	0.3	0.0
Monodominant floodable forest	0.1	0.0
Water	280.3	9.4
Disturbed forest	39.6	1.3
Floodable forest	7.9	0.3
Coastal forest	116.5	3.9
Closed swamp forest	387.8	13.0
Open swamp forest	204.2	6.9
Dryland forest	351.8	11.8
Infrastructure	49.5	1.7
Low mangrove	343.2	11.5
High mangrove	440.9	14.8
Marsh	65.2	2.2
Monodominant dryland forest	68.3	2.3
Raphia stands	79.6	2.7
Humid savanna	186.4	6.3
Dry savanna	353.5	11.9
Total	**2,974.9**	**100**

5.24. The dune ridges located along the lower deltaic plain are home to large sections of littoral forest, which form a tangle of dryland and swamp forests; these alternate with dry savannas and floodable savannas. High mangroves and low mangroves are visible in the background; they adjoin the channels of the Ogooué (Photo: Jean P. Vande weghe).

5.25. Low mangroves located along a tidal river. Along the river, there are small clumps of high mangroves, and islets of dryland forest are present in the savanna. These environments have a high diversity of herbaceous species, which are found across the salinity gradient, from low mangroves to the dry savannas (Photo: Jean P. Vande weghe).

Page 218:
5.26. The lower deltaic plain shelters large expanses almost purely formed by Raphia *that colonize the old high mangroves (isolated trees are visible at the top of the image) (Photo: Jean P. Vande weghe).*

Page 219:
5.27. Vegetation map of the lower section of the Nkomi Lagoon.

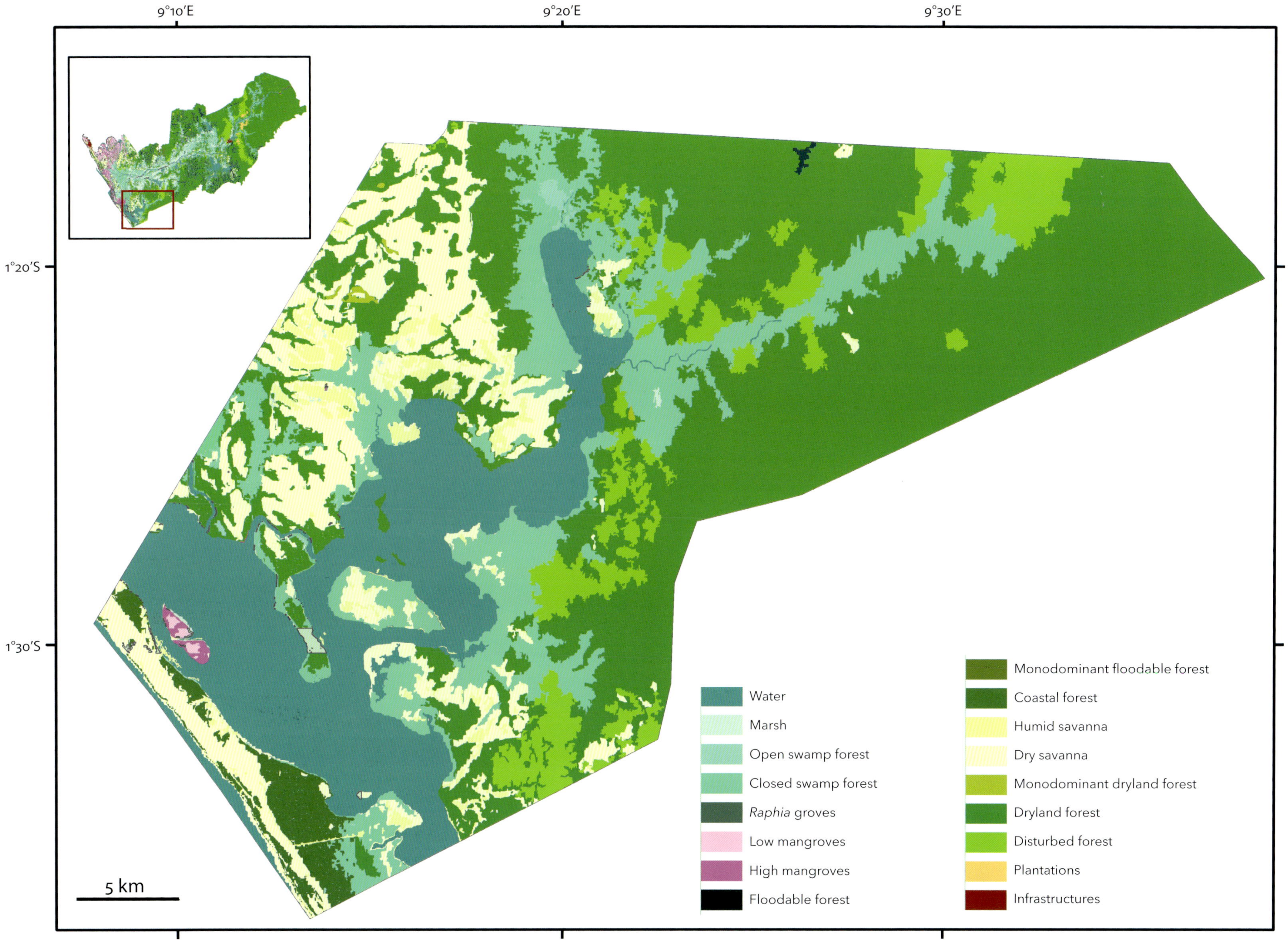
9°10′E
9°20′E
9°30′E
1°20′S
1°30′S
5 km
Water
Marsh
Open swamp forest
Closed swamp forest
Raphia groves
Low mangroves
High mangroves
Floodable forest
Monodominant floodable forest
Coastal forest
Humid savanna
Dry savanna
Monodominant dryland forest
Dryland forest
Disturbed forest
Plantations
Infrastructures

5.28. Swamp forests along the lower section of the Nkomi Lagoon. Some mangroves are visible in the foreground (Photo: Jean P. Vande weghe).

The lower section of the Nkomi Lagoon

This coastal area corresponds to a proposed addition to the Ramsar in order to include the southern part of the Lower Ogooué watershed. It is made up largely of lagoons (18%) (Figures 5.27 and 5.28), but above all by intact (41%) or partially degraded (7%) dryland forest. Wetlands are therefore scarce (around 20%).

Table 5.7. Area covered and proportion of each plant formation in the lower section of the Nkomi Lagoon

Class	**km²**	**%**
Water	208.8	18.4
Disturbed forest	80.0	7.0
Floodable forest	0.6	0.1
Coastal forest	26.2	2.3
Closed swamp forest	158.4	13.9
Open swamp forest	2.9	0.3
Dryland forest	468.8	41.3
Low mangroves	0.9	0.1
High mangroves	1.6	0.1
Marsh	1.0	0.1
Monodominant dryland forest	1.4	0.1
Raphia groves	1.3	0.1
Humid savanna	58.1	5.1
Dry savanna	125.8	11.1
Total	**1,135.7**	**100**

Map validation and reliability

In total, 580 validation polygons were drawn based on field observations and overflights (by airplanes and drones), for a total area of 3,677 ha. These areas of "ground truthing" were used to validate the map and to estimate the reliability of the various classes using a confusion matrix. This matrix consists of a table that, for each class tested on the map, indicates the proportion of classes actually observed in the field. The value of the Kappa coefficient, which assesses the overall classification performance, is 0.76, a value regarded as adequate, although not exceptional, the ideal being to approach a value of 1.

The modest Kappa coefficient obtained is mainly due to the fact that the open swamp forests class is often confused with closed swamp forests and swamp meadows. These swamp forests in fact form a middle ground between the two formations within the colonization dynamics of the Ogooué's banks. The gradual densification of the canopy is relatively well captured by the texture index, but does not lend itself well to categorization into classes. Therefore, in the field, it was quite difficult to assign one of these three classes to the validation areas visited. As a result, quite a few validation zones were captured for the "open swamp forest" category, namely 16 zones for a total of 143 ha.

Similarly, there is significant confusion between the monospecific *Raphia* class and closed swamp forests containing various portions of *Raphia*. In this case, the distinction is extremely subtle (and is notably visible in the illustrations in Figures 5.15 and 5.26) at the resolution of SPOT 6-7 images, and texture data can only differentiate the two in the case of absolutely monospecific *Raphia* formations.

By omitting the *Raphia* and open swamp forests from the validation process, the Kappa coefficient increases to 0.9, which is perfectly satisfactory.

Conclusions

As illustrated by this chapter, mapping the various plant formations of the Lower Ogooué region is a significant challenge due to the size of the region being studied, the complexity and dynamics of the vegetation patchwork, and the area's particularly cloudy climatic conditions. Nevertheless, an objective approach to mapping the plant cover was possible using a combination of layers of satellite information, of which the texture of VHSR images plays a central role. Intensive field surveys by teams of botanists travelling by boat, as well as the use of drones and airplanes, provided a large set of training/validation data covering 580 areas.

We were thus able to map out 16 plant formations, for the most part with a fairly high level of confidence. *Raphia* groves and open swamp forests were retained on the map, despite a lower level of reliability. Each of these formations covers a number of different plant associations that can be observed in the field and are described in the previous chapter. It is, however, impossible to map out all of these combinations across the entire site, although local maps with more details could possibly be produced using Pleiades or drone imagery, for example.

In any case, the proposed map allows for detailed analyses of the position and variation in scale between formations throughout different areas of interest, whether for the study of 1) vegetation dynamics and how this interacts with the dynamics of the delta, 2) water regimes, or 3) the history of human occupation, or in order to define the habitats of fauna, and especially of birds.

Bibliography

BARBIER, N. & P. COUTERON, 2015. Attenuating the bidirectional texture variation of satellite images of tropical forest canopies. *Remote Sens. Environ.*, 171: 245–260.

BLASCHKE, T., 2010. Object based image analysis for remote sensing. *Isprs J. Photogramm. Remote Sens.*, 65: 2–16. doi:10.1016/j.isprsjprs.2009.06.004

Carte de l'Afrique centrale au 1:200.000, 1962. IGN, Libreville.

Carte des formations végétales au 1:200.000, 1998. Ministère des Eaux et Forêts, Libreville.

COUTERON, P., R. PELISSIER, E.A. NICOLINI & D. PAGET, 2005. Predicting tropical forest stand structure parameters from Fourier transform of very high-resolution remotely sensed canopy images. *J. Appl. Ecol,*. 42: 1121–1128. doi:10.1111/j.1365-2664.2005.01097.x

SIMARD, M., G.D. GRANDI, S. SAATCHI & P. MAYAUX, 2002. Mapping tropical coastal vegetation using JERS-1 and ERS-1 radar data with a decision tree classifier. *Int. J. Remote Sens.*, 23: 1461–1474. doi:10.1080/01431160110092984.

6.1. Aphyosemion australe *(Photo: Christophe Aubin).*

6.2. Aphyosemion *sp. (*A. escherichi *group) (Photo: Laurent Chirio).*

6.3. Aphyosemion striatum *near Menguègne (Photo: Laurent Chirio).*

6.4. Aphyosemion gabunense *collected near Lake Zilé (Photo: Laurent Chirio).*

6.5. Chromaphyosemion rubrogaster *of Niengé River (described in 2018) (Photo: Olivier Buisson).*

6.6. Epiplatys sexfasciatus *in the Mabounié River (Photo: Laurent Chirio).*

6.7. Epiplatys *sp. at Wéze River (Photo: Laurent Chirio).*

6.8. Epiplatys multifasciatus *in the Mabounié River drainage (Photo: Laurent Chirio).*

6. Fishes

Yves Fermon, Laurent Chirio, Anton Lamboj, Jean-Félicien Liwouwou & John P. Sullivan

The Lower Ogooué

Situated between 0°30′ and 1°44′ S, 8°50′ and 10°18′ E, the Lower Ogooué basin is located in the central-western part of Gabon. It covers an area of approximately 20,000 km². The upstream limit on the major bed is generally considered to be linked to the area of rapids and falls located downstream of Ndjolé, which marks the transition between the inland relief and the coastal sedimentary basin (page 21). From there, following a center-west/southwest axis to the mouth of the river in the Port-Gentil region, the course of the Ogooué runs for more than 310 km.

The river thus flows through two separate administrative entities: the eastern part of Lower Ogooué is located in the province of Moyen-Ogooué (and the departments of Abanga-Bigné and Ogooué et des Lacs) and the western part is located in the province of Ogooué-Maritime (adapted from Mbega & Teugels, 2003).

Multiple tributaries flow into the lower course of the Ogooué: the Massanga, Abanga (280 km), and Mbiné rivers on its right bank; and the Mboumi, Ngounié (up to the Imperatrice Falls), and Bigouenya rivers upstream from the lakes area. The Olimbe, Mahouaca, Aloué, and Rivière des Coupeurs, among others, flow into the lakes.

Ecological zones and the Ramsar site

In his thesis, Mbega (2004) highlighted three ecological zones:

- The flooded forest zone, which extends from Ndjolé to Lambarené.
- The lake zone, which extends from Lambaréné to near Ngola village. From Lambaréné, the Ogooué splits into two channels, which flow into many lakes and swamps. These lakes constitute the interior delta.
- The deltaic zone extends from the village of Ngola to the mouths of the river.

In practice, delimiting the exact boundaries of these ecological zones is difficult, particularly in the area corresponding to Mbega's "deltaic zone."

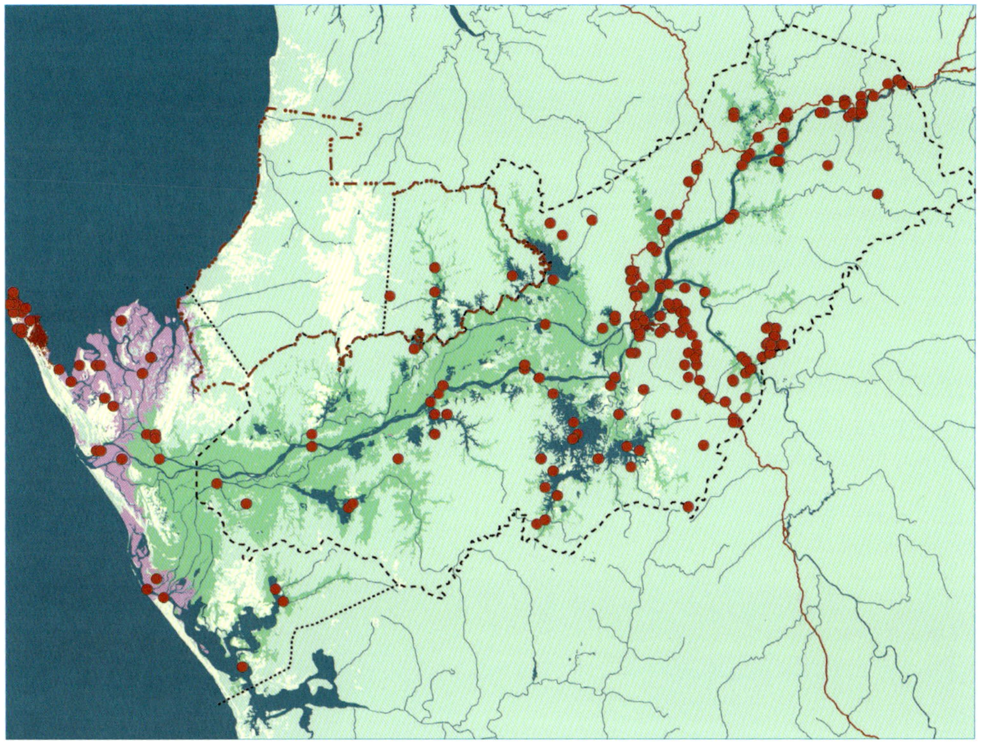

6.9. Distribution of fish collection stations in the Ogooué Delta region.

History of ichthyology in Gabon

based on Josephine W. Hearn & Carl D. Hopkins (Cornell University, Ithaca, New York), supplemented by John Sullivan & Yves Fermon

1859 Auguste Duméril published the first account of fish from Gabon based on a collection made by Aubry-Lecomte. Duméril included several descriptions by Alphonse Guichenot (Duméril, 1859).

1860– 80 French officials and traders began to explore the inland course of the Ogooué. A stretch of rapids over 160 km long through the middle part of the Ogooué was a major impediment to navigation to the upper parts of the river (Günther, 1896).

The earliest explorers of the Ogooué were Du Chaillu, Braouèzec, Serval, Admiral Touchard, Griffon du Bellay, and Admiral Fleuriot de Langle (Sauvage, 1880).

1862 Theodore Gill of the Academy of Natural Sciences of Philadelphia describeds several fish collected by Du Chaillu (Gill, 1862).

1866 Robert Bruce N. Walker of the London Geographic Society made his first trip to Gabon, following the Ogooué to Inenga country (Günther, 1867a; Sauvage, 1880).

1866–1873 Mr. Aymès, "lieutenant de vaisseau" of the French "Canonnière" Pionnier, explored the lower part of the Ogooué up to its tributary, the Ougougne River, for the purpose of map making (Sauvage, 1880).

1867 Albert Günther of the British Museum (Natural History) described the fish collected by Walker from the Lower Ogooué (Günther, 1867b).

1871 Günther described three more fish from Gabon that were "in the possession of the author" (Günther, 1871).

1873 Walker made a second journey to Gabon, this time penetrating as far as the Lopé (Sauvage, 1880; Günther, 1896).

1874–1878 Pierre Savorgnan de Brazza and Noël Balay led an expedition that explored all of the Ogooué, discovering the upper courses of the Licona and Alima Rivers, where the Congo and Ogooué basins are separated by a short distance (Sauvage, 1880). On this expedition, the naturalist Alfred Marche collected fish at Doumé (Ogooué) and Lopé (Sauvage, 1879).

1880 Henry-Émile Sauvage published the first complete account of the known fish of the Ogooué in his "Etude sur la faune ichthyologique de l'Ogôoué" (Sauvage, 1880).

1883 The Mission de l'Ouest Africain, directed by Jacques de Brazza, brother of the famous explorer, collected a large number of fish, mostly from Adouma on the Ogooué.

1886 The large collection made by the Mission de l'Ouest Africain was sent to the Museum of Natural History in Paris. Léon Louis Vaillant very briefly described several of the de Brazza fish in an article on the expedition as a whole (Rivière, 1886; Pellegrin, 1900b; Pellegrin, 1900c).

1894 Mary Kingsley made her first trip to Gabon, following the course of the Ogooué and making collections as far up river as Kondo-Kondo in the Alemba Rapids (Günther, 1896).

1900 Jacques Pellegrin described many more of the fish collected by the de Brazza mission (Pellegrin, 1900a; Pellegrin, 1900d; Pellegrin, 1900b; Pellegrin, 1900c).

1906 Pellegrin received the first of many interesting collections made by pastor Ernest Haug, a correspondent of the Paris Natural History Museum who had established a permanent residence at Ngomo on the Ogooué, 90 km downstream from Lambaréné. The Haug collections made a major contribution to the knowledge of the fishes of Gabon. Furthermore, as a result of his interaction with native peoples, Haug was able to supply the indigenous names of many of the fish he sent to Paris. The names were in three different languages: Galwa, Nkomil, and Pahouin (Fang). In 1916, Haug was posthumously awarded the Prix Secques of the Zoological Society of Paris for his outstanding contributions (Pellegrin, 1906b, 1906c, 1906d, 1907a, 1907b, 1908a, 1908b, 1909a, 1909b, 1911a, 1911b, 1913, 1914a, 1914b).

1914 Pellegrin published a complete list of the known fish of the Ogooué in his "Les Poissons du Bassin de l'Ogooué" (Pellegrin, 1914b).

1919 Pellegrin received the last of the Haug collections, sent from Ngomo by Haug's brother (Pellegrin, 1919).

1964 Gérard Loubens publishes a thesis on the fish and fishing in the Ogooué near the lakes in the Lambaréné region (Loubens, 1964).

1976–80 Huber visited Gabon and reported on a new species of *Aphyosemon* and then published a series of papers on the killifishes of Gabon (Huber, 1976, 1977; Huber & Radda, 1977 ; Huber & Seegers, 1977 ; Huber, 1978, 1980a, 1980b, 1980c, 1981a, 1981b, 1993, 1994).

1976–2002 Carl Hopkins of Cornell University (USA) with his associates and students begin study of Gabon's mormyrid fauna beginning at the Ipassa Research station on the Ivindo River and later including many other parts of Gabon.

1989 Pierre Gilbert *et al.* wrote the first book summarizing the fish of Gabon, which included 103 of the most common species (Gilbert *et al.*, 1989).

1999–2002 This period was marked by the study of fish from the Lower Ogooué region, with a first series of collections (MRAC, AMNH, IRD) made in collaboration with the IRAF, from 60 sites followed by work at nine localities over several seasons.

2000–2012 Several anthropogenic impact studies were carried out by IRAF (Jean-Daniel Mbega, Jean-Hervé Mve Beh, Jean-Félicien Liwouwou, etc.). Several hydrobiological studies were conducted by students from ENEF (National School of Water and Forests).

2003–2004 Teugels and Mbega published the "Guide de détermination des poissons du bassin inférieur de l'Ogooué," which was be followed by J.-D. Mbega's thesis, "Biodiversité des poissons du bassin inférieur de l'Ogooué (Gabon)" at the University of Namur (Belgium).

2007 The two-volume book "Les poissons des eaux douces et saumâtres de Basse Guinée, Afrique Centrale" was published by IRD & MRAC (Stiassny *et al.*, 2007).

2010–2011 Sampling was done in the Ogooué and Ivindo basins by teams from IRAF and Cofrepêche (Yves Fermon) as part of the PSPA project (Projet d'appui au Secteur des Pêches et de l'Aquaculture). IRAF attempt to set up a national collection of fish from Gabon.

2011–2013 Belgian research mission. Revision of the genus *Hepsetus* from Lower Guinea, with the description of a new species endemic to Gabon (Decru *et al.*, 2013).

2013 Synthesis of current knowledge of Gabonese fish by Fermon, thanks to support from WWF.

2014–2015 CBFF project in the Lower Ogooué Ramsar site by WWF, in collaboration with Y. Fermon (Asso. Aimara) and J.-F. Liwouwou (IRAF). ■

Chapter 2 provides a more detailed overview of the ecological zones of the Lower Ogooué (Vande weghe, 2017: page 24, Figure 2.15):

- the more or less flooded lower river valley and adjacent wetlands upstream of Lambaréné;
- the upper deltaic plain, located downstream from Lambaréné, with adjoining lakes and lacustrine deltas;
- the lower or tidal deltaic plain;
- the coastal sand structures;
- the lagoon area.

The number of fish inventory stations (233) varies among the ecological zones (Figure 6.9). The distribution of stations, and therefore knowledge of their fishes, thus appears to be particularly unequal. The Lower Ogooué valley (129) and the upper deltaic plain (65) were much more extensively inventoried than the other zones, with very few stations (just three) in the lagoon zone.

Water chemistry

Water chemistry is of primary importance with regard to the distribution of fish species. It affects all aspects of fish life history (reproduction, migration, egg and larval development, etc.). Chapter 2 presents the general characteristics of water chemistry. Below is a summary of the most important elements.

The physicochemical characteristics and the ionic composition of water in the Lower Ogooué basin indicate:

- The relative poverty of the waters of the Lower Ogooué basin in terms of biogenic elements (ammonium, phosphates, and nitrates);
- The longitudinal succession of two types of environment:
  (1) Strict freshwater environments found in the lower valley and the upper deltaic plain appear to be unaffected by rising seawater. They contain relatively few salts, with almost no salinity, and seasonal physicochemical variation, especially in alkalinity and total hardness, is relatively low;
  (2) Freshwater environments subject to marine influence found in the lower deltaic plain, the lagoon area, and the coastal area, are subject to the influence of seawater intrusion and are relatively rich in dissolved salts. Their salinity is very variable, and seasonal physicochemical variation, especially in alkalinity and total hardness, is relatively high;
- The transverse structuring of two different environments in the flood forest zone and in the lake zone: a lentic environment and a strict lotic environment;
- Throughout all areas, the rainy season is usually accompanied by a decrease in pH, conductivity, salinity, transparency, and total hardness, as well as a simultaneous increase in temperature. The reduction in pH during the rainy season is probably caused by the introduction of humic acids originating from decomposing organic matter that is carried by runoff. Despite the very close proximity to the sea, the pH of the deltaic zone remains acidic during the rainy season, which probably indicates that marine waters have a very weak influence within the area during the rainy season;
- Seasonal fluctuations in conductivity are directly related to seawater intrusion. In the long dry season, the deltaic zone is affected by rising seawater, which brings with it many dissolved elements, leading to significantly increased conductivity during this period. Due to their geographical position relative to the sea, the Lower Ogooué valley and the upper deltaic plain are not affected by this rise in seawater and their conductivity does not appear to be significantly modified;
- The seasonal variation in transparency in both environments is closely linked to the precipitation regime. In the rainy season, soil leaching drains a significant amount of suspended matter and waters become very turbid, decreasing transparency.

The fish population

A bit of history

This section is based on the history established by Josephine W. Hearn and Carl D. Hopkins of Cornell University, Ithaca, New York. More recent information was added by John Sullivan and Yves Fermon (box on page 224).

Biogeographical elements

The fish fauna is not distributed evenly across the African continent. Based on the distribution of the different families and species of fish, ichthyologists currently recognize 10 large ichthyological provinces, each of which hosts characteristic populations (Lévêque & Paugy, 2006).

The rivers of Gabon are part of the Lower Guinea region, which includes the coastal rivers of Cameroon, Equatorial Guinea, and Gabon, up to the mouth of the Congo River.

The Ogooué basin has a large number of Congolese species, which are probably the result of recent colonization, and which cannot be found anywhere else in the Lower Guinea region. Morever, the Ogooué and its main tributaries do not constitute homogeneous entities (Paugy *in* Stiassny *et al.*, 2007). For example, the upper course of the Ivindo (Aïna), the current main tributary of the right bank of the Ogooué, is known to be a "recent" capture. Natural barriers also exist in the Ogooué basin, since *Brycinus bartoni* (Alestidae) is present in the Ngounié (another large tributary, this time from the left bank) and is endemic to this tributary, along with a sister species downstream, which is probably new.

Species composition of the Lower Ogooué ichtyofauna

The ichtyofauna of the Middle Ogooué is one of the best known in Gabon (Fermon, 2013). Old and new data taken together indicate a total of 53 families, 119 genera, and 229 species and subspecies in the area (Appendix). By way of comparison, the Nile basin (without the lake areas) contains a total of 159 species, and the Niger basin has 352 species. This represents a very high level of fish biodiversity in such a short segment of the river. Given that many areas remain poorly documented, this figure may increase. The presence of three of the species included in the count (*Enteromius carens, Epiplatys annulatus,* and *Parachanna africana*) remains doubtful, probably resulting from the erroneous determination of the specimens. Overall, there are 21 families and 32 species exclusive to marine and brackish waters of the Lower Ogooué, and 24 families and 41 species from "salty" water and freshwater.

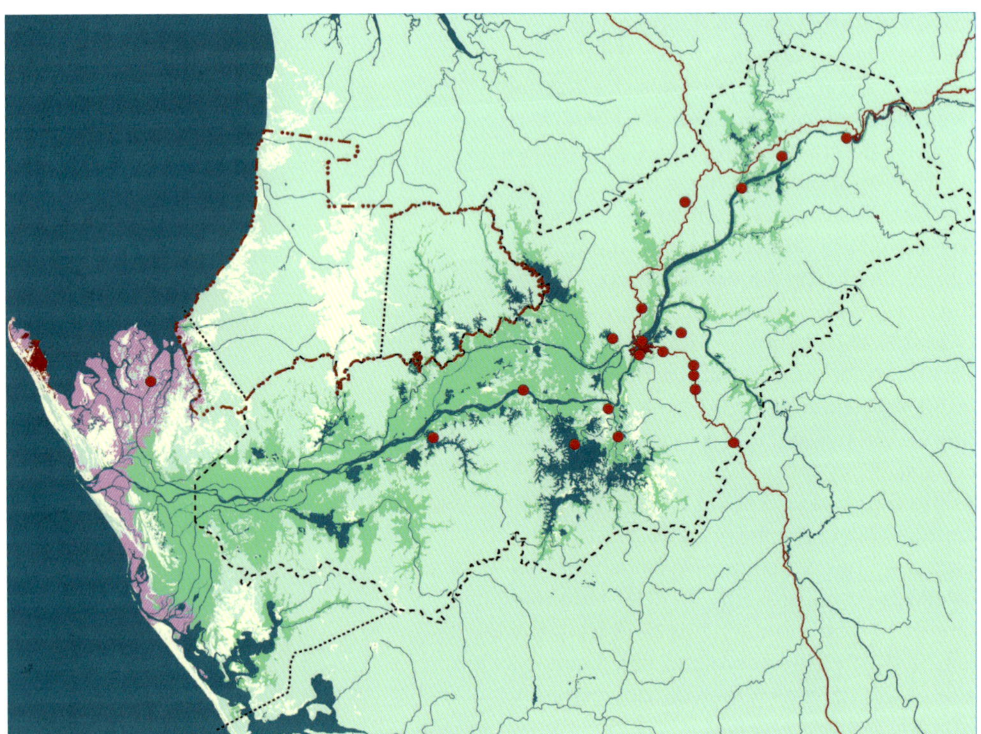

6.10. Distribution of type specimens of fish collected in the Lower Ogooué Delta region.

Table 6.1. Number of families and species (M = salt water, S = brackish water, F = freshwater)

Ecological zones	Stations	Families					Species				
		MS	MSF	SF	F	Total	MS	MSF	SF	F	Total
Lower Ogooué valley	129	2	8	9	21	31	2	9	12	122	145
Upper deltaic plain	65	10	14	9	23	42	10	18	14	114	156
Lower deltaic plain	18	15	18	7	13	38	25	23	7	28	83
Littoral sand structures	18	15	14	5	12	37	26	17	6	17	66
Lagoon area	3	3	8	2	10	21	3	9	2	14	28
Total	**233**	**21**	**19**	**9**	**23**	**53**	**32**	**27**	**14**	**156**	**229**

In all there are 156 species known only from freshwater, which belong to 23 families based on current knowledge (Table 6.1).

General distribution of species

To understand the general distribution of fish species, it is necessary to use previously acquired data based on samples collected during research expeditions on localized sections, referred to as stations. During each visit, fish are caught and sorted by species, providing a list of species. Thanks to the recent contribution of geolocation techniques, geographic data are becoming increasingly precise. By combining all the results into tables, a list of species by stations can be obtained.

The cumulated number of stations described in literature amounts to 233 (Figure 6.9). There are only a few stations along the small tributaries.

For the revision of species known to occur in the Ramsar site, 1,676 species/stations were used (Fermon, 2015). Based on studies conducted since 2014, this number has now increased to 2,245 species/stations.

Twelve families are represented by species present in all zones (Table 6.1), 10 from four zones, 13 from three zones, 12 from two zones, and finally, six have only been only found in a single zone, most of them marine (Appendix). Species diversity is highest in the lower valley and the upper deltaic plain because they contain freshwater and most species can only survive in low-salinity environments (Appendix), although the number of stations is in fact much higher there than in other areas. However, it appears that fewer species occur in the lower valley than in other areas, with the exception of the lagoon area. This can be explained by the presence of marine families and species that are much less well represented in the lower valley. The structure of these populations will be addressed in a later paragraph. The lagoon area, despite a small number of stations (three), includes 28 species, half of which are known only from freshwater environments (14 out of 28).

Type localities

A number of species have their type locality in the Lower Ogooué (Table 6.2). A type is a specimen which was used to describe a given species. It is a bit like the "standard meter" of a species. Data from various international collections account for 29 species which have types that were collected in Gabon. A 30th species (*Chrysichthys walkeri*) proved to be an error, and the classification of this specimen is still doubtful. This represents more than 10% of known species of the Lower Ogooué.

Twenty-one stations, or 9% of the stations, are connected to type specimens (Figure 6.10). Eleven are in the lower valley, one in the lower deltaic plain, and nine in the upper deltaic plain.

Unfortunately, environmental conditions at some of these older stations have deteriorated. When six of them were visited in 2014 (Fermon, 2015), for killifish, only five were found, three of which were badly damaged by urban pollution.

Remarkable species

Heritage and commercial species

The threadfin

The giant African threadfin, *Polydactylus quadrifilis*, also known as the grand captain in order to distinguish it from other captains, notably from the little captain, *Galeoides decadactylus*, is a fish from the Polynemidae family (Figure 6.11). It should not be confused with the freshwater "captain" of West Africa, which is in fact a member of the family Latidae, *Lates niloticus*, the famous Nile perch.

This species is well known by fishermen looking for a challenge. It is a robust fish with silver scales that can reach large sizes, and it has a distinct sort of tactile antennae under its throat that it uses to detect its prey. It also has a very hard, translucent membrane covering its nose and eyes.

The giant African threadfin is a highly valued white fish with fine and much-appreciated flesh.

In the 1960s, Gérard Loubens studied this species in the Lower Ogooué for his doctoral thesis (1966b). It has long been known to reach relatively far up rivers and streams. Specifically, it is abundant throughout the lower basin of the Ogooué throughout the year.

The giant African threadfins of Lower Ogooué are born in oceanic coastal waters during the warm season. The breeding season is very long, since maturing females can be found throughout the warm season, from October to May, but it is from January to March that the breeding season reaches maximum intensity. Maturing females are found in brackish continental waters; on the other hand, they occur exceptionally in freshwater and always a short distance from the ocean.

At the end of their first year, giant African threadfins reach about 25 cm in length, having generally spent the year at sea. Large numbers then invade fresh and brackish waters. Exclusively carnivorous (eating mainly crustaceans and fish), they find suitable feeding conditions in continental waters. Their migration is not, however, the result of searching for specific prey.

They swim downstream toward the ocean at least once for reproduction. The majority of giant African threadfins probably reproduce after two years, at which point they reach an average length of 47 cm.

Sex reversal begins to occur in a few individuals starting at 40 cm long, but the majority only turn into females once they reach sizes between 70 and 95 cm. The likelihood of reversal seems to be increased when individuals stay in brackish waters, and it is in these areas that the majority of hermaphrodites have been caught. In short, brackish waters appear to be both conducive to sexual reversal and the preferred environment for individuals that have undergone sexual inversion. Since sex change usually takes place at larger sizes, most individuals first reproduce as males.

Apart from the invasion of continental waters by year-old individuals, the movement of this species appears to be dispersed over time; no massive departures nor arrivals were observed.

Age and sex change weaken the migratory instinct of the giant African threadfin, which then swim much less far toward the interior of the continent.

Cichlids

The term tilapia is used to refer to species of the family Cichlidae, specifically of the *Tilapia* group, which is widely distributed throughout Africa. In the Lower Ogooué, there are nine species: *Coptodon guineensis* (Figure 6.12), *C. nyongana* (Figure 6.13), *C. tholloni* (Figure 6.14), *C. rendalli* (introduced), *Pelmatolapia cabrae* (Figure 6.15), *Sarotherodon mvogoi* (Figure 6.16), *S. nigripinnis*, *Oreochromis schwebischi* (Figure 6.17), and *O. niloticus* (Figure 6.19).

Cichlids are particularly numerous in fishermen catches, *Oreochromis schwebischi* in particular.

These species exhibit sophisticated parental care. There are three modes of reproduction for tilapia:

(1) In species that lay eggs on substrates of the genera *Coptodon* and *Pelmatolapia cabrae*, the two parents team up during

6.11. The giant African threadfin can reach almost 2 m in length and weigh 75 kg. It was one of the most heavily caught species in the 1960s, accounting for 60% of the fish brought ashore in Lambaréné (Loubens, 1966b).

6.12. Coptodon guineensis *(Photo: Anton Lamboj).*

6.13. Coptodon nyongana *(Photo: Anton Lamboj).*

6.14. Coptodon tholloni *(Photo: Anton Lamboj).*

6.15. Pelmatolapia cabrae *(Photo: Anton Lamboj).*

6.16. Sarotherodon mvogoi *(Photo: Anton Lamboj).*

6.17. Oreochromis schwebischi *(Photo: Anton Lamboj).*

breeding to nest, lay, and take care of the fry by vigilantly standing guard. The adhesive eggs are attached to the substrate.

(2) In two-parent oral incubators of the genus *Sarotherodon*, oral incubation is carried out by both parents involved in laying eggs until the young fish gain independence.

(3) In maternal oral incubators of the genus *Oreochromis*, laying takes place on a substrate and the non-adhesive eggs, laid alone or in small groups, are quickly taken into the female's mouth. The male deposits his sperm when the female collects the eggs or fertilizes them in the mouth. Incubation continues until the young are fully independent. Several species have been introduced to Gabon for aquaculture purposes, including the Nile tilapia, *Oreochromis niloticus*, an invasive species (page 230).

Claroteid catfish

Claroteid catfish are the species in the family Claroteidae, and in Gabon, specifically of the genus *Chrysichthys* (Figure 6.18). Four remarkable species are present in the Lower Ogooué. *Chrysichthys* are rather abundant and easily captured. They have thorns on their pectoral fins and their anterior dorsal, and they are painful if one is stung by them. These fish can reach 50 cm in length.

6.18. *Chrysichthys nigrodigitatus* (Photo: Olivier Lucanus).

The "Yara"

These are catfish of the family Schilbeidae. Two species are remarkable, *Schilbe grenfelli* and *S. multitaeniatus*. *Schilbe grenfelli* is quite distinctive because it can have frugivorous tendencies and can attain significant sizes (the known maximum is 70 cm in length). *Schilbe* are commonly consumed, employing several culinary methods, but are generally eaten fresh or smoked. For example, one of the most common and popular dishes in certain ethnic groups in southern and northern Gabon is "smothered" fresh fish, in which they are wrapped, along with vegetables, in Marantaceae leaves and then cooked over a wood fire.

In terms of abundance in the environment, *Schilbe grenfelli* are more numerous than *S. multitaeniatus*.

In the Lower Ogooué basin, *Schilbe grenfelli* has an omnivorous, phytophagous diet with a strong tendency toward consuming fruit. Its food spectrum mainly comprises fruits from trees or shrubs (*Uapaca heudelotii, Alchornia cordifolia, Elaeis guinensis*), ants, aquatic plant debris, fish, insects, crustaceans, mollusks, and annelids. The diet of juvenile individuals predominantly includes fruits, while adults mostly eat fruit and fish. This is an opportunistic species in the dry season and selective in the rainy season, with continuous feeding activity throughout the year, allowing it to maximally exploit its immediate environment. *Schilbe grenfelli,* with a maximum observed length exceeding 507 mm, is to date the largest documented Schilbeidae. However, in the Lower Ogooué basin the size of the caught individuals of *S. grenfelli* ranged from 123 to 476 mm, with weights ranging from 10.2 g to 824 g.

Schilbe multitaeniatus, on the other hand, is an omnivorous predator species with an ichthyophagous tendency. Its food spectrum mainly comprises fish, insects, crustaceans, animal debris, fruits, and plant debris (Liwouwou *et al.*, 2013). Young individuals are mainly fish-eaters while the adults, in addition to eating fish, seem also to be interested in insects (Liwouwou, 2016). The sizes *S. multitaeniatus* ranged from 100 to 315 mm, and weights ranged from 7 to 282 g (Liwouwou, 2016).

The "eel"

In Gabon, the fish referred to as the eel is somewhat particular in that it has absolutely nothing to do with the eel of temperate waters. Rather, it is a lungfish in the family Protopteridae, *Protopterus dolloi* to be precise, which is distinct in having lungs instead of gills, forcing it to swim to the surface to breathe. This species likes marshy areas with lush vegetation, where it feeds on fish and invertebrates. It can spend the dry season in estivation, buried and curled up in a ball in the mud, surrounded by a kind of cocoon of mucus that hardens as it dries, protecting it from dehydration. It survives this way until the rainy season. The females lay eggs in a nest on the river bottom, which the male defends.

Endemic species

Endemism, for extant species (whether an animal or a plant), means that it has only been found in a given part of the world. The scale of endemism varies, from a continent or a country to a river. At the smaller scale of a country or river, endemism is particularly important with regard to biodiversity and conservation.

Twenty-six species found in the Lower Ogooué are endemic to Gabon, of which 20 are known only from the Ogooué and seven just from the Lower Ogooué

Table 6.2. Numbers of endemic species

Families	Gabon	Ogooué	Gabon/Ogooué	Gab./Lower Ogooué	Total
Alestidae	1		1		2
Cichlidae	1		2		3
Clariidae		1			1
Claroteidae	1				1
Distichodontidae			1	2	3
Hepsetidae			1		1
Kneriidae			1		1
Mochokidae			3	1	4
Mormyridae	3		1		4
Nothobranchiidae	1		2	4	7
Procapotodidae			1		1

6.19. Oreochromis niloticus *(Photo: Anton Lamboj).*

6.20. The "sans-nom" (Photo: Anton Lamboj).

Table 6.3. Number of threatened species

Category	Definition	Number of species	
CR	Critically endangered	1	*Pristis pristis*
EN	Endangered	7	
VU	Vulnerable	7	
NT	Near threatened	1	
LC	Least concern	136	
DD	Data insufficient	8	
NE	Not evaluated	69	

(Appendix). One additional species, *Clariallabes brevibarbis*, is endemic to the Ogooué, but not to Gabon since part of the river's course runs through other countries.

Among these endemic species, the most represented group is that of killifish (Nothobranchiidae and Procapotodidae), with eight species (Table 6.2). This is hardly surprising since these small fish usually live in small, often isolated, ponds or streams. Among the species not described and mentioned in the appendix, there are several killifish, some of which are probably endemic.

Most of these endemic species are present in the lower valley and in the upper deltaic plain.

Threatened species

According to the IUCN list, in the Lower Ogooué area, there are 16 fish species ranging from Vulnerable to Critically Endangered (Table 6.3). Five are mainly marine cartilaginous fish, except for the *Dasyatis ukpam* skate, which is found in brackish water and freshwater. Among them, only one species is Critically Endangered, the sawfish, *Pristis pristis*. It even appears to be extinct in Gabon. Another vulnerable marine species, the tarpon, is found in the Lower Ogooué. Of the freshwater species that are endangered and vulnerable, most are found in small rivers that flow through forests. Three of these species are small killifish in the genus *Plataplochilus*.

Introduced species

Several species have been introduced to Gabon, mainly for aquaculture. Little is known regarding the dates and species introduced. Although Welcomme (1988) published a summary in 1988 (Table 6.4), little information has been published since then regarding species introduced to Gabon. In most cases, these fish have escaped from ponds or cages.

Nile tilapia, *Oreochromis niloticus*

This species, introduced to Gabon for fish farming and which then escaped into the natural environment, is easily recognizable by the regular black vertical bands on its caudal fin (Figure 6.19). It is currently colonizing the lake region, in particular Lake Onangué (Cofrepêche, 2011). It reaches 47 cm long in Gabon, feeds on phytoplankton, and has a detritiphagic tendency; it also consumes zooplankton and invertebrates. It is a very aggressive mouth incubator that often eliminates other species of *Tilapia* and *Oreochromis*, but it hybridizes very easily. During sampling on small streams, several specimens were captured in the Lower Ogooué region (Fermon, pers. comm.).

The catfish, *Clarias gariepinus*

This catfish, also called North African catfish (FAO designation), has a wide natural distribution in Africa. It has been widely

Table 6.4. List of introduced fish species in Gabon (Welcomme, 1988)

Species	Year of introduction	Origin
Coptodon rendalli (Tilapia melanopleura)	1955	Congo, Station of Djoumouna, Brazzaville
Oreochromis macrochir	1955	Congo, Station of Djoumouna, Brazzaville
Oreochromis niloticus	1960?	Congo, Station of Djoumouna, Brazzaville
Heterotis niloticus	1959	Cameroon
Clarias gariepinus (C. lazera)	1977	Central African Republic, Station of the Landjia, Bangui

introduced in other parts of Africa, Europe, and Asia for human consumption, and Gabon has not escaped its introduction, first from the Central African Republic, but more recently also from Nigeria in 2013 or 2014 (Fermon, pers. comm.). Several countries report a negative ecological impact on certain biotopes and habitats after its introduction. It is bred for aquaculture because it is highly recommended for consumption in Africa.

Adults are present mainly in calm lake and pond waters, but can also live at the bottom of fast-flowing rivers. It is a large species that can reach 170 cm long and more than 50 kg. This is an omnivorous, voracious fish that eats everything that passes in front of its mouth.

The successful acclimatization of this species (and its consequences on native diversity) is linked to the presence of an arborescent respiratory organ in its gill cavity that allows this species to breathe atmospheric air when its habitat becomes more dry. When this happens, these catfish remain in the muddy substrates of ponds or water holes and, from time to time, absorb air from their mouths. They are even able to move out of water, using their fins as legs.

The "sans-nom," *Heterotis niloticus*

This species, native to the Nile basin, was not initially known to occur in the Ogooué, where it is called "sans-nom" ("nameless" in French). It was introduced from Cameroon in 1959 for fish farming (Figure 6.20) in the region of Lambaréné and in the fresh waters of the Fernan Vaz Lagoon, and then it colonized the entire Lower Ogooué basin.

It inhabits lakes and forested areas. Reproduction typically occurs during the flooding of grasslands after it first rains. Parental care is provided to the fry.

Many studies have been conducted regarding the feeding habits of this species. Briefly, it is omnivorous with a zoophagous tendency: mollusks, insect larvae, possibly fry, but plankton, seeds, and macrophytes can also be found in its stomach. Its growth is well documented (it reaches 100 to 110 cm in length).

Heterotis niloticus has a modified swim bladder (gas bladder) that allows it to breathe air, when necessary. This is one of the characteristics that allows it to colonize flooded areas and swamps with low oxygenation.

The nests of *Heterotis niloticus* are built in herbaceous vegetation. They are comparable to small hollows measuring approximately 1.2 m in diameter, their slightly excavated center located approximately 30 cm deep (Paugy *et al.*, 2011). The bottom is bare and generally well flattened. The compact edges are 20 cm thick at the top and protrude slightly from the water. They are built with the stems of plants that have been removed from the center of the nest. Parents stay close to the nest when the eggs are laid.

The eggs are quite small (2.5 mm in diameter) and orange in color. They hatch after about two days. The fry have long gill filaments that are dark red in color, which extend outside the operculum. They quickly form a swarm about 30 cm in diameter occupying the center of the nest. On the fifth or sixth day, the fry leave the nest, still forming a dense swarm and under the protection of the parents (Svensson, 1933).

Killifish

Killifish are an order of small fish, the Cyprinodontiformes (Figures 6.1–6.8 and 6.33–6.38). For the most part, they do not exceed 6 cm in length. Their small size places them low down in the food web, and they are generally numerous where present.

The majority of species in the group, in order to flee predation and competition from larger fish, have colonized the upper parts of the watersheds, up to the headwaters. Some *Aphyosemion* even temporarily occupy water holes, and it is not uncommon in forests to find some of them in an elephant's footprints. These fish are also able to tolerate water with the highest levels of acidity in Gabon, down to a pH of 3.5. Killifish remain immobile for long periods of time and have reduced energy needs, but they prefer temperatures that are stable and relatively high: the Lower Ogooué is therefore a region to which they are particularly well suited and in which they are very diverse, with more than 20 recently identified species (many of which have not yet been described).

On the other hand, they are often the only fish to occupy stagnant water holes: the low level of predation in their environment has enabled them to evolve bright colors, and it is the most colorful group of freshwater fish known.

Two families of Cyprinodontiformes are present in the Lower Ogooué region:

(1) Nothobranchiidae, with the genus *Aphyosemion*, which occupy stagnant waters, and the genus *Epiplatys*, which include many surface species specialized in the capture of small insects that land on water.

(2) Procatopodidae, which are more pelagic and include the genus *Plataplochilus*, still poorly known, which move in

The reproduction of some killifish

Most species of the genus *Aphyosemion* generally inhabit temporary pools. They have a very short life cycle that does not exceed a few weeks. The survival of the species is ensured by laying drought-resistant eggs that have a thick membrane, are buried in the ground, and can survive for several years! Once the substrate is again submerged, their development resumes.

This species emits a chemical that can inhibit the reproduction of other killifish species. Thus, when a species arrives in a new environment and develops there, another one is prevented from developing (Paugy *et al.*, 2011).

6.21. Brienomyrus brachyistius *(Photo: John Sullivan).*

6.22. Mormyrops nigricans *(Photo: John Sullivan).*

6.23. Cryptomyrus ona *at Mabounié (Photo: John Sullivan).*

6.24. Mormyrops zanclirostris *(Photo: John Sullivan).*

6.25. Isichthys henryi *(Photo: John Sullivan).*

6.26. Paramormyrops kingsleyae *(Photo: John Sullivan).*

6.27. Marcusenius moori *(Photo: John Sullivan).*

6.28. Paramormyrops ntotom *(Photo: John Sullivan).*

6.29. Petrocephalus simus *(Photo: John Sullivan).*

6.30. Petrocephalus balayi *(Photo: John Sullivan).*

6.31. Petrocephalus microphthalmus *(Photo: John Sullivan).*

6.32. Stomatorhinus walkeri *(Photo: John Sullivan).*

small schools in forest streams. Several dozen species (mostly undescribed) have already been identified in Gabon, including a dozen in the Lower Ogooué region (Chirio, pers. comm.).

Aplocheilichtys spilauchen (Figure 6.33) is the only Gabonese killifish that lives in brackish water; it forms large schools in mangroves.

Their coloration and robustness make them highly valued aquarium fish. The presence of killifish in aquaria is not new, but they are not considered "fashionable" fish and only a few species are regularly imported and bred. However, around the world, associations have hundreds of members, and these groups have discovered many species from Gabon.

These species, also called "mosquito fish," have been used (and may also be used in the future) to fight against diseases transmitted by mosquitoes, given that mosquito larvae are their preferred prey. In fact, their ability to live in residual water holes makes them candidates for introduction into areas of standing water where mosquito larvae proliferate.

Mormyridae or African electric fish

The family Mormyridae is endemic to Africa. Also referred to as "elephantfish," this is one of the most species-rich groups among African freshwater fish, with 228 species in 21 genera (Figures 6.21–6.32).

Known from much of Africa, Mormyridae reach their greatest diversity in the river systems of West and Central Africa.

The fauna of Mormyridae in the Ogooué basin is quite distinctive. In the Ogooué, the genus *Paramormyrops* (Figures 6.26 and 6.28) is particularly diverse (nine described and many undescribed species), while several large genera found in the neighboring Congo basin are absent, such as *Mormyrus, Campylomormyrus, Cyphomyrus,* and *Gnathonemus*.

Mormyridae are valued as edible fish in Gabon. Smaller species are usually prepared in "packets," wrapped in foil or leaves and cooked over a fire. Large species are often smoked and sold in markets.

The Mormyridae of Gabon have, since the 1970s, been the subject of evolutionary, neurobehavioral, and systematic studies at Cornell University (Ithaca, New York, USA) by Professor Emeritus Carl D. Hopkins and his students and associates. They are currently preparing a revision of the genus *Paramormyrops*.

A new genus with two new species was described in 2016 (Sullivan *et al.*, 2016), one of which is found in the Lower Ogooué region (Mabounié River, in the provinces of Moyen-Ogooué and Ngounié).

Thirteen species from nine genera are currently known in the Lower Ogooué region. With the exception *Isichthys henryi* (Figure 6.25) cited by Mbega & Teugels (2003) and *Cryptomyrus ona* (Figure 6.23), recently discovered and described, these figures confirm that the Lower Ogooué region lacks sufficient sampling with regard to this family.

Recordings of the patterns of electric pulse waveforms emitted by these fish are archived in the Macaulay Library at Cornell University, and the associated specimens are deposited in the Museum of Vertebrates at Cornell University.

For detailed descriptions and keys, see Hopkins *et al.* (2007) and the subsequent species descriptions in Sullivan *et al.* (2016) and Rich *et al.* (2017).

Relationship between abiotic parameters and fish fauna

Abiotic parameters (water quality, tide, lunar phase, substrate, current speed, rainfall, habitat, etc.) play a key role in the distribution of fish species or even of individuals of a given species. According to these parameters, whether in combination or individually, juveniles, males, and females will not necessarily be

6.33. Aplocheilichthys spilauchen *(Photo: Christophe Aubin).*

6.34. Plataplochilus *sp. from a small stream near Lake Azingo (Photo: Christophe Aubin).*

6.35. Plataplochilus *sp. from Mabounié drainage (Photo: Olivier Buisson).*

6.36. Plataplochilus *cf.* miltotaenia *at Tchad 3 (Photo: Olivier Buisson).*

6.37. Plataplochilus miltotaenia *from Biweni River upstream (south of Lambaréné) (Photo: Olivier Buisson).*

6.38. Plataplochilus chalcopyrus *from Biweni River downstream (south of Lambaréné) (Photo: Olivier Buisson).*

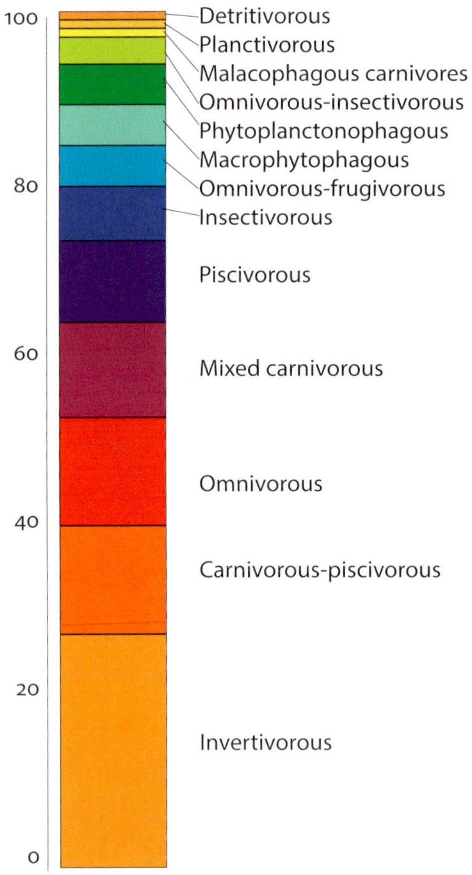

6.39. Relative distribution of different ecological groups.

seen in the same areas. Each species has its own preferred habitat preferences.

Changes in abiotic variables will cause fish to migrate, sometimes seasonally, and they are also linked to migration for reproduction and/or feeding.

Studying the relationship between the fauna and environment for each species, which is mainly a matter of population biology for which it is necessary to study the different ontogenic niches, requires making extremely detailed observations. This has, unfortunately, been done for few species in Gabon.

Ecological groups

Continental aquatic ecosystems in the Ogooué basin function based on a certain number of ecological groups, each with a very specific role in the food chain. These are the main resources:

- organic litter;
- phytoplankton: mainly green and blue algae (Loubens, 1964);
- macrophytes or higher plants: these are mainly found around lakes. Notable species include: *Heteranthoecia guineensis, Cyperus papyrus, Vossia cuspidata,* and *Echinochloa stagnina*. Some fish use their leaves and seeds;
- plant matter: composed of various fruits and seeds that generally come from riparian plants or from plants found in flooded areas. *Alestes macrophthalmus, Xenocharax spilurus*, and *Schilbe grenfelli* are fond of these;
- terrestrial insects in groups such as Hemiptera, Coleoptera, Orthoptera, and Lepidoptera;
- aquatic invertebrates, including chironomids, beetles, Diptera, ostracodes, mayflies, Odonata, and annelids;
- mollusks (it is customary to list them separately);
- crustaceans, including large crabs and shrimps such as *Macrobrachium* spp.;
- fish: those that serve as prey generally belong to small species or are young individuals of larger species.

Fish species can therefore be grouped according to their food preferences:

- pure predators: *Hepsetus* spp. (Figure 6.43), *Parachanna obscura*, *Protopterus dolloi*; or smaller fish such as *Hemichromis elongatus* (Figure 6.44);
- fish-eating carnivores with diets of 75% to 80% fish and crustaceans for the rest: *Ilisha africana, Citharichthys stampfii, Lutjanus dentatus, Polydactylus quadrifilis* (Figure 6.11), *Pseudotolithus elongatus* (Figure 6.46), *Sphyraena afra,* and *Elops lacerta* (Figure 6.45);
- mixed carnivores, which feed on several types of prey (fish, crustaceans, and mollusks in equal proportions): *Caranx hippos* (Figure 6.47), *Pomadasys jubelini* (Figure 6.48), *Trachynotus teraia* (Figure 6.49), and Eleotridae;
- malacophagi (mollusks): only *Cynoglossus senegalensis*;

6.40. Enteromius holotaenia *(Photo: Jean P. Vande weghe).*

6.41. Distichodus notospilus *(Photo: Jean P. Vande weghe).*

6.42. Monodactylus sebae *(Photo: Jean P. Vande weghe).*

6.43. Hepsetus lineata *(Photo: Jean P. Vande weghe).*

6.44. Hemichromis elongatus *(Photo: Jean P. Vande weghe).*

6.45. Elops lacerta *(Photo: Jean P. Vande weghe).*

6.46. Pseudotolithus elongatus *(Photo: Jean P. Vande weghe).*

6.47. Caranx hippos *(Photo: Jean P. Vande weghe).*

6.48. Pomadasys jubelini *(Photo: Jean P. Vande weghe).*

6.49. Trachinotus teraia *(Photo: Jean P. Vande weghe).*

6.50. Plectorhinchus macrolepis *(Photo: Jean P. Vande weghe).*

6.51. Ethmalosa fimbriata *(Photo: Jean P. Vande weghe).*

- insectivores, which prefer to consume aquatic insects and possibly terrestrial insects: Mormyridae of all sizes, *Synodontis* spp.; and some that feed mainly on terrestrial insects: *Bryconalestes longipinnis, Pareutropius debauwi, Xenomystus nigri,* and *Periophthalmus barbarus*;
- omnivorous-insectivorous species with a highly varied diet dominated by insects: *Brycinus taeniurus*;
- omnivores without any specific predominant food sources: *Clarias* spp., *Heterotis niloticus, Schilbe multitaeniatus, Brycinus macrolepidotus, Enteromius holotaenia* (Figure 6.40), and *Monodactylus sebae* (Figure 6.42);
- omnivores with a frugivorous tendency (who have a highly varied diet with a predominance of fruits): *Alestes macrophthalmus, Schilbe grenfelli, Xenocharax spilurus, Parailia occidentalis,* and *Brycinus* spp.;
- species that prefer zooplankton: *Ethmalosa fimbriata* (Figure 6.51), *Pellonula* spp., and *Sardinella* spp.;
- macrophytophagous species: *Tilapia, Pelmatolapia cabrae* (Figure 6.15), *Coptodon tholloni* (Figure 6.14), and *C. guineensis* (Figure 6.12);
- phytoplankton-eaters (species that preferentially consume phytoplankton): *Oreochromis* spp. (Figures 6.17 and 6.19) and *Gobioides ansorgii*.

Reproduction

In tropical and equatorial regions, where temperature and photoperiod vary less, certain species appear to have continuous reproduction, whereas for species with an annual cyclicity, reproduction periods are often linked to rain or flooding (Legendre & Jalabert, 1988). In the Lower Ogooué region, species such as *Alestes macrophthalmus, Xenocharax spilurus, Hepsetus* spp., *Schilbe grenfelli,* and *S. multitaeniatus* have reproductive periods that generally coincide with rainfall or rising waters. For other groups such as Cichlidae (*Pelmatolapia cabrae, Coptodon guineensis, Oreochromis schwebischi*) and *Pseudotholithus elongatus*, breeding appears to be rather continuous throughout the year. However, while these species have the capacity to reproduce several times a year, some of them nevertheless exhibit reproductive seasonality, with periods when reproduction is more intense. In Cichlidae, this occurs when the water level is at its lowest (Mbega, 2004).

Population structure

The study of the structure of fish populations reveals a longitudinal and transverse structure in relation to environmental variables (Mbega, 2004; Fermon, 2015). Unfortunately, too few studies have focused on this phenomenon, which is particularly important for the management of fishing.

Lentic zone species

Lentic environments refer to freshwater areas with slow-flowing or stagnant water (ponds, lakes, ditches, etc.).

In lentic environments, six families are clearly more abundant than others in catches. These include (Mbega, 2004) Alestidae (*Nannopetersius ansorgii* and *Brycinus taeniurus*), followed by Clupeidae, Distichodontidae (*Xenocharax spilurus* and *Distichodus notospilus*), Mormyridae (*Mormyrops zanclirostris*), Cichlidae (*Oreochromis schwebischi, Coptodon tholloni, Coptodon nyongana,* and *Chromidotilapia kingsleyae*), Mochockidae (*Synodontis obsesus, Synodontis polyodon,* and *Synodontis haugi*), Hepsetidae (African pike), as well as small Cyprinidae (*Enteromius holotaenia*).

6.52. *When the water level of the Rivière des Pélicans rises, schools of young African pike, Hepsetus sp., invade flooded forests (Photo: Jean P. Vande weghe).*

Species from swamps and marshes

Although marshes and swamps are lentic environments, they are often treated separately. These areas often contain fry and juvenile individuals of species from open environments, as well as species such as *Nannaethiops unitaeniatus*, although the most remarkable inhabitants are the Protoptera, the eel *Protopterus dolloi*, and—since its introduction—the "sans-nom," *Heterotis niloticus*. Catfish of the genus *Clarias* are also often seen in marshes and swamps.

Lotic zone species

A lotic environment is characterized by rapid water flow. The most common species are Alestidae (*Alestes macrophthamus, Brycinus kingsleyae, B. macrolepidotus, Bryconalestes longipinnis*, and *Bryconaethiops microstoma*), Mormyridae (*Marcusenius moorii, Brienomyrus brachyistius, Petrocephalus microphthalmus, P. balayi, P. simus, Mormyrops nigricans*, and *Ivindomyrus marchei*), Schilbeidae (*Schilbe multitaeniatus, S. grenfelli, Parailia occidentalis,* and *Pareutropius debauwi*), Cyprinidae (*Raiamas buccholzi*), and Claroteidae (*Chrysichthys ogooensis, C. thysi*, and *C. nigrodigitatus*).

However, there are a number of small species typically found in small forest undergrowth streams that are not found in larger rivers: Distichodontidae (*Nannocharax* spp. and *Neolebias* spp.), *Mastacembelus* spp., *Plataplochilus* spp., the ubiquitous predator *Hemichromis elongatus*, catfish, *Microsynodontis* spp., *Notoglanidium* spp., and *Amphilius* spp., as well as Mormyridae including *Paramormyrops* spp. and *Chiloglanis* spp. Small Cyprinidae are typically found in slower waters.

Migrations

Seasonal changes in the composition of fish populations in relation to the hydrological regime of the river can be seen by comparing the seasonal evolution of the structure of populations in fresh and brackish water environments in the Lower Ogooué basin (Mbega, 2004; Fermon, 2015). With regard to the families represented in fresh and brackish water environments, the hydrological regime influences the presence of certain families. Cichlidae, Distichodontidae, and Hepsetidae are numerically abundant only in freshwater catches during the long dry season, when the river is at its lowest. Conversely, Mormyridae are numerically abundant only in freshwater catches during the long rainy season, when water levels are high. In brackish water, the presence of Monodactylidae seems to coincide with the long dry season.

Variations in the results from analyses based on abiotic parameters mainly show a contrast between more marine or brackish water species (*Polydactylus quadrifilis, Aplocheilichthys spilauchen, Chelon dumerili*, and *Awaous lateristriga*) in relation to the physical and climatological parameters of other species (Fermon, 2015).

Surprisingly, the phases of the moon also have an influence. This highlights differences between species: for example, the moon has little influence on genera such as *Petrocephalus* and other members of Mormyridae. The moon also has little influence on Schilbeidae.

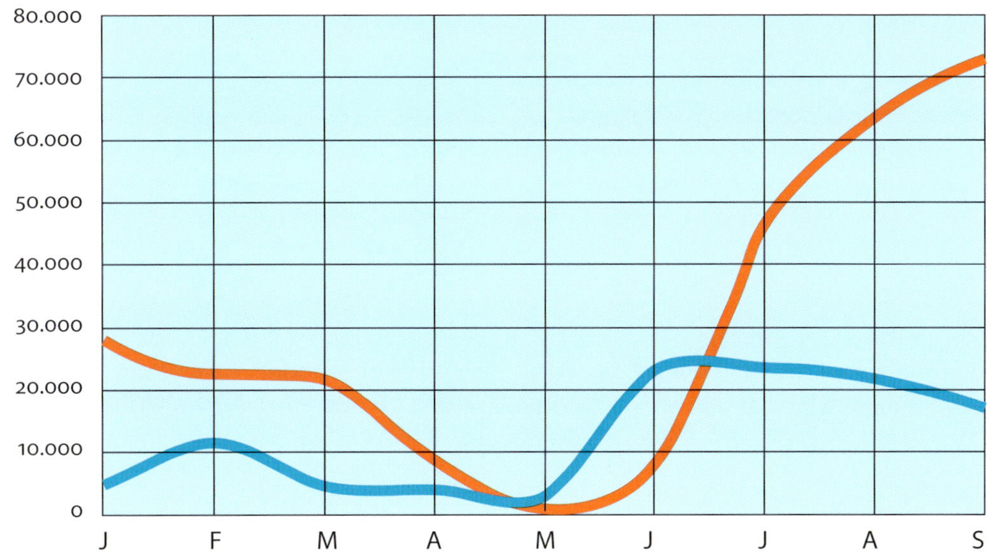

6.53. The quantity of fish caught per month in Ndjolé (in blue) and Lambaréné (in red), in kilograms, according to statistics from IPPAMO (DGPA).

6.54. The typical environment of several species of Aphyosemion *(Photo: Laurent Chirio).*

Water color influences other parameters. Species influenced by water color include those found in marine or brackish waters as well as Distichodontidae. The presence of marine species that do not travel far, including of juvenile individuals, near Lambaréné and Ndjolé remains difficult to explain without more precise knowledge of the hydrology of the area.

Longitudinal migrations

There is a longitudinal contrast between species restricted to brackish waters (*Neochelon falcipinnis, Ilisha africana, Plectorynchus macrolepis*, and *Monodactylus sebae*) and freshwater species (*Oreochromis schwebischi, Xenocharax spilurus, Schilbe grenfelli, Parachanna obscura*, and *Alestes macrophthalmus*). Diversity indicators suggest that this contrast is between the undiversified and relatively unstable brackish water environment and the highly diverse and more balanced freshwater environment.

Lateral migrations

Regarding lateral migrations, the available data seem to indicate that they mainly occur in lentic environments (Mbega, 2004). The lateral migrations that have been defined are probably related to the greater availability of food in places where the river expands laterally.

Species restricted to lentic environments in areas of forest and in lakes (*Oreochromis schwebischi, Pelmatolapia cabrae, Xenocharax spilurus*, and *Parachanna obscura*), which typically have slow currents, are transversally opposed to lotic environments in these same areas (*Schilbe grenfelli, Schilbe multitaeniatus, Ivindomyrus marchei*, and *Petrocephalus simus*) with stronger currents. Unstable lentic environments with high diversity therefore stand in contrast to balanced lotic environments, which also have relatively high diversity.

Fisheries

Inland river-lagoon fishing is practiced mainly by Gabonese fishermen (numbering 4,400 in 2008) in rivers, the Lower Ogooué basin, coastal lagoons, estuaries, dams, and reservoirs (Cofrepêche, 2011). This activity is more intensive during the long dry season. Catches total close to 10,000 tons per year, according to the DGPA. The total number of wooden canoes involved is said to be more than 2,200, 23% of which are motorized, or up to 40% of the total in the province of Ogooué-Maritime (Cofrepêche, 2011).

Airbreathing catfish (*Clarias* spp.), claroteid catfishes (*Chrysichthys* spp.), cichlids (*Coptodon* spp., *Pelmatolapia cabrae,* and *Oreochromis* spp.), "yara" (*Schilbe* spp.), pikes (*Hepsetus* spp.), and "sans-nom" constitute the main species being harvested.

Those involved in fishing are divided into three groups: permanent, seasonal, and occasional fishermen and women, the latter two of which also practice agricultural activities. Fishing is practiced by both men (82%) and women (18%). In general, men use different techniques from those used by women.

Men mostly fish as an individual activity. They primarily use nets, harpoons, cast-nets, trammels, trolling lines, or hooks. Most angling is done using trolling lines, which are either held by fishermen themselves (the hook) or affixed to a support on the raft or canoe.

Women, on the other hand, fish in groups in small forest rivers. They use landing nets, traps, dams, and poison. Most of them practice subsistence fishing, but some fish for commercial purposes with their husbands. Besides fishing activities, women are strongly involved in the processing and marketing of the products.

Most fishing communities engage in seasonal migrations, often at a small scale. In the Lower Ogooué, fishermen and women set up temporary camps on the sandbanks during the dry season to fish along the river.

Fishing is highly seasonal, with maximum activity occurring during the long dry season (May to September). This is the period when the waters are lowest, both in lakes and rivers, and some rivers are even dried up (which is also the case for ponds and marshes). When the water level goes down, fish exit flooded forests and marshy fringes where they were feeding and reproducing during periods of higher water level and return to open waters, where they are more easily caught. In fact, during this period, both depth and current decrease significantly, revealing sandbanks and facilitating the movement of fishermen and women, allowing them to set up fishing camps. A second fishing period, although less important than the first, takes place between mid-December and mid-March. This corresponds to the short dry season when the level of rivers and lakes drops significantly.

IPPAMO agents (DGPA) in Ndjolé and Lambaréné have provided the results of surveys they conducted to collect data from fishermen and women in the region of Lambaréné and Ndjolé (Figure 6.53). These results differ between the two station sites, and it is likely that not all species were recorded. It should also be noted that fish caught near Lambaréné mostly came from the lakes region. The species being studied are, therefore, not the same in these two regions, and the numbers of fishermen are also very different. The catches are much more significant in low waters where fish populations are more concentrated, and therefore easier to catch. It is also possible that fishing activities depend on the season and are more intense when the water level is low. Marine fish are also mostly caught in March and April, and also in July and August. However, the curves representing gillnet sampling differ slightly: the increase in catch begins around July or August.

Fish typically found in swamps and grasslands (Arapaimidae, Protopteridae,

Clariidae, and Channidae) are caught in large numbers in January and sometimes in February, where water levels are low, which may indicate those marshes that are favored fishing grounds for fishermen.

Threats

A degradation in the quality of inland waters can be attributed to several factors. They can cause a deterioration of both the physical and chemical qualities of water. Most of the causes responsible for a deterioration in the quality of fish are anthropogenic. In Gabon, factors responsible for the potential degradation of inland waters can be linked to the following human activities: logging, oil exploitation, mining and aggregate extraction, agriculture, and urban expansion (waste and excrement discharge). The Gabonese economy relies on some of these activities, which contribute to a sharp deterioration in the quality of water and watersheds. Overfishing must be added to this list, as well as the ecological damage caused by the introduction of invasive species.

Towns and cities

Uncontrolled urbanization results in increased pollution of urban environments. According to a report on the state of pollution in Gabon (ANCR-GEMN, 2008), the consequences of urbanization include an increase in uncontrolled land occupation leading to degradation of the urban environment:

- inadequate housing and the proliferation of spontaneous construction;
- pollution of rivers, streams, and the sea by industrial discharge has a significant impact;
- pollution by the discharge of atmospheric particles by the logging industry;
- the practice of small trades (whether mechanical or nonmechanical) and overall commerce also factor into environmental degradation by generating both solid and liquid waste;
- the establishment of unauthorized areas of aggregate extraction, including of sand, gravel, etc., which mostly takes place in natural areas and with often little control;
- household products such as the phosphate found in detergents also pose a similar problem.

Eutrophication can result in the following possible consequences for watercourses:

- an increase in nitrate concentration;
- an increase in phosphate concentration;
- degradation of organic matter in deeper layers, resulting in a drop in the water's oxygen concentration;
- an increase in primary production, i.e., a strong increase in phytoplankton, which causes an increase in turbidity;
- the spread of toxic products such as H_2S and NH_3, which cause putrid odors.

Sampling carried out in 2014 resulted in the capture of significantly fewer fish specimens in urban areas than in non-urban zones (Fermon, 2015).

Forest exploitation

The country's vast forest cover helps protect the soil and limits the risk of erosion. Indeed, erosion increases when plant density decreases. Residues and vegetation protect the soil from the impact of raindrops and from splashing, and they tend to slow down the speed of runoff and allow for better infiltration. Logging, if not properly controlled, can accelerate erosion. Furthermore, treating wood with pesticides, especially fungicides, presents an additional risk of pollution, especially when treatment is done near rivers or even directly in them.

Oil exploitation, mining, and industrial activities

Pollution caused by oil exploration comes from extraction, transportation, and the use of oil products (lubricants and fuels). Hydrocarbons are very harmful to fauna and flora, and cause significant and lasting physical and chemical damage. Through erosion, petroleum residues can end up in streams and lakes, causing them to be polluted, depending on the quantity of residues involved. The greatest risks probably come from accidents (ruptured pipelines, drilling accidents, etc.), which can cause massive petroleum spills in natural environments.

The impact of mining on the environment can be observed at three levels: the destruction of vegetation, the modification of physical environments, and the pollution of water and soils through the use of chemicals such as cyanide, mercury, etc.

Agriculture

Agriculture typically involves agro-industrial complexes organized by subsector (rubber, palm oil, cocoa, coffee, etc.).

The impact of agriculture on the environment can be seen by the removal of the plant cover, which plays an important role in protecting the soil against erosion. The disappearance of the vegetation cover leads to an acceleration of water erosion, which in turn fills in watercourses. Furthermore, phytosanitary products and fertilizers are used for agriculture, which, in high doses, can end up in watercourses, causing pollution.

Fishing

Pressure from fishing is very high, especially in the lakes of Lower Ogooué, where monoxyle canoes–with oars or motors–and plastic canoes are often equipped with two outboard motors.

The rare fishermen who have motorized plastic canoes combine fishing activities with marketing their own products and fish they have caught, selling them in villages and camps along the lakes. The boats used by these collectors are equipped with old domestic freezers for onboard cold storage of caught/collected products.

Native fishermen criticize nonnative fishermen who work in the same areas, "which are supposed to belong to them," for not respecting official regulations—especially the mesh size of the nets—even though there are in fact no regulations with regard to inland fishing. This gear typically used probably captures juvenile fish, which considerably impacts the available resources. In addition, theft of catches and anchor nets often occurs due to the lack of an appropriate system for control and surveillance, which in turn results in income loss and a significant reduction in the amount of fish sold at local markets.

Conflicts also arise between fishermen using passive fishing gear and those favoring active gear. Fishermen and women who beat the surface of the water scare away fish, which then escape the nets set by other fishermen and women operating on the same fishing grounds.

It is difficult to confirm whether overexploitation of existing stocks is already a problem, since data regarding fishery are unfortunately sparse and not standardized.

Introduced species

The previous section described several species introduced to the Lower Ogooué region (the Nile tilapia; *Oreochromis niloticus*, the "sans-nom"; and the catfish, *Clarias gariepinus*). These three species are well established. It is difficult to know what the long-term consequences will be since little to no monitoring and follow-up of regular fish communities in the Lower Ogooué region has been done. However, various possible consequences should be a cause for concern:

- The disappearance, through competition, of species with similar ecological niches;
- Possible hybridization with local species (*Oreochromis*, for example);
- Predation on juveniles of other species, leading to a decline in population for several species;
- Epidemics linked to the introduction of fish diseases previously absent from the area.

Taken together, these could result in deep and lasting ecological changes and an imbalance that could take years to stabilize.

Synthesis

Fishery management

The Lower Ogooué system exhibits a high level of complexity, which is linked to the hydrological regime. The movement of species is varied and involves factors that are barely understood. The following should, however, be noted:

- The presence of exotic species—increasingly numerous in catches—is a cause for concern, indicating increased ecological imbalance;
- The presence of small marine species and juvenile individuals of marine fish;
- The presence of marine species mostly found in coastal waters or mangroves;
- Variation resulting from the rise and fall of water levels.

The fish fauna of the Lower Ogooué region is one of the best known in Gabon. However, studies on small tributaries remain inadequate. About a quarter of the species known to Gabon, both in marine and continental waters, are found in the region. Few studies, except those of Mbega (2004) and Fermon (2015), have examined the temporal variability of fish communities and the biology of the species present.

The system for collecting data on fishery is still insufficient, with too few agents to monitor a very large area and who often lack the means necessary to conduct quality work. Fishing regulations are, therefore, poorly monitored and there are clear signs of overfishing.

The Ramsar site

The current boundaries of the Lower Ogooué Ramsar site include both geographic and administrative boundaries.

- Since certain marine fish have a clear influence on the Lower Ogooué, it seems appropriate to extend the site's borders all the way to the coast, even including the northern area of the Fernan Vaz Lagoon, since it is connected to the Ogooué basin.
- The Lower Ogooué Ramsar site should include the southeastern part of the Wonga-Wongué Reserve, especially the Rivière des Coupeurs; this would not affect any of the reserve's borders.
- The remaining borders should follow the limits of the Ogooué basin, stopping at tributaries along the first biogeographic limits, such as Imperatrice Falls on the Ngounié.
- Upstream along the river, near Ndjolé, an administrative limit would be appropriate. It is, indeed, difficult to find a biogeographical limit in this area since the first rapids encountered upstream can be crossed by numerous species.

Bibliography

6.55. *The fishing camps are a source of plastic pollution (Photo: Jean P. Vande weghe).*

ANCR-GEMN, 2008. Étude sur l'état de la pollution au Gabon. GEF-PNUD.

AUGUSTE, H.M., 2004. *Poissons du Gabon : les beautés miniatures.* Éditions Elf Gabon. Total Fina Elf: 64 pp.

COFREPÊCHE (multi-auteurs), 2011. *Rapport de tâche i. Les eaux gabonaises et leurs ressources halieutiques : le point des connaissances actuelles. Évaluation des Stocks, Gestion des Ressources Halieutiques et Mise en place d'un Observatoire des Pêches* ESOP. GAB1R03 et GAB1R04. 161 p.

COFREPÊCHE (multi-auteurs), 2011. *Rapport de tâche ii. Résultats des campagnes d'évaluation des stocks en milieu continental. Évaluation des Stocks, Gestion des Ressources Halieutiques et Mise en place d'un Observatoire des Pêches* ESOP. GAB1R08. 118 p.

FERMON, Y., 2013. *Caractérisation de la faune ichtyologique du Gabon.* Rapport WWF, 213 p.

FERMON, Y., 2015. *Caractérisation ichtyologique dans le massif forestier d'Evaro, Bas Ogooué, Gabon.* WWF Gabon-Aimara. 132 p.

FROESE, R. & D. PAULY (eds.), 2017. Fishbase. World Wide Web electronic publication. www.fishbase.org, version (02/2017).

GALLANT, J.R., ARNEGARD, M.E., SULLIVAN, J.P., CARLSON, B.A., HOPKINS, C.D., 2011. Signal variation and its morphological correlates in Paramormyrops kingsleyae provide insight into the evolution of electrogenic signal diversity in mormyrid electric fish. *Journal of Comparative Physiology A.*, 197(8): 799-817. doi:0.1007/s00359-011-0643-8

HOPKINS, C.D., LAVOUÉ, S., & SULLIVAN, J.P., 2007. Mormyridae. Ch. 12 in *The Fresh and Brackish Water Fishes of Lower Guinea, West-Central Africa.* IRD Editions, Paris. 115 pp. electronic access

LÉVÊQUE, C. & PAUGY, D., 2006. *Les poissons des eaux continentales africaines: diversité, écologie, utilisation par l'homme.* IRD, 2006, 564 p.

LIWOUWOU, J.-F., AHOUANSOU, M S., CHIKOU, A., ADANDEDJAN, D., MBEGA, J.-D. & LALÈYÈ, A.P., 2013. Le régime alimentaire de *S. multitaeniatus* (Pellegrin, 1913) (Siluriforme, Schilbeidae) de la Rivière Rembo Bongo au Gabon (Afrique Centrale). *Int. J. Biol. Chem. Sci.* 7(5): 1853-1865.

LIWOUWOU, J.-F., 2016. Biologie, Ecologie et exploitation de deux espèces de Schibe (poissons Siluriforme, Schilbeidae) des cours d'eau d'eau Rembo Bongo, Ogooué et Nyanga au Gabon. Thèse de doctorat de l'Université d'Abomey -Calavi, Bénin. 219 p.

LOUBENS, G., 1964. *Rapport technique sur les travaux effectués en 1963.* Lambaréné: Eaux et Forêts, 1964, 38 p.

LOUBENS, G., 1966a. *Travaux en vue du développement de la pêche dans le bassin inférieur de l'Ogôoué.* Publication No. 27. Centre Technique Forestier Tropical, Nogent sur Marne, France, 151 p.

LOUBENS, G., 1966b. *Biologie de* Polydactylus quadrifilis *dans le bas Ogooué (Gabon).* Paris: Université de Paris, 1966, 142 p.

LOUBENS, G., 1966c. *Peuplement en poissons des eaux saumâtres et son origine.* Paris: Faculté des Sciences, 1966, 29 p.

MBEGA, J.-D., 2004. *Biodiversité des poissons du bassin inférieur de l'Ogooué (Gabon).* Volume 1 et Volume 2. Thèse de Doctorat en Sciences Faculté des Sciences. Université de Namur. 614 p.

MBEGA, J.D. & TEUGELS, G.G., 2003. *Guide de détermination des poissons du bassin inférieur de l'Ogooué.* IRAFG: 165 pp.

PAUGY, D., LÉVÊQUE, C. & MOUAS, I., 2011. *Poissons d'Afrique et peuples de l'eau.* IRD, 2011, 320 p. (Focus). ISBN 978-2-7099-1711-7.

RICH, M., SULLIVAN, J.P., HOPKINS, C.D., 2017. Rediscovery and description of *Paramormyrops sphekodes* (Sauvage, 1879) and a new cryptic *Paramormyrops* (Mormyridae, Osteoglossiformes) from the Ogooué River of Gabon using morphometrics, DNA sequencing, and electrophysiology. *Zoological Journal of the Linnean Society*, 177.

SEEGERS, D.L., 1997. *Killifishes of the World. Old World Killis.* Verlag A.C.S. GmbH (1): 160 pp.

SEEGERS, D.L., 1997. *Killifishes of the World. Old World Killis.* Verlag A.C.S. GmbH (2): 112 pp.

STIASSNY, L.J.M., TEUGELS, G.G. & HOPKINS, D.C., 2007. *Poissons d'eaux douces et saumâtres de la basse Guinée, ouest de l'Afrique Centrale*, Vol. 1 & 2. Institut de recherche pour le développement (IRD) (Paris, France), Musée national d'histoire naturelle (MNHN) (Paris, France) & Musée Royal de l'Afrique Centrale (MRAC) (Tervuren, Belgium), 804 + 622 p.

SULLIVAN, J.P., LAVOUÉ, S., ARNEGARD, M.E. & HOPKINS, C.D., 2004. AFLPs resolve phylogeny and reveal mitochondrial introgression within a species flock of African electric fish (Mormyroidea: Teleostei). *Evolution*, 58 (4): 825-841.

SULLIVAN, J.P., LAVOUÉ, S. & HOPKINS, C.D., 2016. *Cryptomyrus*: a new genus of Mormyridae (Teleostei, Osteoglossomorpha) with two new species from Gabon, West-Central Africa. *ZooKeys*, 561: 117-150. doi: 10.3897/zookeys.561.7137

SULLIVAN, J.P., LAVOUÉ, S. & HOPKINS, C.D., 2002. Discovery and phylogenetic analysis of a riverine species flock of African electric fishes (Mormyridae: Teleostei). *Evolution*, 56: 597-616. doi:10.1111/j.0014-3820.2002.tb01370.x

WILDEKAMP, R.H., 2012. *A World of Killies: Atlas of the Oviparous Cyprinodontiform Fish of the World.* American Killifish Association (tomes 1 à 4).

Appendix: List of the fish of the Ogooué Delta

LT: maximum total lengh in cm;
IUCN status: LC = Least Concern; DD = Data Deficient; NE = Not Evaluated; CR = Critically Endangered; EN = Endangered; VU = Vulnerable
Biogeography: G = endemic to Gabon; O = endemic to the Ogooué basin; BO = endemic to the Lower Ogooué; I = introduced; BG = endemic to Lower Guinea
Ecology: M = salt water; S = brackish water; D = freshwater
Distribution: BV = lower valley of the Ogooué; PDS = upper deltaic plain; PDI = lower deltaic plain; SL = coastal sands; Lg = lagoon

The assignment to family and orders does not follow the most recent taxonomy for fishes.

	LT	Type	IUCN Status	Biogeogr.	M	S	D	Migrations	BV	PDS	PDI	SL	Lg
Elasmobranchii													
CARCHARHINIFORMES													
CARCHARHINIDAE													
Carcharhinus leucas (Müller & Henle, 1839) [1]	400		NT		■	■	■	Amphidrome		■	■		■
Rhizoprionodon acutus (Rüppell, 1837)	175		LC		■	■	■	Amphidrome		■		■	
SPHYRNIDAE													
Sphyrna lewini (Griffith & Smith, 1834)	430		EN		■	■		Oceanodrome		■		■	
PRISTIFORMES													
PRISTIDAE													
Pristis pristis (Linnaeus, 1758) [2]	750		CR		■	■	■	Oceanodrome		■	■		■
RAJIFORMES													
RHINOBATIDAE													
Rhinobatinae													
Rhinobatos rhinobatos (Linnaeus, 1758)	147		EN		■	■				■		■	
MYLIOBATIFORMES													
GYMNURIDAE													
Gymnura altavela (Linnaeus, 1758)	400		VU		■	■						■	
DASYATIDAE													
Dasyatis ukpam (Smith, 1863) [3]	65		EN			■	■		■	■	■	■	■
Actinopterygii													
OSTEOGLOSSIFORMES													
ARAPAIMIDAE													
Heterotis niloticus (Cuvier, 1829)	122		LC	I			■				■	■	■
MORMYRIDAE													
Mormyrinae													
Brienomyrus brachyistius (Gill, 1862)	15.1		LC			■	■		■	■	■		
Cryptomyrus ona Sullivan, Lavoué & Hopkins, 2016	12		NE	G			■		■				
Isichthys henryi Gill, 1863	35		LC				■				■		
Ivindomyrus marchei (Sauvage, 1879)	12		LC	G			■		■	■	■		
Marcusenius moorii (Günther, 1867)	26.1	1	LC				■		■	■	■		■
Mormyrops nigricans Boulenger, 1899	34		LC				■		■		■		
Mormyrops zanclirostris (Günther, 1867)	27		LC				■		■	■	■		
Paramormyrops curvifrons (Taverne, Thys van den Audenaerde, Heymer & Géry, 1977)	17.5		LC				■				■		

1. Species recorded from Port-Gentil; also present in the Ogooué (BMNH 1913.7.12.1).
2. Known from Portugal to Angola, but very rare since 20-30 years and probably extinct in Gabon. A fisher from the Fernan Vaz Lagoon told us in August 2014 to have caught a specimen in 2011, but this could not be verified.
3. Occurrences in Gabon: USNM 219780, CAS 42761 "Lake Ezanga, S. of Lambarene," MNHN 1979.244 "Bouée," MRAC P-95004.0001 "Réserve de la Lopé."
4. Known from the Ogooué and Nyanga rivers. One specimen collected in Cameroon, close to the border.
5. Known from the Ivindo and Nyong rivers in Cameroon.

	LT	Type	IUCN status	Biogeogr.	M	S	E	Migrations	BV	PDS	PDI	SL	Lg
Paramormyrops gabonensis Taverne, Thys van den Audenaerde & Heymer, 1977 [6]	19.9		VU				■		■		■		
Paramormyrops kingsleyae (Günther, 1896)	16.5		DD	BG			■		■				
Paramormyrops ntotom Rich, Sullivan & Hopkins, in press [7]			NE	G/O			■				■		
Paramormyrops sp. [8]							■				■		
Paramormyrops sp. "magnotipes" [8]							■				■		
Paramormyrops sphekodes (Sauvage, 1879)	13.9		LC				■				■		
Stomatorhinus walkeri (Günther, 1867)	9	1	LC				■		■		■		
Petrocephalinae													
Petrocephalus balayi Sauvage, 1883	11	1	LC				■		■	■	■		
Petrocephalus microphthalmus Pellegrin, 1908	5.2		LC				■		■		■		
Petrocephalus simus Sauvage, 1879	12		LC				■	Potamodrome	■		■	■	
Petrocephalus sullivani Lavoué, Hopkins & Kamdem Toham, 2004	11		LC	G			■		■		■		
NOTOPTERIDAE													
Xenomystus nigri (Günther, 1868)	36.6		LC				■		■		■		
ELOPIFORMES													
ELOPIDAE													
Elops lacerta Valenciennes, 1847 [9]	100		LC		■	■	■		■	■	■	■	■
Elops senegalensis Regan, 1909 [10]	90		DD		■	■				■	■		
MEGALOPIDAE													
Megalops atlanticus Valenciennes, 1847	250		VU		■	■	■	Amphidrome		■		■	
ANGUILLIFORMES													
OPHICHTHIDAE													
Ophichthinae													
Dalophis cephalopeltis (Bleeker, 1863) [11]	53.5		NE		■	■					■		
CLUPEIFORMES													
CLUPEIDAE													
Dorosomatinae													
Ethmalosa fimbriata (Bowdich, 1825)	45		LC		■	■	■	Catadrome		■		■	■
Laeviscutella dekimpei Poll, Whitehead & Hopson, 1965	4.9		LC			■	■		■		■		
Odaxothrissa ansorgii Boulenger, 1910	15.9		LC			■	■		■	■	■		■
Pellonula leonensis Boulenger, 1916	10		LC		■	■	■	Anadrome	■		■		
Pellonula vorax Günther, 1868	14.6		LC		■	■	■	Anadrome	■	■	■		■
Sardinella aurita Valenciennes, 1847	31		NE		■	■		Oceanodrome		■		■	
Sardinella maderensis (Lowe, 1838)	41.4		NE		■	■		Oceanodrome		■		■	
PRISTIGASTERIDAE													
Pelloninae													
Ilisha africana (Bloch, 1795)	36.6		NE		■	■				■		■	
GONORHYNCHIFORMES													
KNERIIDAE													
Grasseichthys gabonensis Géry, 1964	2.5		VU	G/O			■				■		
CYPRINIFORMES													
CYPRINIDAE													
Barbinae													
Enteromius brazzaii Pellegrin, 1901	13.4		LC				■		■		■		
Enteromius brichardi Poll & Lambert, 1959	7.8		LC				■		■				
Enteromius camptacanthus (Bleeker, 1863)	16		LC				■		■		■		■
(*Enteromius carens* Boulenger, 1912) [12]	3		LC	?			■		■				

6. Species known from the Ivindo, Ntem, and Woleu rivers.
7. Publication in press.
8. Undescribed species.
9. Often confused with *E. senegalensis*.
10. Often confused with *E. lacerta*.
11. Known from Liberia to the Congo Estuary. Several occurrences in Gabon: BMNH 1909.7.27.33, MNHN 1909-0018/0020, "Ngomo, Ogooué."
12. Species known from Kouilou and Chiloango rivers. One occurrence in Gabon: MRAC P-80027.00538/0059 1962 "PK10, route Lambaréné-Mouila" Lambert. Possibly an identification error.

	LT	Type	IUCN status	Biogeogr.	M	S	E	Migrations	BV	PDS	PDI	SL	Lg
Enteromius diamouanganai Teugels & Mamonekene, 1992	11		LC				■		■				
Enteromius holotaenia Boulenger, 1904	12		LC				■		■	■	■	■	■
Enteromius martorelli Roman, 1971	12.2		LC				■		■				
Enteromius rubrostigma Poll & Lambert, 1964	13.8	1	LC				■		■				
Enteromius trispilomimus Boulenger, 1907	4.5		LC				■		■		■		
Labeobarbus batesii (Boulenger, 1903)	43.5		LC				■		■		■		
Labeobarbus compiniei (Sauvage, 1879) [13]	73		LC				■		■				
Danioninae													
Opsaridium ubangiense (Pellegrin, 1901)	12		LC				■		■	■	■		
Raiamas buchholzi (Peters, 1876) [14]	14		LC				■		■	■	■		■
Labeoninae													
Labeo annectens Boulenger, 1903	48.5		LC				■		■		■		
Labeo batesii Boulenger, 1911	21.4		LC				■				■		
Labeo cyclorhynchus Boulenger, 1899	16	1	LC				■				■		
Labeo lukulae Boulenger, 1902	32.3		LC				■				■		
CHARACIFORMES													
ALESTIDAE													
Alestes macrophthalmus Günther, 1867	73		LC				■		■	■	■	■	
Brachypetersius gabonensis Poll, 1967	8.5	1	LC	G			■		■				
Brycinus bartoni (Nichols & La Monte, 1953)	11.2		EN	G/O			■		■		■		
Brycinus intermedius (Peters, 1852)	9.8		LC				■				■		
Brycinus kingsleyae (Günther, 1896)	20.2		LC				■		■	■	■	■	
Brycinus macrolepidotus Valenciennes, 1850	64.7		LC			■	■	Potamodrome	■		■	■	
Brycinus opisthotaenia (Boulenger, 1903)	15.6		LC				■		■		■		
Brycinus taeniurus (Günther, 1867)	12		LC				■		■		■		
Bryconaethiops microstoma Günther, 1873	21	1	LC				■		■		■		
Bryconalestes longipinnis (Günther, 1864)	14.6		LC			■	■	Potamodrome	■	■	■	■	
Nannopetersius ansorgii (Boulenger, 1910)	7.5		LC				■		■		■		
Phenacogrammus aurantiacus (Pellegrin, 1930)	10		LC				■		■		■		
Phenacogrammus urotaenia (Boulenger, 1909)	7		LC				■		■				
DISTICHODONTIDAE													
Distichodus hypostomatus Pellegrin, 1900	23.8	1	LC				■		■		■		
Distichodus notospilus Günther, 1867	19.5		LC				■		■	■	■		■
Monostichodus elongatus Vaillant, 1886	11.2	1	LC	G/O			■		■		■		
Nannaethiops unitaeniatus Günther, 1872	6.2		LC				■		■		■		
Nannocharax fasciatus Günther, 1867 [16]	8.1		LC				■	Potamodrome	■		■		
Nannocharax intermedius Boulenger, 1903	6.8		LC				■				■		
Nannocharax parvus Pellegrin, 1906	6.5	1	LC				■		■		■		
Neolebias ansorgii Boulenger, 1912	3.2		LC				■	Potamodrome	■		■	■	
Neolebias gossei Poll & Lambert, 1964	3.6	1	LC	BO			■		■		■		
Neolebias kerguennae Daget, 1980	3.5		EN	BO			■		■		■		
Neolebias unifasciatus Steindachner, 1894	5.3		LC				■		■		■		
Xenocharax spilurus Günther, 1867	26		NE				■		■		■		
HEPSETIDAE													
Hepsetus kingsleyae Vreven, Decru & Snoeks, 2013	26.8	1	NE	G/O			■	Potamodrome	■		■		
Hepsetus lineata (Pellegrin, 1926)	34.5		NE				■	Potamodrome			■		
Hepsetus sp. [17]				–			■		■	■	■		

13. Type locality of holotype: "*Barbus compiniei*" MNHN A-2845 1876 "Ogôoué."
14. The holotype was collected in the "Ogowe." At that time it could be only the Lower Ogooué.
15. According to FishBase, two occurrences: MNHN 1886-0392/393 "Adoumas, Ogooué," collected by Savorgnan de Brazza. These specimens should be verified.
16. This species is known from Senegal to Cameroon, including the Lake Tchad basin and the Upper Cross River. The holotype was collected in "Gabon." Several specimens from Gabon mentioned by Faunafri should be verified.
17. All specimens of *Hepsetus* have not been reverified during the revision of 2013.

	LT	Type	IUCN status	Biogeogr.	M	S	E	Migrations	BV	PDS	PDI	SL	Lg
SILURIFORMES													
AMPHILIIDAE													
Amphiliinae													
Amphilius longirostris (Boulenger, 1901) [18]	4		LC				■		■				
Amphilius nigricaudatus Pellegrin, 1909	8	1	NE				■		■		■		
Amphilius pulcher Pellegrin, 1929	8		LC				■						■
Amphilius sp. "lamani"							■		■				
Doumeinae													
Phractura brevicauda Boulenger, 1911	7		LC				■		■				
ARIIDAE													
Ariinae													
Arius latiscutatus Günther, 1864	9		NE		■	■	■			■	■	■	
CLARIIDAE													
Channallabes apus (Günther, 1873) [19]	32.7		LC				■		■				
Clariallabes brevibarbis Pellegrin, 1913 [20]	29	1	DD	0			■				■		
Clarias angolensis Steindachner, 1866 [21]	35		LC				■				■		
Clarias buthupogon Sauvage, 1879	109.8		LC				■	Potamodrome	■		■		
Clarias camerunensis Lönnberg, 1895	46.6	1	LC				■		■	■	■		
Clarias gabonensis Günther, 1867	36		LC				■		■	■	■		
Clarias gariepinus (Burchell, 1822)	170		NE	I			■	Potamodrome			■	■	
Clarias jaensis Boulenger, 1909	48.3		LC				■				■		
Clarias pachynema Boulenger, 1903	35.6		LC				■		■		■		
Heterobranchus longifilis Valenciennes, 1840	183		LC				■	Potamodrome	■				
CLAROTEIDAE													
Auchenoglanidinae													
Notoglanidium boutchangai (Thys van den Audenaerde, 1965) [22]	22.5		LC	BG			■		■				
Notoglanidium macrostoma (Pellegrin, 1909)	24	1	LC				■		■		■		
Parauchenoglanis balayi (Sauvage, 1879)	39		LC				■				■		
Parauchenoglanis punctatus (Boulenger, 1902)	50		LC				■		■		■		
Claroteinae													
Chrysichthys auratus (Geoffroy Saint-Hilaire, 1809)	57		LC				■	Potamodrome	■	■	■	■	■
Chrysichthys nigrodigitatus (Lacepède, 1803)	79.3		LC				■	Potamodrome	■	■	■	■	■
Chrysichthys ogooensis (Pellegrin, 1900)	25	1	LC	G			■		■		■	■	
Chrysichthys thysi Risch, 1985 [23]	30	1	LC				■				■		
MALAPTERURIDAE													
Malapterurus beninensis Murray, 1855	27.2		LC				■		■		■		
Malapterurus oguensis Sauvage, 1879	26.2		LC				■		■		■		
MOCHOKIDAE													
Chiloglanidinae													
Chiloglanis cameronensis Boulenger, 1904	5.5		LC			■	■		■		■		
Mochokinae													
Microsynodontis batesii Boulenger, 1903 [24]	12.2		DD				■		■		■		
Microsynodontis notata Ng, 2004 [25]	6	1	DD	G/O			■				■		
Microsynodontis vigilis Ng, 2004 [26]	6	1	DD	0			■		■		■		
Synodontis batesii Boulenger, 1907	12.6			LC	■			1		■		■	
Synodontis haugi Pellegrin, 1906	30.2	1	LC	G/O			■		■		■		
Synodontis obesus Boulenger, 1898	37.5		LC				■		■		■		■

18. Several occurrences from the Okano River.
19. This species is mainly known from Angola and the Kouilou River in Congo; a few specimens were collected in Gabon in the Woleu and Lower Ogooué rivers.
20. This species, considered to be endemic to the Ogooué River, was collected in the Democratic Republic of the Congo.
21. One occurence in Gabon: MRAC P-A0048-1853/1856 "Rivière Verié, affluent du lac Avanga, Ogooué" (Lower Ogooué).
22. Known in the Ogooué and Kouilou rivers.
23. This species is considered endemic, but was also collected in Cameroon.
24. Known from Gabon in the Ntem and Ivindo basins.
25. Known from only the Lower Ogooué. Type locality: MRAC 80-51P-839 "Rivière Ezanga, environ à mi-chemin entre le lac Ezanga et le lit majeur de Ogooué, Gabon."
26. Species known from only the Lower Ogooué. Type locality: "Gabon: Province du bas Ogooué, 12 km au nord de Lambaréné, branche de la crique Nzorbang près du village du même nom."

27. Species known from Guinea, Liberia, and Sierra Leone. Possibly an identification error.
28. Undescribed species.
29. Species known from the Ngounié basin, Lebamba road.
30. Known in Lower Ogooué and northwestern Gabon.
31. In Gabon known from Lower Ogooué, Kango, Nyanga River, and other coastal river systems, such as Mayumba.
32. Known from the coastal rivers, up to Cape Esterias in the north.
33. Described from the Ngounié basin: MRAC P-77017.0472 1976 "Dans une rivière de montagne à débit rapide près de Yéno, Mogambi, Ougoudou-Ogoulou sous-système de la Ngounié" (Radda & Huber).
34. The name of this species is subject to discussion. Following Huber (1998) *Aphyosemion escherichi* could be a synonym of *Aphyosemion striatum*. The correct name of the species, named here *A. escherichi*, could then be *A. microphtalmum* Lambert & Géry, 1968. Since no material of *A. escherichi* from the type locality was collected recently, we follow the revision of Seegers (1988).
35. Known from between Mandilou and Ngoudoufala, west Gabon, in the middle Ngounié region: MNHN 1978-0140 1977, 1981-0861 1976.
36. This subspecies is described from small forest rivers in the humid forest along the Ngounié River. Type localité: NMW 90751 "Ruisseau sur la route de Bigouenia à Mora."
37. *A. gabunense* is represented by three forms in Gabon (Stiassny *et al.*, 2007): *gabunense* in the lower Ngounié, *boehmi* in the upper Ngounié (above the waterfalls of Fougamou) and *marginatum* in the area of Bifoun.
38. Subspecies known from only its type locality near Bifoun, north of the Ogooué River: NMW 90750 "mare résiduelle à 9 km sud-ouest de Bifoun."
39. Type locality: MHNG 2590.64 "Gabon, 45 km nord-est de Lambaréné (point de départ le pont sur l'Ogooué) sur la route de Bifoun, près de Benguié."
40. Found in the small rivers of the humid forest in the upper Ivindo basin, formerly a subspecies of *A. wildekampi*.

	LT	Type	IUCN status	Biogeogr.	M	S	E	Migrations	BV	PDS	PDI	SL	Lg
Synodontis polyodon Vaillant, 1895	31.4	1	LC	G/O			■		■		■		
SCHILBEIDAE													
Parailia occidentalis (Pellegrin, 1901)	8.5		LC				■		■		■		
Pareutropius debauwi (Boulenger, 1900)	12.7		LC				■		■	■	■		■
Schilbe grenfelli (Boulenger, 1900)	61		LC				■		■	■	■	■	■
Schilbe multitaeniatus (Pellegrin, 1913)	32.1	1	LC				■		■	■	■		
CYPRINODONTIFORMES													
NOTHOBRANCHIIDAE													
Epiplatyinae													
(*Epiplatys annulatus* (Pellegrin, 1913)) [27]	4		LC	?			■		■				
Epiplatys ansorgii (Boulenger, 1911)	8		LC				■		■		■		
Epiplatys cf. *ansorgii* [28]							■		■	■			
Epiplatys cf. *huberi* [28]							■		■				
Epiplatys cf. *multifasciatus* [28]							■		■				
Epiplatys cf. *singa* [28]							■		■				
Epiplatys huberi (Radda & Pürzl, 1981) [29]	6		LC	G/O			■		■				
Epiplatys multifasciatus (Boulenger, 1913)	6		DD				■		■				
Epiplatys sexfasciatus sexfasciatus Gill, 1862 [30]	12.2		LC				■	Potamodrome	■		■		
Epiplatys singa (Boulenger, 1899) [31]	6		LC				■		■	■	■	■	
Nothobranchiinae													
Aphyosemion australe (Rachow, 1921) [32]	6		LC				■			■	■	■	
Aphyosemion citrineipinnis Huber & Radda, 1977 [33]	5		LC	G			■		■				■
Aphyosemion escherichi (Ahl, 1924) [34]	5		LC				■		■		■		
Aphyosemion exigoideum Radda & Huber, 1977 [35]	6		LC	G/O			■		■				
Aphyosemion gabunense boehmi Radda & Huber, 1977 [36]	5		NE	BO			■				■		
Aphyosemion gabunense gabunense Radda, 1975 [37]	5		NE	BO			■		■				
Aphyosemion gabunense marginatum Radda & Huber, 1977 [38]	5		NE	BO			■		■				
Aphyosemion hera Huber, 1998 [39]	4	1	DD	BO			■		■				
Aphyosemion punctatum Radda & Pürzl, 1977 [40]	5		LC				■		■				
Aphyosemion striatum (Boulenger, 1911) [41]	6		LC				■		■		■		
PROCAPOTODIDAE													
Aplocheilichthyinae													
Aplocheilichthys spilauchen (Duméril, 1861)	70		LC			■	■		■	■	■	■	
Procatopodinae													
Plataplochilus cabindae (Boulenger, 1911)	5		LC				■		■				
Plataplochilus chalcopyrus Lambert, 1963	4	1	EN				■		■				
Plataplochilus miltotaenia Lambert, 1963 [42]	5	1	VU				■		■				
Plataplochilus ngaensis (Ahl, 1924)	5		LC				■		■			■	
Plataplochilus sp. *albomarginatus* [43]							■		■				
Plataplochilus sp. *cyanolineatus* [43]							■		■		■		
Plataplochilus sp. *minimus* [43]							■		■				
Plataplochilus sp. *omoueyi* [43]							■				■		
Plataplochilus sp. *supermiltotenia* [43]							■				■		
Plataplochilus sp. *zamatae* [43]							■		■				
Plataplochilus terveri (Huber, 1981) [44]	5.5		EN	G/O			■		■				
Poropanchax stigmatopygus Wildekamp & Malumbres, 2004 [45]	3		LC				■			■			

	LT	Type	IUCN status	Biogeogr.	M	S	E	Migrations	BV	PDS	PDI	SL	Lg
BELONIFORMES													
BELONIDAE													
Strongylura senegalensis (Valenciennes, 1846) [46]	183		NE		■	■	■			■		■	
SYNGNATIFORMES													
SYNGNATHIDAE													
Syngnathinae													
Enneacampus ansorgii (Boulenger, 1910)	16.8		LC			■	■		■	■	■		
Enneacampus kaupi (Bleeker, 1863)	20		LC			■	■		■		■		
Microphis aculeatus (Kaup, 1856) [47]	24.4		NE		■	■	■		■		■		
SYNBRANCHIFORMES													
MASTACEMBELIDAE													
Mastacembelus marchei Sauvage, 1879	32.9		NE				■				■		
Mastacembelus niger Sauvage, 1879	45.8		LC				■		■	■	■		
MUGILIFORMES													
MUGILIDAE													
Liza dumerili (Steindachner, 1870) [48]	40		NE		■	■	■	Catadrome			■		
Mugil curema Valenciennes, 1836	90		NE		■	■	■	Catadrome				■	
Neochelon falcipinnis (Valenciennes, 1836)	45.5		NE		■	■	■	Catadrome	■	■	■	■	
Parachelon grandisquamis (Valenciennes, 1836)	40		NE		■	■	■	Catadrome		■	■	■	
PERCIFORMES													
ANABANTIDAE													
Ctenopoma gabonense Günther, 1896	20.1		NE				■				■		
Ctenopoma kingsleyae Günther, 1896	24.5		LC				■		■		■		
Ctenopoma nigropannosum Reichenow, 1875	17		LC				■				■		
Microctenopoma nanum (Günther, 1896)	8		LC				■		■		■		
CARANGIDAE													
Caranginae													
Alectis alexandrina (Geoffroy Saint-Hilaire, 1817)	100		NE		■	■				■		■	
Caranx hippos (Linnaeus, 1766)	124		NE		■	■				■		■	
Caranx senegallus Cuvier, 1833	100		NE		■	■				■		■	
Chloroscombrus chrysurus (Linnaeus, 1766)	65		NE		■	■				■		■	
Selene dorsalis (Gill, 1863)	38		NE		■	■				■		■	
Trachinotinae													
Trachinotus ovatus (Linnaeus, 1758)	70		NE		■	■				■	■	■	
Trachinotus teraia Cuvier, 1832	68		NE		■	■	■			■	■	■	
CHANNIDAE													
[*Parachanna africana* (Steindachner, 1879)] [49]	32		LC	?			■				■		
Parachanna obscura (Günther, 1861) [50]	61		NE				■	Potamodrome	■		■		
CICHLIDAE													
Pseudocrenilabrinae													
Chromidotilapia kingsleyae Boulenger, 1898	15.9	1	LC				■		■	■	■		
Chromidotilapia melaniae Lamboj, 2003 [51]	9.7	1	LC	G			■		■				
Chromidotilapia mrac Lamboj, 2002 [52]	11.3	1	LC	G/O			■		■		■		
Chromidotilapia regani (Pellegrin, 1906)	15.5		VU	G/O			■		■		■		
Coptodon guineensis (Günther, 1862)	36.6		LC		■	■	■		■	■	■	■	■
Coptodon nyongana Thys van den Audenaerde, 1971	21		LC				■				■		
Coptodon rendalli (Boulenger, 1897) [53]	45		LC	I		■	■				■		

41. Described from Equatorial Guinea, but also collected in the Mbei (or Mbé), Komo, Abanga, and Lower Ogooué rivers in Gabon. Several subspecies introduce some confusion.
42. Species present in the lower reaches of the Ogooué River and the Komo River in northwest Gabon.
43. Species not yet described.
44. Species found in small rivers and streams in the upper course of the Ogooué River in southeastern Gabon (Mpassa River, Haut-Ogooué).
45. Species found between the Sanaga estuary (Cameroon) and Port-Gentil in northwestern Gabon.
46. Species found ranging from Senegal and Guinea (Conakry) to the Gulf of Guinea and Angola. One occurrence in Gabon: MRAC P-73002-270373002-2705 1963 "Nangue Boni, estuaire de l'Ogooué" (Thys van den Audenaerde). It is in fact very common in mangroves throughout the country.
47. Formerly a subspecies of *Microphis brachyurus.*
48. Known from Mauritania to South Africa. One occurrence in Gabon: MNHN 1997-3914.
49. Known from the Lower Cross River and the Niger Delta in Nigeria to southern Benin (Ouémé River). One occurrence in Gabon (Lambaréné, Ogooué River). But this seems erroneous. Collected and identified by G. Teugels.
50. Present in the Ogooué and Nyanga rivers in Gabon.
51. Mainly in south and southwestern Gabon.
52. Known in Gabon from the Lower Ogooué, the Nyanga and Ngounié rivers, and the coastal regions of northern Gabon.
53. Introduced in several areas (e.g., Libreville, Makoukou, Lebamba, and Franceville). One occurrence in Makokou, Ivindo (MRAC P-154570).

54. Known in Gabon from the Ogooué River.
55. Known from the Ivindo, Ntem, and Ogooué rivers (Ndjolé region).
56. Found in the Ivindo and Ntem rivers.
57. Following Froese & Pauli (2014) and Stiassny *et al.* (2007), *Kribia nana* is limited to the Nile, Tchad, and Congo basins, and from Upper Guinea to the Cross River in Nigeria. Fermon (2013) mentions occurrences in the Lower Ngounié and Upper Ivindo basins, but these data have to be verified.
58. *Porogobius schlegelii* is known from Senegal to RDC. In Gabon this species was collected in the lower deltaic plain of the Ogooué and in Lake Onangué (upper deltaic plain) (Fermon, 2013).

	LT	Type	IUCN staus	Biogeogr.	M	S	E	Migrations	BV	PDS	PDI	SL	Lg
Coptodon tholloni (Sauvage, 1884)	22		LC				■		■		■	■	
Hemichromis elongatus (Guichenot, 1861)	15		LC				■		■	■	■	■	
Hemichromis stellifer Loiselle, 1979 [54]	10		LC				■			■	■		
Oreochromis niloticus niloticus (Linnaeus, 1758)	73.2		NE	I		■	■	Potamodrome			■		
Oreochromis schwebischi (Sauvage, 1884)	36.6		LC				■		■		■		
Parananochromis brevirostris Lamboj & Stiassny, 2003 [55]	6.3		VU				■		■				
Pelmatochromis nigrofasciatus (Pellegrin, 1900)	14.1		LC				■				■		
Pelmatolapia cabrae Boulenger, 1899	37		LC				■		■	■	■		■
Sarotherodon mvogoi (Thys van den Audenaerde, 1965) [56]	24.6		LC				■				■		
Sarotherodon nigripinnis nigripinnis (Guichenot, 1861)	22		NE			■	■		■	■	■	■	
ELEOTRIDAE													
Butinae													
Kribia kribensis (Boulenger, 1907)	5.7		LC				■		■	■	■		
Kribia nana (Boulenger, 1901) [57]	4.9		LC				■		■				
Eleotris daganensis Steindachner, 1870	12.1		LC			■	■		■		■	■	
Eleotrinae													
Eleotris senegalensis Steindachner, 1870	21.2		LC		■	■	■	Amphidrome		■	■		
EPHIPPIDAE													
Ephippus goreensis Cuvier, 1831	30		NE		■	■						■	
GERREIDAE													
Eucinostomus melanopterus (Bleeker, 1863)	30		NE		■	■	■	Amphidrome		■	■	■	
Gerres nigri Günther, 1859	20		NE		■	■				■		■	
GOBIIDAE													
Gobiinae													
Porogobius schlegelii (Günther, 1861) [58]	14.9		NE		■	■	■	Amphidrome		■	■	■	
Gobionellinae													
Awaous lateristriga (Duméril, 1861)	5		NE		■	■	■	Amphidrome	■	■	■		
Ctenogobius lepturus (Pfaff, 1933)	6		NE		■	■					■		
Oxudercinae													
Periophthalmus barbarus (Linnaeus 1766)	25		LC		■	■	■	Amphidrome		■		■	■
HAEMULIDAE													
Haemulinae													
Pomadasys jubelini (Cuvier, 1830)	60		NE		■	■	■		■	■	■	■	■
Pomadasys perotaei (Cuvier, 1830)	36		NE		■	■				■	■	■	
Pomadasys rogerii (Cuvier, 1830)	60		NE		■	■				■		■	
Plectorhinchinae													
Plectorhinchus macrolepis (Boulenger, 1899)	45		NE		■	■				■		■	
LUTJANIDAE													
Lutjaninae													
Lutjanus agennes Bleeker, 1863	139		NE		■	■						■	
Lutjanus dentatus (Duméril, 1861)	150		NE		■	■				■	■	■	
Lutjanus goreensis (Valenciennes, 1830)	80		NE		■	■	■			■			■
MONODACTYLIDAE													
Monodactylus sebae (Cuvier, 1829)	25		NE		■	■	■			■	■	■	■
POLYCENTRIDAE													
Polycentropsis abbreviata Boulenger, 1901	8		LC				■		■		■		

	LT	Type	IUCN status	Biogeogr.	M	S	E	Migrations	BV	PDS	PDI	SL	Lg
POLYNEMIDAE													
Galeoides decadactylus (Bloch, 1795)	50		NE		■	■				■		■	
Pentanemus quinquarius (Linnaeus, 1758)	35		NE		■	■				■		■	
Polydactylus quadrifilis (Cuvier, 1829)	200		NE		■	■			■	■	■	■	■
SCIAENIDAE													
Pseudotolithus elongatus (Bowdich, 1825)	47		LC		■	■			■	■	■		■
Pseudotolithus typus Bleeker, 1863	140		NE		■	■	■			■			
SCOMBRIDAE													
Scombrinae													
Scomberomorus tritor (Cuvier, 1832)	100		LC		■	■				■		■	
SERRANIDAE													
Epinephelinae													
Cephalopholis nigri (Günther, 1859)	30		LC		■	■					■		
SPHYRAENIDAE													
Sphyraena afra Peters, 1844	205		NE		■	■				■		■	
PLEURONECTIFORMES													
CYNOGLOSSIDAE													
Cynoglossinae													
Cynoglossus senegalensis (Kaup, 1858)	66		NE		■	■				■	■		■
PARALICHTHYIDAE													
Citharichthys stampflii (Steindachner, 1894)	16		NE		■	■	■		■	■		■	
PSETTODIDAE													
Psettodes belcheri Bennett, 1831	80		NE		■	■				■		■	
TETRAODONTIFORMES													
TETRAODONTIDAE													
Tetraodontinae													
Lagocephalus laevigatus (Linnaeus, 1766)	100		NE		■	■						■	
Sarcopterygii													
LEPIDOSIRENIFORMES													
PROTOPTERIDAE													
Protopterus dolloi Boulenger, 1900	130		LC				■		■		■	■	

7.1. Flood forest with Anthostema aubryanum *near Lake Evaro. This type of pond is typically inhabited by* Xenopus *and* Hymenochirus, *and the surrounding litter is frequented by* Arthroleptis *and* Phrynobatrachus *(Photo: Jean P. Vande weghe, November 2005).*

7.2. Pond with water lilies, Nymphaea lotus, *and* Ceratopteris thalictroides *ferns on Mandji Island. These ponds are very important for the reproduction of many amphibians, notably* Afrixalus *and* Hyperolius *(Photo: Jean P. Vande weghe, June 2016).*

7. Amphibians

Olivier S. G. Pauwels

The state of research on amphibians

Knowledge about the amphibians of the Lower Ogooué region is mainly based on the result of opportunistic collections of specimens, often in association with the collection of reptile material by people who do not make the same level of effort and thoroughness (see Chapter 8 on reptiles by Pauwels). In the area of study, batrachology thus remains the lesser branch of herpetology. However, knowledge of the area's batrachofauna has gradually increased thanks in particular to the collections made by the English explorer Mary Kingsley, the Italian explorer Leonardo Fea, and the French Protestant missionary Ernest Haug, conducted in certain localities–such as the Fernan Vaz Lagoon, Lambaréné, and Ndjolé–at the end of the 19th century, as well as some much more recent publications. While many species remain to be documented, the currently established list is sufficient for making comparisons and determining research priorities.

Diversity and species richness

The table presented in the appendix summarizes bibliographic research conducted in two distinct areas: the Lower Ogooué Ramsar site, encompassing the lower river valley and upper deltaic plain, and the marine delta, including the lower deltaic plain and coastal sand structures. Together these two zones extend along the Ogooué River from the vicinity of Ndjolé to the beaches of Ogooué-Maritime province. They are adjacent to one another, overlapping only for about 10 kilometers along the course of the Ogooué. Their boundaries were given by Vande weghe *et al.* (2016: Figures 771 and 779) and are detailed in the first chapter of this book (pages 24 and 25). The main bibliographic references for the area are those of Peters (1876), Boulenger (1900, 1906), Jongsma *et al.* (2016), Lötters *et al.* (2001), Loveridge (1936), Mocquard (1897a-b, 1902), and Pauwels (2016). Some older works simply reference "Ogooué" (without further precision) as a locality for various species, but since it is unclear what portion of Ogooué is being referenced, they have not been included here. Certain reference materials have also been ignored because they relate to groups whose systematics are complex, and which require re-examination of the material in museum collections on which they are based.

The list in the table given in the appendix replaces Pauwels's earlier list (2016) for the Lower Ogooué region, which does not include the data that became available after its publication. The best-known localities from a batrachological perspective are the Nkomi Lagoon (Fernan Vaz), Lambaréné and its surroundings, and Ndjolé and surroundings. The combined list of species from the Ramsar site and the

7.3. Afrixalus osorioi *frog atop an individual of* Amaranthus *sp. in a vegetable garden near Lambaréné (Photo: Cyrille Mvele).*

7.4. Hyperolius ocellatus *at the edge of the gallery forest (00°55′22″ S, 09°11′28.2″ E) near Kendié (Photo: Ehoarn Bidault).*

7.5. *A sprinkled long reed frog,* Hyperolius adspersus, *in a flood forest (01°07′40.8″ S, 09°25′31.2″ E, November 2016) near Lake Anengué (Photo: Ehoarn Bidault).*

marine delta includes 44 amphibian taxa in 11 families, including one *Gymnophiona* (legless amphibians) and 43 frogs.

Jongsma *et al.* (2016) mentioned a species of *Leptodactylodon* in Mittoné, near Lambaréné, without further details. It was later identified as the Gabon endemic *L. blanci* by Dewynter & Frétey (2019). The genus *Leptodactylodon* is poorly known and rare in Gabon. The same authors (loc. cit.) also made reference to a "*Xenopus amieti* sp. group" from Carivenville, near Ndjolé, regarded as probably *X. parafraseri* by Dewynter & Frétey (2019). Based on current knowledge, the Lower Ogooué is inhabited by more than 40% of the amphibians known to Gabon (42% of a total of 104 confirmed species, based on Dewynter *et al.* (2017) and Dewynter & Frétey (2019), as well as the references cited therein). By comparison, this represents more species than recorded in Loango National Park (32 species), which is nearby geographically, but fewer than in Moukalaba National Park, which holds the current record for the number of amphibians among the national parks of Gabon (57 species) (Pauwels, 2016).

7.6. Phrynobatrachus auritus *at the edge of Lake Evaro (Photo: Jean P. Vande weghe).*

7.7. Phrynobatrachus auritus *on the shore of Lake Oguémoué (Photo: Heather Arrowood).*

Ecological groups

The Seraphin's caecilian (*Geotrypetes seraphini*), which lives in mud, as well as *Aubria subsigillata, Leptodactylodon blanci, Conraua crassipes,* and Pipidae, are almost exclusively aquatic species. Although there are currently no data on the ecology of *Hymenochirus feae*, its morphology points to a strictly aquatic habitat, just like *H. boettgeri.*

Arthroleptis and *Phrynobatrachus* (probably including *Phrynobatrachus ogoensis*, whose natural history is currently unknown), as well as *Cardioglossa* and *Scotobleps*, are frogs found in undergrowth and litter. *Nectophryne* and *Leptopelis* comprise arboreal species. Depending on their ecology, these species are varyingly able to cope with forest degradation, but most of them still show a certain level of ecological plasticity (Blanc & Frétey, 2004). *Afrixalus* (Figure 7.3) and *Hyperolius* from the Lower Ogooué region generally live in tall herbaceous growth in marshy areas, in clearings, or in open areas, and also in savannas, depending on the species.

Endemism

According to our current knowledge, the Gabon dwarf clawed frog *Hymenochirus feae* (Figure 7.11) is only known from the six specimens that were used to describe the species, collected in the Fernan Vaz Lagoon. As its author, the Belgian zoologist Georges A. Boulenger, explained in 1906, this species is identical in all respects to *H. boettgeri* except that its fingers and toes are webbed to the end. *Hymenochirus boettgeri* is also found in the Lower Ogooué region, and is common throughout Gabon.

The Ogooue puddle frog, *Phrynobatrachus ogoensis* (Figures 7.9 and 7.10), also described by Boulenger in 1906, was known only from its type locality, Lambaréné, until it was reported very recently in the Lopé National Park buffer zone, about 175 km further east, based on the capture of an adult female (Larson & Zimkus, 2015). However, the identification of the Lopé specimen as *P. ogoensis* was immediately rejected (Rödel *et al.*, 2015). Even more recently, Dewynter *et al.* (2017) identified a *Phrynobatrachus* collected at the foot of Mt. Koumounabouali (Tsamba-Magotsi department, Ngounié) as *P.* cf. *ogoensis*, expressing doubt about its identification. Until further notice, this species of frog can therefore still be considered endemic to the Lower Ogooué Ramsar site. *Phrynobatrachus ogoensis* has been regarded by some authors as a synonym of *P. brongersmai*, which is found in Liberia, but this synonymy has not been documented and is unlikely from a zoogeographic point of view (Pauwels & Rödel, 2007a; Frétey, 2008).

Apart from these two latter species and *Leptodactylodon blanci*, the Lower Ogooué does not appear to harbor any other species endemic to Gabon, just a subendemic species from Gabon, *Amnirana amnicola* (Pauwels & Rödel, 2007a-b).

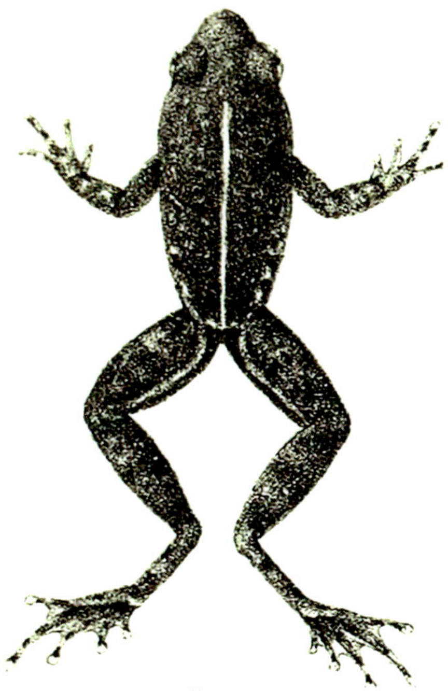

Conservation

The International Union for Conservation of Nature (IUCN; http:// www.iucnredlist.org) has assessed the global conservation status of each of the amphibian species found in the Lower Ogooué region (including the recently discovered *Xenopus mellotropicalis* and *X. parafraseri*, which were described by Evans et al. in 2015), except two. According to the IUCN, all are of minor concern (cf. Table), except for *Hymenochirus feae* and *Phrynobatrachus ogoensis,* the two species endemic to the region, for which data were deemed insufficient for proper classification, and for *Leptodactylodon blanci*, evaluated as Endangered. There is no local systematic hunting of amphibians in the area, though it is possible that *Xenopus* are occasionally eaten, as they are in other parts of Gabon. The main threat to the conservation of amphibians in the Lower Ogooué region is undoubtedly forest degradation, which affects strictly forest-dwelling species.

Jongsma *et al.* (2016) tested the presence of the chytrid *Batrachochytrium dendrobatidis*—a vector for chytridiomycosis in amphibians—among others, in two localities within the Lower Ogooué region, Mittoné near Lambaréné and Carivenville near Ndjolé. The prevalence observed in these two localities was 26.6% and 26.8% respectively among tested amphibians. However, they did not observe individuals who appeared to be in poor health, and it is possible that some amphibians have developed some resistance to chytrid.

7.8. A Gaboon forest tree frog, Leptopelis aubryi, *at the edge of a track (01°06′45.2″ S, 09°25′39.5″ E, alt. 12 m, November 2016) along Lake Anengué (Photo: Ehoarn Bidault).*

7.9. Illustration of a male Ogooué puddle frog, Phrynobatrachus ogoensis, *published as part of the original description in 1906.*

7.10. Illustration of a female Ogooué puddle frog, Phrynobatrachus ogoensis, *published as part of the original description in 1906.*

7.11. Illustration of a male Gabon dwarf clawed frog, Hymenochirus feae, *published as part of the original description. This species, described based on a specimen from the Fernan Vaz Lagoon, has never been collected since its description in 1906.*

Gaps and future research

Important steps for the study of amphibians would include the rediscovery of the two taxa endemic to the Lower Ogooué region, establishing their conservation status, and clarifying the taxonomic status of *Phrynobatrachus ogoensis* (which has direct implications on its distribution, and therefore its conservation status). The ecology of most species and the potential impact of *Batrachochytrium* should also be addressed.

On the basis of their global distribution and ecology (notably using the Ogooué-Maritime inventories of Burger *et al.*, 2006), one could expect to find a certain number of amphibians in the Lower Ogooué region during future inventories, especially the Gabon caecilian, *Herpele squalostoma* (Herpelidae), *Arthroleptis adelphus* and *A. taeniatus*, *Astylosternus batesi*, *Cardioglossa gratiosa*, *Leptopelis aubryioides* (Arthroleptidae), *Nectophryne batesii*, *Sclerophrys superciliaris*, and *S. tuberosa* (Bufonidae), *Hoplobatrachus occipitalis* (Dicroglossidae), *Acanthixalus spinosus*, *Alexteroon obstetricans*, *Hyperolius kuligae*, and *H. pardalis* (Hyperoliidae), *Petropedetes* sp. (Petropedetidae), *Phrynobatrachus cornutus* (Phrynobatrachidae), and *Amnirana lepus* (Ranidae). The most northerly point of the known distribution of Perret's snout burrower frog, *Hemisus peretti* (Hemisotidae), is in Loango National Park (Pauwels, 2007); this species might very well reach as far north as the Ogooué marine delta region. The available information thus suggests that the overall diversity of amphibians in the Lower Ogooué region includes at least 60 species.

Acknowledgments

The author is very grateful to Ehoarn Bidault (Missouri Botanical Garden) and Jean Pierre Vande weghe, as well as Heather Arrowood and Cyrille Mvele of the Organisation Ecotouristique du Lac Oguémoué (OELO) for making available their photographs of amphibians in Lower Ogooué.

Bibliography

BLANC, C.P. & FRÉTEY, T., 2004. Répartition écologique des amphibiens dans la Réserve de Faune de la Lopé et la Station biologique de la Makandé (Gabon). *Bulletin de la Société zoologique de France*, 129 (3): 297–315.

BOULENGER, G.A., 1900. A list of the batrachians and reptiles of the Gaboon (French Congo), with descriptions of new genera and species. *Proceedings of the Zoological Society of London*: 433–456.

BOULENGER, G.A., 1906. Report on the batrachians collected by the late L. Fea in West-Africa. *Annali del Museo Civico di Storia Naturale di Genova*, 3 (2): 157–172 + 2 pl.

BURGER, M., PAUWELS, O.S.G., BRANCH, W.R., TOBI, E., YOGA, J.-A. & MIKOLO, E.-N., 2006. Inventaire des amphibiens du Complexe d'Aires Protégées de Gamba, Gabon. Pp. 79–90 in: ALONSO, A., LEE, M.E., CAMPBELL, P., PAUWELS, O.S.G. & DALLMEIER, F. (éds). Gamba, Gabon: biodiversité d'une forêt équatoriale africaine / Gamba, Gabon: biodiversity of an equatorial African rainforest. *Bulletin of the Biological Society of Washington*, 12: i-xii + 436 pp. + 32 pl.

DEWYNTER M., CHIRIO, L., MELKI, F., CORDIER, J. & FRÉTEY, T., 2017. Premières données herpétologiques (amphibiens et reptiles) sur le mont Koumouna-Bouali (Gabon). *Les cahiers de la fondation Biotope*, 11: 1–42.

DEWYNTER, M. & FRÉTEY, T. 2019. *Liste taxonomique commentée et catalogue illustré des amphibiens du Gabon*. Les cahiers de la fondation, 27: 84 pp.

EVANS, B.J., CARTER, T.F., GREENBAUM, E., GVOŽDÍK, V., KELLEY, D.B., MCLAUGHLIN, P.J., PAUWELS, O.S.G., PORTIK, D.M., STANLEY, E.L., TINSLEY, R.C., TOBIAS, M.L. & BLACKBURN, D.C., 2015. Genetics, morphology, advertisement calls, and historical records distinguish six new polyploid species of African clawed frog (*Xenopus*, Pipidae) from West and Central Africa. *PLoS ONE*, 10 (12): e0142823.

FRÉTEY, T., 2008. Revue des genres africains *Arthroleptis* Smith, 1849 et *Phrynobatrachus* Günther, 1862 (Amphibia, Anura). *Alytes*, 25 (3–4): 99–172.

JONGSMA, G.F.M., BAMBA KAYA, A., YOGA, J.-A., MBEGA, J.-D., MVE BEH, J.-H., TOBI, E., EMRICH, A.M., DIXONMACCALLUM, G.P., DAVIS, D.R., KERBY, J.L. & BLACKBURN, D.C., 2016. Widespread presence and high prevalence of *Batrachochytrium dendrobatidis* in Gabon. *Herpetological Review*, 47 (2): 227–230.

LARSON, J.G. & ZIMKUS, B.M., 2015. Geographic distribution. *Phrynobatrachus ogoensis* (Ogooué River Frog). *Herpetological Review*, 46 (1): 57.

LÖTTERS, S., GOSSMANN, V., OBAME, F. & BÖHME, W., 2001. Zur Herpetofauna Gabuns. Teil I: Einleitung, Untersuchungsgebiet und Methodik, kommentierte Artenliste der gefundenen Froschlurche. *Herpetofauna*, 23 (133): 19–34.

LOVERIDGE, A., 1936. African reptiles and amphibians in Field Museum of Natural History. *Zoological Series, Field Museum of Natural History*, 22 (1): 1–111.

MOCQUARD, F., 1897a. Note préliminaire sur une collection de reptiles recueillie par M. Haug à Lambaréné. *Bulletin du Muséum d'Histoire naturelle*, 3: 54–55.

MOCQUARD, F., 1897b. Sur une collection de reptiles recueillie par M. Haug à Lambaréné. *Bulletin de la Société Philomatique de Paris*, 9: 5–20.

MOCQUARD, F., 1902. Sur des reptiles et batraciens de l'Afrique orientale anglaise, du Gabon et de la Guinée française (région de Kouroussa). *Bulletin du Muséum d'Histoire naturelle*, 6: 404–417.

PAUWELS, O.S.G., 2007. La grenouille fouisseuse de Perret. P. 250 in: VANDE WEGHE, J.P. *Les Parcs Nationaux du Gabon. Loango, Mayumba et le bas Ogooué*. Wildlife Conservation Society, Libreville: 320 pp.

PAUWELS, O.S.G., 2016. Liste des amphibiens. Pp. 362–363 in: VANDE WEGHE, J.P., CHRISTY, P., DUCROCQ, M., LEE, M., VANDE WEGHE, G.R. & PAUWELS, O.S.G. *Biodiversité des parcs nationaux et réserves du Gabon. 2. Espèces, écosystèmes et populations*. Agence Nationale des Parcs Nationaux, Libreville: 384 pp.

PAUWELS, O.S.G., 2017. Les reptiles. Pp. 257–265 in: T. STÉVART & J.P. VANDE WEGHE, Le delta de l'Ogooué. ANPN, Libreville.

PAUWELS, O.S.G. & RÖDEL, M.-O., 2007a. Amphibians and national parks in Gabon, western central Africa. *Herpetozoa*, 19 (3/4): 135–148.

PAUWELS, O.S.G. & RÖDEL, M.-O., 2007b. An update on the amphibian richness of Gabon and its representation in national parks. *Froglog*, 79: 1–2.

PETERS, W. 1876. Eine zweite Mittheilung über die von Hrn. Professor Dr. R. Buchholz in Westafrica gesammelten Amphibien. *Monatsberichte der Königlich Preußischen Akademie der Wissenschaffen zu Berlin*: 117–123 + pl.

RÖDEL, M.-O., BURGER, M., ZASSIBOULOU, A.G., EMMRICH, M., PENNER, J. & BAREJ, M.F., 2015. Two new *Phrynobatrachus* species (Amphibia: Anura: Phrynobatrachidae) from the Republic of Congo. *Zootaxa*, 4032 (1): 55–80.

Appendix: List of amphibians of the Ogooué Delta

Distribution: A + B = valley upstream from Lambaréné and upper deltaic plain; C + D = lower deltaic plain, coastal sands, and Nkomi Lagoon
IUCN status: LC = Least Concern; DD = Data Deficient; NE = Not Evaluated

	A + B	C + D	IUCN status
GYMNOPHIONES			
DERMOPHIIDAE			
Séraphin's caecilian, *Geotrypetes seraphini*	■		LC
ANURA			
ARTHROLEPTIDAE			
Arthroleptinae			
Long-fingered squeaker, *Arthroleptis macrodactylus*	■	■	NE
Forest squeaker, *Arthroleptis sylvaticus*	■		LC
Buea squeaker, *Arthroleptis variabilis*	■		LC
Rio Benito long-fingered frog, *Cardioglossa gracilis*	■		LC
Silver long-fingered frog, *Cardioglossa leucomystax*	■		LC
Astylosterninae			
Blanc's egg frog, *Leptodactylodon blanci*	■		EN
Gabon forest frog, *Scotobleps gabonicus*	■		LC
Leptopelinae			
Aubry's tree frog, *Leptopelis aubryi*	■		LC
Boulenger's tree frog, *Leptopelis boulengeri*	■		LC
Millson's tree frog, *Leptopelis millsoni*	■		LC
Soccer tree frog, *Leptopelis notatus*	■	■	LC
Ocellated tree frog, *Leptopelis ocellatus*	■		LC
BUFONIDAE			
Common dwarf tree toad, *Nectophryne afra*		■	LC
Cameroon toad, *Sclerophrys camerunensis*	■		LC
Angola toad, *Sclerophrys funerea*		■	LC
Slender-foot toad, *Sclerophrys gracilipes*	■		LC
African savanna toad, *Sclerophrys regularis*	■	■	LC
CONRAUIDAE			
Abo slippery frog, *Conraua crassipes*	■		LC
HYPEROLIIDAE			
Striped banana frog, *Afrixalus dorsalis*	■	■	LC
Osorio's banana frog, *Afrixalus osorioi*	■	■	LC
Foulassi banana frog, *Afrixalus paradorsalis*	■		LC
Sprinkled reed frog, *Hyperolius adspersus*	■		LC
Dotted reed frog, *Hyperolius guttulatus*	■		LC
Ocellated reed frog, *Hyperolius ocellatus*	■	■	LC
Olive reed frog, *Hyperolius olivaceus*	■	■	NE
Fantastic reed frog, *Hyperolius phantasticus*		■	LC

	A + B	C + D	IUCN status
Benito river reed frog, *Hyperolius platyceps*	■	■	LC
Rough reed frog, *Hyperolius tuberculatus*	■		LC
Grey-eyed frog, *Opisthothylax immaculatus*		■	LC
Leonard's tree frog, *Phlyctimantis leonardi*		■	LC
PHRYNOBATRACHIDAE			
African puddle frog, *Phrynobatrachus africanus*	■	■	LC
Golden puddle frog, *Phrynobatrachus auritus*	■	■	LC
Ogooue puddle frog, *Phrynobatrachus ogoensis*	■		DD
PIPIDAE			
Boettger's dwarf clawed frog, *Hymenochirus boettgeri*	■		LC
Gabon dwarf clawed frog, *Hymenochirus feae*		■	DD
Gabon clawed frog, *Xenopus mellotropicalis*	■	■	LC
Mfoundi clawed frog, *Xenopus parafraseri*	■		LC
PTYCHADENIDAE			
Mascarene grass frog, *Ptychadena* aff. *mascareniensis*	■	■	LC
Perret's grass frog, *Ptychadena perreti*	■		LC
PYXICEPHALIDAE			
Brown fishing frog, *Aubria subsigillata*	■		LC
RANIDAE			
Forest white-lipped frog, *Amnirana albolabris*	■	■	LC
Ilanga white-lipped frog, *Amnirana amnicola*	■		LC
RHACOPHORIDAE			
African foam-nest tree frog, *Chiromantis rufescens*	■		LC

8. Reptiles

Olivier S. G. Pauwels

The state of herpetological research

The first attempt to catalog Gabon's reptile species, published in 1900 by the Belgian zoologist Georges A. Boulenger, notably included a certain number of specimens captured in the Lambaréné area by the English explorer Mary Kingsley during her 1895 expedition, and by the French Protestant missionary Ernest Haug, who was stationed in Ngomo. Within the Ramsar site, the locality that has to date been best studied for its herpetofauna is Lambaréné (notably through the works of W. Peters in 1876, of Albert Günther in 1896, those of François Mocquard in 1897 and 1902, and those of Dieter Gramentz cited in Pauwels & Vande weghe, 2008). Although the famous Alsatian doctor Albert Schweitzer (1950) devoted an entire chapter to the snakes of Lambaréné, his descriptions were, unfortunately, so imprecise that few species are identifiable. During the same period, mention of reptiles from the coastal region was made in several publications, notably from the Fernan Vaz Lagoon and Port-Gentil, i.e., just west of the Ramsar site, in the marine part of the Ogooué Delta. Older references to reptiles from the Fernan Vaz Lagoon are notably based on the collections of the Italian explorer Leonardo Fea and on the accounts of the explorer Paul Du Chaillu (1868). While the reptile fauna of the Lower Ogooué Ramsar site has not been the subject of systematic research and some species certainly remain to be documented, combining the data accumulated for more than a century has nevertheless resulted in a clear idea of what lives there.

Diversity and species richness

Extensive bibliographic research was carried out regarding two areas adjacent to the Lower Ogooué region, which only took into account precise indications coming from these respective areas. The first area strictly corresponds to the Lower Ogooué Ramsar site, as illustrated by Vande weghe *et al.* (2016: Figure 771). The second includes the part of the Ogooué closest to the river's mouth as well as the mouth itself, downstream from the Ramsar site, as delimited by Vande weghe *et al.* (2016: Figures 771 and 779, in which the area is referred to as the "marine delta"). Unlike the marine delta, the Ramsar site does not include environments under marine influence. The city of Port-Gentil is included in the northwestern part of the second area, and the Fernan Vaz Lagoon marks its southern limit. The two areas overlap for a portion of approximately 10 km of the Ogooué.

The complete list of reptiles that have currently been identified within these two areas is provided in the table given in the appendix. It includes aquatic snakes of the genus *Grayia* mentioned by Pauwels (1995), the synthesis on Gabonese reptiles published by Pauwels & Vande weghe (2008), and certain more recent works that provide data on this area (Bell &

8.1. Young slender-snouted crocodile sunbathing on a branch overlooking the Nkomi Lagoon (Photo: Gaël R. Vande weghe).

8.2. Adult Pelusios niger *in the village of Tsam Tsam on Lake Oguemoué. This turtle is eaten when trapped in fishing nets (Photo and comment: Heather Arrowood).*

8.3. Green bush viper on branches overlooking the Abanga River, near the village of Ebel Abanga (Photo: Quentin Meunier).

8.4. Adult Trachylepis albilabris *on Lake Oguémoué near the village of Saint Louis (Photo: Cyrille Mvele).*

8.5. A couple of Agama agama *(male on the right) in the village of Odimba on Lake Oguémoué. Village elders claim that this species was introduced in the 1960s via barges from Port-Gentil (Photo and comment: Heather Arrowood).*

Stuart, undated; Pauwels *et al.*, 2011, 2016, 2017; Pauwels & Vande weghe, 2011; and Pauwels, 2016–which includes occasional personal field observations made by the author in Port-Gentil, Lambaréné, Ndjolé, and their surroundings from 2001 to 2011). Certain citations of species from the "Ogooué" and "Lower Ogooué" in older herpetological literature are too imprecise to determine whether they were made in the Ramsar site or the marine part of the delta, or even further upstream along the Ogooué. They almost exclusively concern species already listed in one of the two areas, and would therefore not significantly change the list of overall diversity for these two areas. A mention of *Pelusios carinatus* from Port-Gentil is present in the literature, but this species was not retained here since its presence appears to be erroneous (see Maran & Pauwels' discussion of this record, 2005).

The list presented here includes a total of 72 reptiles found strictly within the Ramsar site: five freshwater and terrestrial turtles, three crocodiles, 24 lizards, two amphisbaenians, and 38 snakes. It supersedes the incomplete list presented by Pauwels (2016). These 72 reptiles represent well over half of the total number of species currently known in Gabon, which stands at 130 (Pauwels, 2016; Pauwels *et al.*, 2019). By comparison, it represents a higher total than those obtained for Ivindo National Park (65 species), which nevertheless has been the focus of significant attention from herpetologists, especially in recent years (Pauwels *et al.*, 2016), as well as Loango Park (45 species), or any other protected area in Gabon (Pauwels, 2016; Pauwels *et al.*, 2004, 2006, 2018, 2020).

The marine part of the delta includes, among others, 13 species not listed in the Ramsar site (cf. Table). These include four sea turtles that are certainly not present in the Ramsar site, and several other species that will certainly be found there in future field studies. In total, the two areas house at least 85 reptile species.

Ecological groups

The Ramsar site includes 11 aquatic species (15% of the site's total number of reptiles), including four freshwater turtles, three crocodiles, and four snakes (two species of *Grayia*, the black water snake, *Hydraethiops melanogaster*, and the banded water cobra, *Naja a. annulata*). Although these species are capable of living out of water, they spend most of their time in an aquatic environment; their morphology is in fact adapted to aquatic conditions. There is, therefore, a relatively high proportion of aquatic forms among the snakes in the Ramsar site compared to the group at large, about 5% of which are aquatic (Pauwels *et al*., 2008). There are also several species that are not strictly aquatic, but which always live at the edge of water and regularly enter it to move or hunt: the ornate monitor, *Varanus ornatus*, the Seba's rock python, *Python sebae*, and two species of *Natriciteres*. For these four species, the freshwater environments of the Ramsar site are therefore almost as vital as they are for the first 11 taxa.

Another peculiarity of the fauna of the Ramsar site is that it includes a relatively large proportion of burrowing reptiles (12 species, or 17% of the total), which are quite diversified taxonomically: three skinks of the genus *Feylinia*, two amphisbaenians belonging to separate genera, two Typhlopidae belonging to two genera (*Afrotyphlops* and *Letheobia*), one Boidae of the genus *Calabaria*, and four Atractaspidinae belonging to three genera (*Aparallactus, Atractaspis,* and *Polemon*). Burrowing reptiles are rarely encountered because they hardly move on the surface and are often neglected in environmental studies, although they contain species with restricted geographic distributions–generally more so than other ecological groups–which are therefore important for conservation.

The Ramsar site is home to at least two savanna species: the lizard *Gerrhosaurus nigrolineatus* and the snake *Dasypeltis confusa*, the latter feeding exclusively on bird eggs. The savanna-dwelling snake, *Psammophis mossambicus*, is known to occur in the marine delta, which forms a patchwork of savannas and mangroves, but it will certainly soon be found within the Ramsar site, in pockets of savanna, especially those surrounding several large lakes (as visible for example in the vegetation map of Lakes Onangué and Ezanga presented in Vande weghe *et al.*, 2016: 217).

Among the remaining species, some are anthropophilic, such as *Agama agama* and *Hemidactylus mabouia*, which are found on all types of human infrastructure in the area. Others are arboreal or live in the forest litter, some of which are quite sensitive to forest quality (such as the Cameroon stumptail chameleon, the Gabon tail-pad gecko, the spiny-tailed lizard, and the Rohde's skink), while others are much less demanding or ubiquitous (such as the crested chameleon, the ornate monitor, the blue-tailed tree snake, Blanding's tree snake, Jameson's mamba, and the forest cobra). Haug's professional activities (see below) explain why his collections included an individual of *Gastropholis echinata*, a lizard very rarely encountered because it inhabits the canopy, and which he probably collected while felling trees; more than a hundred years have passed since the publication of the specimen collected by Haug and the latest mention of this species in Gabon.

Some species show a marked pref-erence for open or degraded environments, such as the flap-necked chameleon, the Owen's chameleon, the brown-flanked skink, the striped grass snake, and the green grass snake, as well as the Gabon viper (specific ecological data have been summarized by Pauwels & Vande weghe, 2008).

8.6. Kinixys erosa *near the village of Saint Louis on Lake Oguémoué (Photo: Heather Arrowood).*

8.7. An adult male of the Gabon Gerrhosaurus *on the shore of Lake Evaro. Its bright orange throat indicates that it is in nuptial livery (Photo: Jean P. Vande weghe).*

8.8. An adult male Agama picticauda *on a tree in Port-Gentil. Males often perch up high to monitor their territory and their small harem of females (Photo: Jean P. Vande weghe).*

8.9. An adult female Agama picticauda *in Port-Gentil; it is part of the harem photographed in Figure 8.8 (Photo: Jean P. Vande weghe).*

Endemism

The first known specimen of the amphisbaenian *Cynisca bifrontalis* was collected by L. Fea near the Fernan Vaz Lagoon. This was the only known material of this species until it was rediscovered, about a century later, by Branch *et al.* (2003), some 90 km further to the southeast, although still within Ogooué-Maritime province. *Cynisca haugi* is, to this day, still only known from its type specimen, which was used to describe the species. It was collected by Ernest Haug, as were the types used to describe the snakes *Atractaspis boulengeri* and *Polemon bocourti*. These three species burrow and are as a result difficult to find. Haug must have discovered them because, although he was a missionary, he was also director of the Ngomo sawmill, which must have led him frequently to be involved in the felling of trees, giving him an opportunity to capture animals uncovered when trees were uprooted. In fact, it was while uprooting a tree that *C. bifrontalis* was recently rediscovered and specimens of *Monopeltis* were collected (Branch *et al.*, 2003) during the construction of a new road running through the forest. The morphological differences between the six specimens of *C. bifrontalis* and the only currently known specimen of *C. haugi* are subtle, but there is currently no evidence indicating that they are the same species. If more specimens were found and morphological or genetic studies were to indicate that the differences between the two are due only to intraspecific variation, then the oldest name should be maintained, *C. haugi,* which was described in 1904, rather than *C. bifrontalis,* which was described two years later. In the meantime, particularly with regard to conservation, it is better to consider them as separate species, one of which, *C. haugi,* is therefore presumed to be endemic to the Lower Ogooué Ramsar site.

Only two specimens of the legless burrowing skink, *Feylinia boulengeri,* are known to this day, one collected by the French adventurer Maurice Patry around

1947 in the vicinity of the Mbilagone River in Estuaire, and the other in 1913 by the Protestant missionary Victor Ellenberger at Ngomo in the Lower Ogooué Ramsar site. This species differs from other *Feylinia* in certain aspects of its scaling; it was once placed in the genus *Chabanaudia*, established especially for it but no longer recognized by specialists. It is quite possible that the distribution of this species is limited to the Lower Ogooué region and its immediate surroundings.

In addition to these three species endemic to Gabon, the Ramsar site includes several taxa endemic to Lower Guinea: the Gabon tail-pad gecko (which has its type locality within the Ramsar site itself), the Rohde's skink, and the Gabon helmeted worm lizard.

Conservation issues

Forty-three of the species found in the Ogooué Delta have had their conservation status evaluated by experts from the International Union for Conservation of Nature (IUCN; http://www.iucnredlist.org; Chirio et al., 2017; Pauwels & Schmitz, 2010). All the sea turtles, which frequent the beaches of the marine delta, including those found on the "Barre des Portugais," have been evaluated by the IUCN: *Chelonia mydas* (Endangered), *Eretmochelys imbricata* (Critically Endangered), *Lepidochelys olivacea* (Vulnerable), and *Dermochelys coriacea* (Vulnerable). These classifications accurately reflect the situation of the sea turtles in the Lower Ogooué region, where serious efforts must be made quickly in order to safeguard these species. Fortunately, sea turtles have all been added to Gabon's list of wholly protected animals. All the other species which have been evaluated are of Least Concern, except *Pelusios niger* (Near Threatened), *Kinixys erosa* (Data Deficient), *Cycloderma aubryi* (Vulnerable), *Trionyx triunguis* (Vulnerable), *Mecistops cataphractus* (Critically Endangered), *Osteolaemus tetraspis* (Vulnerable), and *Monopeltis galeata* (Data Deficient). Of the 85 species known from the Ogooué Delta, 51% have therefore been assessed by the IUCN. It should be noted that the IUCN does not necessarily assess species based on prior risk estimates regarding their conservation, and that a conservation status concerns an entire species, even though conditions and threats may vary locally. Consequently, the analysis of these species is still fragmented and does not properly represent the conservation status of the herpetofauna inhabiting the Ramsar site.

Turtles of the genus *Pelusios* are among the species for which a local assessment would be important. The brown mud terrapin is not systematically hunted because of its small size, but it is still consumed locally, notably in Port-Gentil and Lambaréné (Maran & Pauwels, 2005). The larger black mud terrapin is hunted for its flesh and eggs, particularly in the lakes near Lambaréné. The two local species of softshell turtles are actively sought after and their flesh and eggs sold as preferred foods in the markets of Lambaréné. This hunting has already seriously impacted the populations of these two species in certain localities along the Ogooué (Maran & Pauwels, 2005; Pauwels & Maran, 2007). The eroded hinge-back tortoise is regularly consumed and sold in markets, particularly in Lambaréné, but it is ubiquitous in Gabon, ecologically very flexible, and seems to present a certain resilience to the regular hunting it faces.

As described in the 19th century accounts of the explorer Paul Du Chaillu (1863), the Lower Ogooué and surrounding lakes were swarming with Nile crocodiles; he specifically mentioned gatherings of large numbers on the banks of the Anengué River, which flows into the lake of the same name. Du Chaillu described his encounter with crocodiles in this locality as follows [our translation]: "There were several hundred of these disgusting monsters, who warmed themselves in the sun while kneading mud [...]. Some were at least twenty feet long [...]. I maneuvered my canoe into the midst of them in order to isolate the largest individual from the rest of the troop, and I shot him in the body, [...] he sank into the mud. [...] I shot a second one, which also sank; and since my men didn't bother to venture out in the black mud to look for them, we couldn't pick bring in up." This first encounter between a European explorer and crocodiles in the area of the Lower Ogooué Ramsar site is a reflection of the interactions between humans and crocodiles that have reigned in the area ever since. Postcards from the early 20th century depicting massacres

8.10. An Indotyphlops braminus *found in a garden in the SEEG district of Port-Gentil in 2008. This tiny, invasive, exotic snake of Asian origin probably arrived in the city well before 2008 via international maritime traffic (Photo: Olivier S. G. Pauwels).*

of fairly large slender-snouted crocodiles in the Lower Ogooué region suggest that it was a frequent and popular activity. Crocodile hunting campaigns in this area reached their peak in the 1960s and 1970s when the trade in crocodile skins was officially and nationally organized for massive export (Pauwels, 2006). These hunting activities ceased on their own in the 1970s due to insufficient natural stocks for large scale exploitation to continue. Since then, hunting has continued, although on a smaller scale, and is now driven by local food consumption. This hunting, which has been carried out for many years on the three species of crocodiles present in the Lower Ogooué region, has drastically reduced the numbers of each, and gatherings such as those previously described by Du Chaillu can no longer be seen today. More so than other Gabonese crocodiles, the slender-snouted crocodile can easily be drowned by fishing nets, even though they are not specifically targeted.

On a more optimistic note, it is worth mentioning that, in 2011, three crocodile species were added by decree to Gabon's list of wholly protected animals, and it is also worth noting the existence of a conservation project specifically designed to stop the hunting and consumption of slender-snouted crocodiles, the "Projet Faux Gavial," which was launched very recently in Lambaréné. Set up by the American biologist Heather Arrowood and the Gabonese naturalist Cyrille Mvele, it is the first conservation project in Gabon dedicated to reptiles other than sea turtles. The first results appear to be positive (Arrowood & Mvele, 2017). Of the three Gabonese crocodiles, the slender-snouted crocodile faces the worst situation and truly requires the urgent implementation of such conservation programs. The fact that a recent genetic study (Shirley *et al.*, 2013) indicates that the slender-snouted crocodile is actually a complex involving two species, one of which is an undescribed species that inhabits Central Africa and the Ogooué region, strongly reinforces the importance of the population living within the Ramsar site, which offers ideal biotopes for all three species and could therefore play a major role in ensuring their survival.

Apart from turtles and crocodiles, few reptile species are eaten on a regular basis in the area, with the exception of the African rock python. Besides hunting, the main threat to reptiles comes in the form of environment degradation in forests, which affects forest-dwelling species.

It is currently impossible to determine whether the presence of the invasive exotic snake *Indotyphlops braminus*, native to Asia and parthenogenetic, poses a threat to Typhlopidae or other native burrowing reptiles with which it could compete. This snake was reported for the first time from Gabon in 2004 and is still only known from Libreville and Port-Gentil (Pauwels *et al.*, 2011), where its presence probably results from both cities being ports. In any case, there is a significant risk that this species will spread widely across the country, as it has done in dozens of other tropical countries where it has colonized. It is uncertain whether the gecko *Hemidactylus angulatus*, is indigenous to the region; this anthropophilic lizard has currently only been seen within Gabon in Port-Gentil and Gamba, two coastal cities in Ogooué-Maritime province, where freight regularly arrives from abroad.

Gaps and future research

Without any doubt, the most urgent activities concerning the herpetofauna of the Lower Ogooué Ramsar site and the marine delta concern the conservation of crocodiles, especially the slender-snouted crocodile and the Nile crocodile, as well as sea turtles. It is also important to complete the list of species present in the area by organizing field expeditions, especially in the pockets of savanna where several not-yet-identified species surely live. A more complete list of species will allow the area to be better positioned in relation to other protected areas in terms of specific richness and representativeness. Being present when large trees are felled in forested areas would allow access to more specimens of *Cynisca*, which would help with deciding whether there are one or two species, or even with confirming their validity, as well as with discovering new ones, which is very likely. More generally, a targeted search for burrowing reptiles, particularly using pitfall traps, which are commonly used by herpetologists but have never been used in this region, would probably lead to an increase in their already very high known proportion within the local herpetofauna. It would also help to study the possible spread of *Indotyphlops* from Port-Gentil, if they have not already progressed further inland, or have not already been introduced to Lambaréné through transporting building materials and seedlings.

A number of reptile species are known to exist in areas close to the Lower Ogooué region and live in the biotopes that are present within the region. These species, which will more than likely soon be documented within these areas, include: *Holaspis guentheri* (Lacertidae), *Trachylepis p. polytropis* (Scincidae), *Afrotyphlops angolensis* (Typhlopidae), *Crotaphopeltis hotamboeia*, *Rhamnophis batesii*, *Thelotornis kirtlandii*, and *Thrasops flavigularis* (all Colubridae), *Pseudohaje goldii* (Elapidae), *Atractaspis c. corpulenta*, *A. reticulata*, *Bothrophthalmus brunneus*, and *Limaformosa savorgnani* (all Lamprophiidae), and *Causus maculatus* (Viperidae). It is very reasonable to assume that the total diversity of reptiles at the Lower Ogooué Ramsar site actually exceeds 90 species.

Acknowledgments

The author warmly thanks Heather Arrowood, Cyrille Mvele (Ecotouristic Organization of Lake Oguemoué), Quentin Meunier (OLAM Gabon), and Jean P. Vande weghe for allowing the use of their photographs.

Bibliography

ARROWOOD, H. & MVELE, C., 2017. Projet Faux Gavial reduces commerce of slender-snouted crocodile in Gabon. *Oryx,* 51 (1): 13–14.

BELL, R. & STUART, B. [undated] *Quelques reptiles du Lac Oguemoué*. Poster avec photos-couleurs.

BOULENGER, G.A. 1900. A list of the batrachians and reptiles of the Gaboon (French Congo), with descriptions of new genera and species. *Proceedings of the Zoological Society of London*: 433–456.

BRANCH, W.R., PAUWELS, O.S.G. & BURGER, M., 2003. Rediscovery of *Cynisca bifrontalis* in Gabon, with additional notes on *Monopeltis galeata* (Reptilia: Amphisbaenia). *African Journal of Herpetology*, 52 (2): 93–100.

CHIRIO, L., DIAGNE, T. & PAUWELS, O.S.G. 2017. *Cycloderma aubryi. The IUCN Red List of Threatened Species* 2017: e.T163448A1009393. https://dx.doi.org/10.2305

DU CHAILLU, P.B., 1863. *Voyages et aventures dans l'Afrique équatoriale*. Michel Lévy Frères, Libraires éditeurs, Paris: xiii + 546 pp. + pl. + carte.

DU CHAILLU, P.B., 1868. *L'Afrique sauvage. Nouvelles excursions au pays des Ashangos*. Michel Lévy Frères, Libraires éditeurs, Paris: 411 pp. + carte.

GÜNTHER, A., 1896. Report on a collection of reptiles and fishes made by Miss M.H. Kingsley during her travels on the Ogowe River and in Old Calabar. *Annals and Magazine of Natural History*, 6 (17): 261–285 + pl. 13–15.

MARAN, J. & PAUWELS, O.S.G., 2005. Etat des connaissances sur les tortues continentales du Gabon: distribution, écologie et conservation. *Bulletin de l'Institut Royal des Sciences naturelles de Belgique*, 75: 47–60.

MOCQUARD, F., 1897a. Note préliminaire sur une collection de reptiles recueillie par M. Haug à Lambaréné. *Bulletin du Muséum d'Histoire naturelle,* 3: 54–55.

MOCQUARD, F., 1897b. Sur une collection de reptiles recueillie par M. Haug à Lambaréné. *Bulletin de la Société Philomatique de Paris*, 9: 5–20.

MOCQUARD, F., 1902. Sur des reptiles et batraciens de l'Afrique orientale anglaise, du Gabon et de la Guinée française (région de Kouroussa). *Bulletin du Muséum d'Histoire naturelle*, 6: 404–417.

PAUWELS, O.S.G., 1995. *Etude biométrique des couleuvres aquatiques africaines du genre* Grayia *Günther (Serpentes: Colubridae).* Mémoire présenté pour l'obtention du Grade de Licencié en Sciences zoologiques. Université Libre de Bruxelles, Bruxelles: 61 pp. + Annexes.

PAUWELS, O.S.G., 2006. Crocodiles and national parks in Gabon. *Crocodile Specialist Group Newsletter*, 25 (1): 12–14.

PAUWELS, O.S.G., 2016. Liste des reptiles. Pp. 364–366 in: VANDE WEGHE, J.P., CHRISTY, P., DUCROCQ, M., LEE, M., VANDE WEGHE, G.R. & PAUWELS, O.S.G. *Biodiversité des parcs nationaux et réserves du Gabon. 2. Espèces, écosystèmes et populations*. Agence Nationale des Parcs Nationaux, Libreville: 384 pp.

PAUWELS, O.S.G., BAHAA-EL-DIN, L., ALBERT, J.-L., CARLINO, P., GIANNUZZI, F., CHIRIO, L., GILLET, J.-F., POIRIER, E. & STÉVART, T. 2018. Miscellanea Herpetologica Gabonica XIV. *Bulletin of the Chicago Herpetological Society*, 53 (9): 185-190.

PAUWELS, O.S.G., BRANCH, W.R. & BURGER, M., 2004. Reptiles of Loango National Park, Ogooué-Maritime Province, south-western Gabon. *Hamadryad*, 29 (1): 115–127.

PAUWELS, O.S.G., CARLINO, P., CHIRIO, L., MEUNIER, Q., OKOUYI OKOUYI, J.V., ORBELL, C., ROUSSEAUX, D. & TESTA, O., 2017. Miscellanea Herpetologica Gabonica IX. *Bulletin of the Chicago Herpetological Society*, 52 (6): 97–102.

PAUWELS, O.S.G., CHRISTY, P. & HONOREZ, A. 2006. Reptiles and national parks in Gabon, western central Africa. *Hamadryad,* 30 (1–2): 181–196.

PAUWELS, O.S.G., LE GARFF, B., INEICH, I., CARLINO, P., MELCORE, I., BOUNDENGA, L., VIGNA, C., STÉVART, T., JEFFERY, K., ORBELL, C., SQUARCINI, J.-B., VANDE WEGHE, J.P. & WHITE, L.J.T., 2016. Miscellanea Herpetologica Gabonica V & VI. *Bulletin of the Chicago Herpetological Society*, 51 (11): 177–185.

PAUWELS, O.S.G. & MARAN, J., 2007. Occurrence of tortoises and freshwater turtles (Pelomedusidae, Testudinidae, and Trionychidae) in the national parks of Gabon – 2006 status of knowledge. *Turtle and Tortoise Newsletter*, 10: 21–26.

PAUWELS, O.S.G., MORELLE, S., ALBERT, J.-L., CARLINO, P., RAHOLA, N. & TRAPE, J.-F. 2019. New reptile records from Lékédi Park and Haut-Ogooué Province, southeastern Gabon. *Amphibian & Reptile Conservation*, 13 (1): 143-161.

PAUWELS, O.S.G., PAULY, A., ARALDI, A., NDONDA MAKEMBA, R., FONTEYN, D., OSLISLY, R. & WHITTAKER, A. M. 2020. Miscellanea Herpetologica Gabonica XVI. *Bulletin of the Chicago Herpetological Society*, 55 (5): 93-99.

PAUWELS, O.S.G. & SCHMITZ, A., 2010. *Cynisca bifrontalis.* The IUCN Red List of Threatened Species 2010: e.T176236A7202113. http://dx.doi.org/10.2305/IUCN.UK.2010-4.RLTS.T176236A7202113.en.

PAUWELS, O.S.G., TOBI, E. & SARRAZIN, P., 2011. Geographic distribution. *Ramphotyphlops braminus* (Brahminy Blind Snake). *Herpetological Review*, 42 (1): 115–116.

PAUWELS, O.S.G. & VANDE WEGHE, J.P., 2008. *Reptiles du Gabon*. Smithsonian Institution, Washington: 272 pp.

PAUWELS, O.S.G. & VANDE WEGHE, J.P. ,2011. Les espèces invasives. Pp. 253–254 in: VANDE WEGHE, J.P. *Les parcs nationaux du Gabon. Akanda et Pongara. Plages et mangroves*. Wildlife Conservation Society & Agence Nationale des Parcs Nationaux, Libreville: 272 pp.

PAUWELS, O.S.G., WALLACH, V. & DAVID, P. ,2008. Global diversity of snakes (Serpentes: Reptilia) in freshwater. Pp. 599–605 in: BALIAN, E.V., LÉVÊQUE, C., SEGERS, H. & MARTENS, K. (éds). Freshwater animal diversity assessment. *Hydrobiologia*, 595: 1–637.

PETERS, W. 1876. Eine zweite Mittheilung über die von Hrn. Professor Dr. R. Buchholz in Westafrica gesammelten Amphibien. *Monatsberichte der Königlich Preuβischen Akademie der Wissenschaffen zu Berlin*: 117-123 + pl.

SCHWEITZER, A., 1950. *The animal world of Albert Schweitzer. Jungle insights into reverence for life.* Translated and edited, with an introduction, by Charles R. Joy. The Ecco Press, Hopewell.

SHIRLEY, M.H., VLIET, K.A., CARR, A.N. & AUSTIN, J.D., 2013. Rigorous approaches to species delimitation have significant implications for African crocodilian systematics and conservation. *Proceedings of the Royal Society B*, 281: 20132483.

VANDE WEGHE, J.P., CHRISTY, P., DUCROCQ, M., LEE, M., VANDE WEGHE, G.R. & PAUWELS, O.S.G., 2016. *Biodiversité des parcs nationaux et réserves du Gabon. 2. Espèces, écosystèmes et populations*. Agence Nationale des Parcs Nationaux, Libreville: 384 pp.

Appendix: List of reptiles of the Ogooué Delta

Distribution: A + B = Lower Ogooué valley and upper deltaic plain; C + D = lower deltaic plain, coastal sands, and Nkomi Lagoon (or Fernan Vaz Lagoon)
IUCN status: CR = Critically Endangered; EN = Endangered; VU = Vulnerable; LC = Least Concern; DD = Data Deficient; NE = Not Evaluated

	A + B	C + D	IUCN status
CHELONII			
PELOMEDUSIDAE			
Brown mud terrapin, *Pelusios castaneus*	■	■	NE
Black mud terrapin, *Pelusios niger*	■	■	NT
CHELONIIDAE			
Green sea turtle, *Chelonia mydas*		■	EN
Hawksbill sea turtle, *Eretmochelys imbricata*		■	CR
Olive ridley sea turtle, *Lepidochelys olivacea*		■	VU
DERMOCHELYIDAE			
Leatherback sea turtle, *Dermochelys coriacea*		■	VU
TESTUDINIDAE			
Eroded hinge-back tortoise, *Kinixys erosa*	■	■	DD
TRIONYCHIDAE			
Aubry's flapshell turtle, *Cycloderma aubryi*	■	■	VU
Nile softshell turtle, *Trionyx triunguis*	■	■	VU
CROCODYLIA			
CROCODYLIDAE			
Nile crocodile, *Crocodylus niloticus*	■	?	LC
Slender-snouted crocodile, *Mecistops cataphractus*	■	■	CR
Dwarf crocodile, *Osteolaemus tetraspis*	■	■	VU
SQUAMATA			
AGAMIDAE			
Red-headed agama, *Agama agama*	■	■	NE
CHAMAELEONIDAE			
Flap-necked chameleon, *Chamaeleo dilepis*	■	■	LC
Cameroon stumptail chameleon, *Rhampholeon spectrum*	■		LC
Crested chameleon, *Trioceros cristatus*	■		LC
Owen's chameleon, *Trioceros oweni*	■		LC
GEKKONIDAE			
Gabon half-toed gecko, *Hemidactylus angulatus*		■	NE
Banded half-toed gecko, *Hemidactylus fasciatus*	■	■	LC
House half-toed gecko, *Hemidactylus mabouia*	■	■	LC
Ghana half-toed gecko, *Hemidactylus muriceus*	■	■	LC
Richardson's half-toed gecko, *Hemidactylus richardsonii*	■		NE
Fischer's dwarf gecko, *Lygodactylus fischeri*	■		NE
Gabon tail-pad gecko, *Urocotyledon palmata*	■		NE

	A + B	C + D	IUCN status
GERRHOSAURIDAE			
Gabon plated lizard, *Gerrhosaurus nigrolineatus*	■	■	LC
LACERTIDAE			
Spiny-tailed lizard, *Gastropholis echinata*	■		NE
Ford's striped lizard, *Poromera fordii*	■		LC
SCINCIDAE			
Boulenger's limbless skink, *Feylinia boulengeri*	■		LC
Curror's limbless skink, *Feylinia currori*	■	■	LC
Large-scaled limbless skink, *Feylinia grandisquamis*	■	■	LC
Reichenow's skink, *Lacertaspis reichenowi*	■		LC
Rohde's skink, *Lacertaspis rohdei*	■		LC
Striated fire skink, *Lepidothyris striatus*	■		NE
Short-headed skink, *Panaspis breviceps*	■		NE
Brown-flanked skink, *Trachylepis affinis*	■	■	NE
White-lipped skink, *Trachylepis albilabris*	■	■	LC
VARANIDAE			
Ornate monitor, *Varanus ornatus*	■	■	NE
AMPHISBAENIDAE			
Omboué worm lizard, *Cynisca bifrontalis*		■	LC
Haug's worm lizard, *Cynisca haugi*	■		NE
Gabon helmeted worm lizard, *Monopeltis galeata*	■	■	LC
TYPHLOPIDAE			
Afrotyphlopinae			
Marbled blind snake, *Afrotyphlops congestus*	■	■	NE
Gabon beaked blind snake, *Letheobia caeca*	■		LC
Asiatyphlopinae			
Flowerpot blind snake, *Indotyphlops braminus*		■	NE
BOIDAE			
Calabar boa, *Calabaria reinhardtii*	■		NE
PYTHONIDAE			
Seba's python, *Python sebae*	■	■	NE
COLUBRIDAE			
Colubrinae			
Confusing egg-eating snake, *Dasypeltis confusa*	■	■	NE
Blue-tailed tree snake, *Dipsadoboa duchesnii*	■		LC
Underwood's tree snake, *Dipsadoboa underwoodi*	■		LC
Green tree snake, *Dipsadoboa viridis*	■		NE
Weiler's tree snake, *Dipsadoboa weileri*	■		LC

	A + B	C + D	IUCN status
Black-lined emerald snake, *Hapsidophrys lineatus*	■		NE
Common emerald snake, *Hapsidophrys smaragdinus*	■	■	NE
Thirteen-scaled grass snake, *Philothamnus carinatus*	■		NE
Striped grass snake, *Philothamnus dorsalis*	■		LC
Green grass snake, *Philothamnus nitidus*	■		NE
Large-eyed green tree snake, *Rhamnophis aethiopissa*	■		NE
Blanding's tree snake, *Toxicodryas blandingii*	■		NE
Powdered tree snake, *Toxicodryas adamantea*	■	■	NE
Grayiinae			
Long-tailed water snake, *Grayia caesar*	■		LC
Ornate water snake, *Grayia ornata*	■	■	LC
ELAPIDAE			
Jameson's mamba, *Dendroaspis jamesoni jamesoni*	■		NE
Banded water cobra, *Naja annulata annulata*	■		LC
Forest cobra, *Naja melanoleuca*	■		LC
LAMPROPHIIDAE			
Aparallactinae			
Western forest centipede-eater, *Aparallactus modestus*	■	■	NE
Bocourt's snake-eater, *Polemon bocourti*	■		LC
Collared snake-eater, *Polemon collaris*	■		NE
Cameroon snake-eater, *Polemon notatus*		■	LC
Atractaspidinae			
Boulenger's burrowing asp, *Atractaspis boulengeri*	■		LC
Lamprophiinae			
Olive house snake, *Boaedon olivaceus*	■		NE
Dark-bellied house snake, *Boaedon virgatus*	■		NE
Banded dwarf snake, *Chamaelycus fasciatus*		■	NE
Brussaux's file snake, *Gonionotophis brussauxi*	■		NE
Brown file snake, *Hormonotus modestus*	■		NE
Guiral's file snake, *Limaformosa guirali*	■		NE
Spotted wolf snake, *Lycophidion laterale*	■		NE
Bioko file snake, *Mehelya poensis*		■	NE
Cap Lopez file snake, *Mehelya stenophthalmus*		■	NE
Psammophiinae			
Mozambique sand snake, *Psammophis mossambicus*		■	NE
NATRICIDAE			
Black water snake, *Hydraethiops melanogaster*	■	■	LC
Brown marsh snake, *Natriciteres fuliginoides*	■		LC
Olive marsh snake, *Natriciteres olivacea*	■		LC
VIPERIDAE			
Viperinae			
Green bush viper, *Atheris squamigera*	■	■	NE
Gabon viper, *Bitis gabonica*	■		NE
Rhinoceros viper, *Bitis nasicornis*	■		NE
Lichtenstein's night adder, *Causus lichtensteini*		■	NE

9.1. Common tern in flight over the lake delta of the Rivière des Pélicans in July 2014 (Photo: Jean P. Vande weghe).

9.2 and 9.3. The "Ogooué" tern in flight over the Ozouri sandbar. It has a very dark underside that rises very high, and its beak is all black (Photo: Jean P. Vande weghe).

9.4. A gathering of common terns on the Ozouri sandbar. These birds are all in "winter" plumage. In the foreground and a little to the left, there is a much smaller Damara tern. In the foreground on the right, there is an "Ogooué" tern in breeding plumage, identifiable by its smaller size and very dark underside (Photo: Jean P. Vande weghe).

9.5. A couple of "Ogooué" terns on a sandbank at the Ozouri sandbar in August 2014. In the background and on the left, a black tern in winter plumage can be seen (Photo: Jean P. Vande weghe).

9.6. Three common terns in breeding plumage and in full display on the Ozouri sandbar in August 2014. The thicker and longer red beak is visible, as well as the lighter undersides and long black feathers of their heads, which continue down their necks. In the "Ogooué" tern, these feathers are much less developed (Photo: Jean P. Vande weghe).

9. Birds

Patrice Christy & Jean P. Vande weghe

Historical knowledge

The avifauna of the Lower Ogooué region is now well known, but what is most remarkable is that the region is among the first areas to be studied from an ornithological point of view in Gabon. The reason for this is simple: at a time when boats were a main mode of transport, its coasts were the first accessible areas, while the rivers were used to navigate further inland. From the founding of Libreville, the first bird collector in Gabon was Paul Belloni Du Chaillu, who first stayed there from 1848 to 1852. Du Chaillu had been trained in taxidermy by a French taxidermist shop–the Paris-based Verreaux brothers; who were well reputed at the time–before he was 20 years old. His biography also states that a quarrel arose between him and his sponsors due to the loss of a cargo shipment of taxidermied animals (one of his first Gabonese collections) when his boat sank, at which point his sponsors proved to be unreasonable in demanding that Du Chaillu reimburse the incurred costs.

His first trip to the United States dates back to 1852 and allowed him to contact the natural history museums of Boston, Philadelphia, and New York. On his return to Gabon, one of his expeditions led him–from April to July 1856–to the current Wonga-Wongué region and, from there, to Apomandé in the northern Ogooué Delta. The specimens he collected and skinned were loaded onto a boat, probably at Cape Lopez Bay, even if they were not from immediate nearby areas. They were shipped to the Museum of Natural History in Philadelphia, and studied by John Cassin, the museum's curator. In his description of this collection, published in 1859 in the Museum's Proceedings, Cassin mentions Cape Lopez Bay as a collection place or typical locality for new species. This collection included 49 species, four of which were new to science: the scaly francolin, *Pternistis squamatus*; the black guineafowl, *Agelastes niger*; the plumed guineafowl, *Guttera plumifera*; and the red-billed dwarf hornbill, *Lophocerus camurus*. From Du Chaillu's account, we know that these birds were not collected in Cape Lopez Bay, but east of Sangatanga, in the lands of the current Wonga-Wongué Reserve. This collection does not include any species of shorebird. In his following voyages, Du Chaillu did not travel any further upstream along the Ogooué than Lake Anengué.

The first travel accounts published by Du Chaillu had a strong impact in scientific circles at the time, mainly because of his descriptions of the behaviors of the gorilla, a species which had just been scientifically described, but especially for its adventures hunting large primates. The bird collectors who followed Du Chaillu had indeed read his adventures and landed in Gabon shortly after, in 1873. The objectives of the Marquis de Compiègne and Alfred Marche were twofold: to collect specimens for a merchant based in Paris, A. Bouvier, and partially finance their trip by supplying bird skins intended for the feather market, a flourishing trade linked to feminine fashion using brightly colored bird ornaments (or immaculate white plumage like that of the egrets) to adorn the hats of sophisticated women. Their first trip took them further upstream than Du Chaillu along the Ogooué since they reached Lambaréné and the region of the Great Lakes, specifically Lake Onangué. A study of this collection was published by Bouvier in 1875 under the title "Catalogue géographique des oiseaux recueillis par A. Marche et le marquis de Compiègne." It mentions 65 bird species from the "confluence of the Ogooué" and 21 species from "Lake Onangué." It was during this trip that a new bird species was discovered, though further upstream along the river, at the confluence of the Ogooué and the Lopé rivers, the vermiculated fishing owl, *Scotopelia bouvieri*.

The third historic collector, an English physician named Ansorge, made a much shorter trip to the Ogooué and appears to have collected only bird eggs for the British Museum in July and August 1907.

In 1949, all of the knowledge on the mammals and birds of Gabon was gathered into two books published by René Malbrant and Alain Maclatchy. Maclatchy resided in Gabon as a colonial administrator and was assigned to several posts but did not carry out any missions in Lambaréné. Their work therefore provides few personal observations within the Lower Ogooué region. A few years later, in 1951 and 1952, the Chicago Museum of Natural History sent one of its collectors, Harry Beatty, to Gabon and to what was then called the Middle Congo, now the Republic of Congo. He collected birds in Omboué, in the Nkomi Lagoon (or Fernan Vaz Lagoon), and briefly in Port-Gentil, but he did not visit the region of Lambaréné nor the Great Lakes region.

Recent data

Recent research on the Lower Ogooué avifauna reaches back about 40 years. Regular visits to the region of Lambaréné and the lake region of the lower delta were carried out starting in 1978 by Patrice Christy in order to improve knowledge of the avifauna of this vast region. In Port-Gentil, most of the observations were conducted by Peter Alexander-Marrack—a geologist working for Shell. During his stay from 1986 to 1990, he provided precious information on migratory birds and on the species of savannas and coastal swamps. In the first quarter of 1992, a study was conducted mainly focusing on resident and migratory waterbirds of the Gabonese coast, organized by the Netherlands WIWO, a group dedicated to research on shorebirds and waterbirds, which stayed at Port-Gentil during February. The results of the study were published under the title: Coastal waterbirds in Gabon, winter 1992 (Schepers & Marteijn, 1993).

Within the limits of the current Lower Ogooué Ramsar site, environmental impact studies for oil companies, mining companies, and agro-industrial companies then helped improve knowledge of the species and their distribution. These include the Ezanga and Onal sites (oil sites south of Lambaréné); the Bifoun and Ebel-Abanga zones, including Nguène Lakes (which justifies their inclusion in the Ramsar zone) in the far northeast; the Makouké and Lake Zilé areas, east of Lambaréné (for projects related to agro-industrial development); and the Mabounié sector, a mining site in the extreme southeast of the Ramsar site.

In addition to these environmental impact studies, there are ornithological observations and inventories more directly linked to conservation. These include the numerous observations made at various seasons by Jean P. Vande weghe from 1999 to 2017 throughout the complex, those made in the Lake Gomé region in December 2011 by Patrice Christy, or the census of nesting birds of the Ogooué sandbanks during the 2014 dry season, by Patrice Christy and Jean P. Vande weghe, as part of a multidisciplinary study organized by WWF.

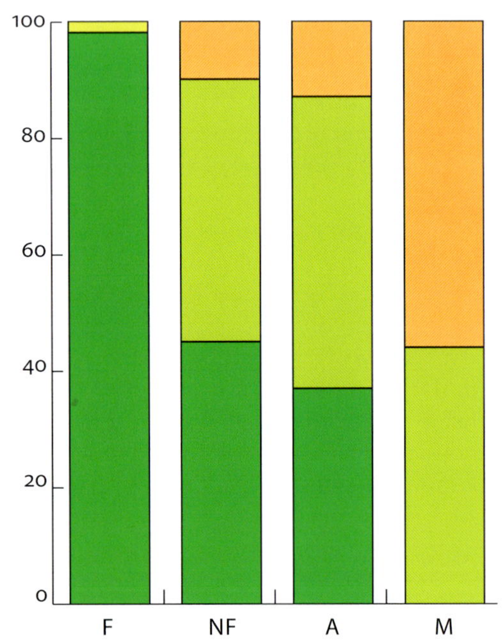

9.7. Composition of the bird fauna of the Lower Ogooué wetland complex, by ecological category:
(F) Forest birds
(NF) Nonforest birds from dryland environments
(A) Aquatic birds
(M) Marine birds
Resident species
Regular migratory species
Rare or occasional species

The avifauna

Given the diversity of environments and the extent of the Lower Ogooué deltaic complex, it is obvious that the above listed ornithological observations are incomplete or fragmentary, both from a spatial and a temporal point of view. What can be observed of the avifauna when traveling on the region's rivers and lakes in a dugout canoe remains very incomplete and biased, and available observations are too spread out in time to allow a detailed understanding of the seasonal phenomena or interannual variations. Large parts of the ecosystem are also completely invisible, particularly the large marshes of the Mbiné river and the upper deltaic plain, as well as most of the lower deltaic plain. Nevertheless, available observations allow for the identification of this vast site's avifauna's essential characteristics.

According to the latest censuses, the avifauna of the Lower Ogooué deltaic complex counts 465 species (Appendix 2), or two-thirds of the known species in Gabon. This makes it the richest site in Gabon, though this is due both to its large size and the great diversity of habitats it encompasses.

Overall, this avifauna counts 179 (38%) forest species, 137 (29%) nonforest species, 39 (8.3%) forest and nonforest species, 98 (21%) aquatic species, and 12 (2.6%) marine species (Figure 9.7).

Forest species

The forest avifauna is for 98% sedentary (Figure 9.8), and of the three migratory species, two are Afrotropical and one is of Palearctic origin: the European honey buzzard, *Pernis apivorus*.

Throughout most of the deltaic complex, this avifauna is typical of the northern part of the Gabon coastal basin. The genus *Stiphrornis* (Muscicapidae) for example includes the Gabon forest robin, *S. gabonensis*, in this northern part, while further south, in the region of the Ndogo

Lagoon, there is the olive-backed forest robin, *S. pyrrholaemus*. However, the border between these two species is not well known, and it may well be formed by the wetland complex (Vande weghe *et al.*, 2016). The Omboué region is also the northern distribution limit of the red-capped robin-chat, *Cossypha natalensis*, within Gabon.

Throughout most of the Lower Ogooué complex, the forest avifauna is therefore less diverse than that of the plateaus and reliefs of the interior of the country. It is only in the area of Mabounié and to the far east of the current Ramsar site that the avifauna is more typical of the forests of the interior of the country.

Overall, the forest avifauna of the deltaic complex is therefore not paramount in terms of conservation and will not be described any further in this book

Nonforest dryland species

These are mainly opportunistic species that colonize anthropized spaces among others. True savanna species are few in number, probably since savannas cover only small areas and are almost always devoid of any shrubby vegetation.

Resident birds represent 45% of the species, as do regular migrants. The remaining 10% are occasional migrants, which are rare or lost species, observed mainly on Mandji Island (Figure 9.8).

Among the 79 nonresident species, regular or occasional, 58% are of Afrotropical origin, 38% of Palearctic origin, and 2.5% may be of either Afrotropical or Palearctic origin. An occasional species is of Nearctic origin: the upland sandpiper, *Bartramia longicauda*, observed on Mandji Island.

Similar to the rest of the country, and to sub-Saharan Africa in general, there are few birds which are indifferent to forest and nonforest environments. They are all widely distributed birds, mostly sedentary. The three migratory species are of Afrotropical origin.

Aquatic species

Aquatic birds include species that depend on water—whether it is fresh, brackish, or salty—but which also need dry land, not just for nesting, but also for resting or feeding. The majority of waterfowl are also distinguished by the fact that they often make use of very different environments depending on the season and time of the day. Coastal birds, although they live in marine environments, will hereby be considered aquatic species.

Overall, these include 37% species which are more or less sedentary, some of which nevertheless travel locally depending on the state of their prefered environments, 50% migratory species, and 13% very infrequent or occasional species. This attests to the need for a specific approach to conservation (Figure 9.8).

Among the 63 nonsedentary species, 38% are of Afrotropical origin, 53% of Palearctic origin, 7.3% of either Afrotropical or Palearctic origin, and 2.9% of Nearctic origin (Figure 9.8). These are the buff-breasted sandpiper, *Calidris subruficollis,* and the Wilson's phalarope, *Phalaropus tricolor* (page 290).

Among these species, there are two "forms" of common tern. One, *Sterna hirundo,* is mainly of Palearctic origin. The other, which nests in the region, is very different and its taxonomic status is uncertain: perhaps it is a tropical form of the common tern, perhaps it is a separate species, already described in the 19th century but which has since been forgotten (page 293).

Marine species

These are all species that live on the open sea, outside of nesting, and are therefore completely independent from dry land, which some species do not even approach.

These are all very mobile species, 44% of which are regular migrants and 56% are rare or occasional species (Figure 9.8). However, given the limited number of observations made at sea, it is likely that some of them are more frequent than the few current observations suggest.

Their origins are very diverse: 29% are of Palearctic origin, 7% are of Nearctic origin, and 64% come from the South Atlantic or islands in the Gulf of Guinea (Figure 9.8).

In reality, these birds are not part of the avifauna of the Lower Ogooué deltaic complex, but they can occasionally be observed along the beaches or at sea, not far from the coast. As for mammals, this is also the case for certain dolphins and the humpback whale, *Megaptera novaeangliae*.

The distribution of species within the delta

Finally, it is important to note that of the 465 species observed in the Lower Ogooué region, 110 (24%) are not present in the Ramsar site as it is currently defined. These 110 species include 0.8% forest species, 47% nonforest species, 1.7% forest and nonforest species, 37% aquatic species, and 14% marine species that have been seen only in the area of the Mandji and Ozouri islands, Cape Lopez Bay, or the lower deltaic plain.

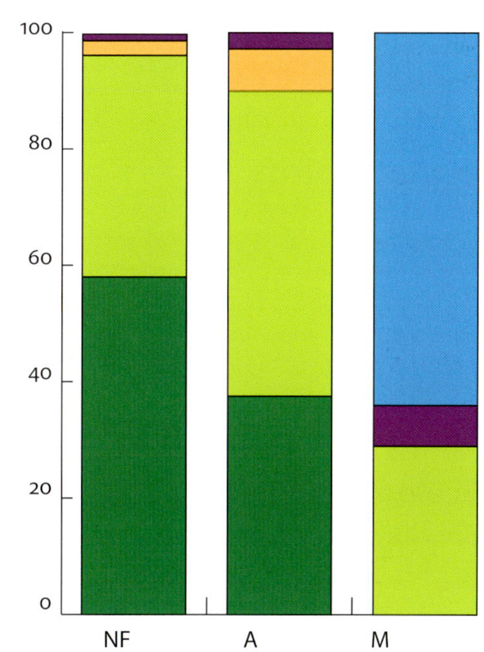

9.8. Composition of the bird fauna of the Lower Ogooué wetland complex by biogeographic category:
(NF) Nonforest birds in dryland environments
(A) Aquatic birds
(M) Marine birds
Afrotropical species
Palearctic species
Species that are either Afrotropical or Palearctic
Nearctic species
Species from the South Atlantic and the Gulf of Guinea

Table 9.1 . Estimation of the numbers of breeding birds on the sandbanks in August 2014. Nonbreeding birds have not been included in the counts. A = Lower Ogooué valley; B = upper deltaic plain; C = lower deltaic plain; D = littoral sands, including the sandbars; E = lagoon region

Species	A	B	C	D	E	Total counted (*estimated*)
White-fronted plover	6	10	0	180-280	0	*200-300*
White-headed lapwing	10	14	0	0	0	24 (*40-60*)
Gray pratincole	104	215	0	0	0	320 (*400-600*)
Skimmer	68	411	0	209	0	688 (*780-840*)
Caspian tern	0	0	0	460	0	460 (*500-550*)
"Common" tern	50	40	0	220	0	310 (*400-500*)
Little tern	22	43	4	8	0	73 (*110-150*)
Gray-rumped swallow	460	985	0	0	0	1,445 (*2,500-3,500*)
Number of species	7	7	1	5	0	8
Total numer of individuals without the swallows	260	733	4	1,126	0	2,119

Their absence from the Ramsar site is partly due to the fact that many sightings have occurred in the Port-Gentil region, and some nonforest or aquatic species may be present further east. For most species, however, this situation is linked to their ecological preferences. These are in fact mainly birds that prefer beaches, mudflats and mangroves, environments that are not found further east, or wide open environments, as are found on the Mandji and Ozouri islands or in the Animba and Inguéssi plains.

Aquatic birds

Regardless of their taxonomic affiliation, aquatic birds can be divided into various groups according to the environments they frequent, their food, and their way of nesting. These various aspects of their biology often bring about anatomical and behavioral specializations.

Ecological groups

Sandbank species

This group of birds constitutes one of the main conservation interests of the Lower Ogooué complex and one of the most important elements to justify the extension of the current Ramsar site in order for it to reach the coast.

These birds mainly belong to the families Burhinidae, Charadriidae, Glareolidae, and Laridae, all from the order of Charadriiformes. They have in common that they nest directly on the ground atop exposed sandbanks, by depositing their eggs in depressions in the bare sand, sometimes surrounded by twigs, dry herbs, or shells. Other species belong to the Meropidae, Alcedinidae, and Hirundinidae families. These are swallows, kingfishers, and bee-eaters that nest in galleries dug in the sand.

The nesting of birds of this group is conditioned by the appearance of large sandbanks in the valley upstream of Lambaréné and in the upper deltaic plain caused by the recession of water in the dry season. These sandbanks are most often islands made of sand in the widest parts of the river, and more rarely sandbanks which form alongside the river banks and remain in contact with the mainland—these therefore remain accessible to predatory mammals, and birds tend to avoid them.

Upstream from Lambaréné, sandbanks are present up to the town of Ndjolé, in Lake Nguène on the Abanga, and on the lower course of the Ngounié River, probably as far as Fougamou. These same birds, however, also nest on the sandbars that form at sea in front of the central delta in the sector of Ozouri or in front of the southern delta, namely the "Barre de l'Arabe" and the "Barre des Portugais." They are therefore absent only from the lower deltaic plain, where there are practically no sandbanks which remain dry for long periods.

Since the river's large sandbanks can move from year to year, the nesting colonies also move according to available sites. These birds are therefore very opportunistic and adapted to exploit changing sites.

Seasonality

All these birds settle from the end of May or the beginning of June, as the sandbanks progressively fall dry. The first layings take place in the second half of June, but most take place in July or even early August. These birds generally nest earlier upstream in the Lambaréné region than they do downstream in the Mandji Island region.

When the water levels rise, these birds gradually leave the region. Only the "Ogooué" tern, *Sterna "senegalensis"*—generally associated with the white-winged tern, *Chlidonias leucopterus* of Palearctic, origin—the white-headed lapwing, *Vanellus albiceps*, and the water thick-knee, *Burhinus vermiculatus*, remain in the region. All the other species leave for the coast, some perhaps even beyond the borders of Gabon.

The brevity of the short dry season in January or February does not allow for the nesting of these species, which therefore takes place only during the long dry season.

Distribution

Most of the breeding populations of white-fronted plover, *Ochthodromus marginatus*, and Caspian tern, *Hydroprogne caspia*, are found on sandbars. The white-headed lapwing, *Vanellus albiceps*, and the white-fronted tern, *Sternula albifrons*, are mostly found in the eastern upper deltaic plain and the Lower Ogooué valley. The gray pratincole, *Glareola cinerea*, is distributed throughout the upper deltaic plain and along the Ogooué upstream of Lambaréné. The local form of common tern nests on every sandbanks, but is more abundant upstream of Lambaréné and on sandbars. As for the African skimmer, *Rynchops flavirostris*, 47% of its population nests on the banks of the upper deltaic plain and 30% on sandbars. The gray-rumped swallow, *Pseudhirundo griseopyga*, nests in very large numbers on all the sand- banks of the deltaic complex except for the sandbars.

Tendencies

Partial counts suggest that populations of Caspian terns are roughly stable, but that the numbers of "Ogooué" terns and little terns have declined in the Lambaréné region over the past 10–15 years. Gray-rumped swallow numbers also appear to have declined significantly since 1999.

Threats

(1) These birds are particularly vulnerable and suffer greatly from human disturbance. In many countries, their nesting sites are set up as reserves to allow them to breed. Despite this, many species are declining steadily. In Gabon, they are disturbed by children, who are on school leave precisely during their nesting periods, and by fishermen and women and hunters who roam the sandbanks, with or without dogs. Thus, on numerous occasions, we have seen groups of children, armed with small shovels and buckets, harvesting eggs and chicks atop the sandbanks of the Lambaréné region. The fact that certain species are strictly protected by law does not seem to concern them.

(2) These birds are also very sensitive to pollution.

(3) They are very sensitive to anthropogenic changes in their environment resulting from development projects, which influence the variations in the river's water level and can make sandbanks disappear or considerably reduce water productivity.

The fish-eating birds of open waters

This group includes four species of fish-eating birds, which either catch fish by swimming or by diving down from the surface of the water. Two of them are pelican species (Pelecanidae), one a cormorant (Phalacrocoracidae), and the last one a drafter (Anhingidae). These birds have a fairly wide distribution in Africa. The pink-backed pelican, *Pelecanus rufescens*, the reed cormorant, *Microcarbo africanus*, and the African darter, *Anhinga rufa*, are resident species that breed in the Lower Ogooué region. The great white pelican, *Pelecanus onocrotalus*, is an Afrotropical migratory species which can be fairly regularly observed in the region.

Seasonality

Apart from the great white pelican, which is a large migratory bird which seeks temporary environments, the three other species are sedentary. They nest so as to raise their young in the dry season when the water is low and the fish are more easily accessible. During the rest of the year, they scatter throughout the Lower Ogooué complex. Like all pelicans, pink-backed pelicans can fish up to 100 or 150 km from their settlements.

Distribution

Although some colonies have probably gone unobserved, the highest concentrations of nesters in this ecological group have settled in the upper deltaic plain, and especially in the lake deltas (Figure 9.10).

Tendencies

Pink-backed pelican numbers may have decreased, but reed cormorant populations appear to have grown in the past 10–15 years.

Threats

(1) Cormorants and darters do not appear to be heavily targeted by hunters. Pink-backed pelicans, on the other hand, are regularly slaughtered, and the individuals observed in the northern delta and the Bay of Nazaré were all very shy. The colonies of pink-backed pelicans settled on large kapok trees are also very visible, which makes them especially susceptible to rifle hunting, even though the species is protected. As pelicans are long-lived birds with slow and late reproduction, their populations are very sensitive to an increase in mortality caused by hunting.

(2) All species in this category are very sensitive to various types of pollution. As they are predators at the top of the food chain, they accumulate toxic substances ingested through their prey.

Flamingos

These birds belong to a very particular ecological group which specializes in the consumption of microscopic animals that live in brackish or shallow coastal saltwater environments, as well as inland salt lakes. They constitute the Phoenicopteridae family, which counts two species in Africa. Both are found in Gabon, but only the lesser flamingo, *Phoeniconaias minor,* has been seen in the Ogooué Delta. The significantly larger greater flamingo, *Phoenicopterus roseus,* has only been observed in Akanda National Park, for the first time in 2013.

These birds occasionally form inrushes, sometimes massive ones, into unexpected places, possibly due to disturbances in the areas they normally inhabit. The lesser flamingo has been observed in very small numbers in the Port-Gentil region in 1981, 1987, 1988, and 2013 (page 280).

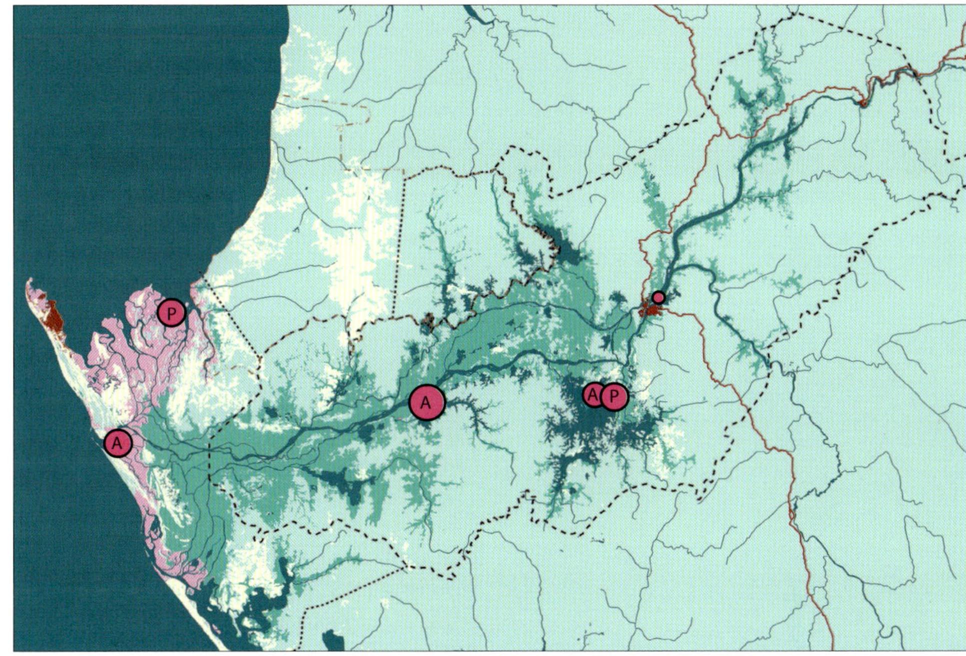

9.9. Distribution of open-water fish-eating bird colonies: anhinga (A), reed cormorant and pink-backed pelican (P). Cormorants are still rare.

9.10. Distribution of the main colonies of large wading birds (herons, egrets, etc.).

9.11. Main feeding areas for large wading birds during nesting periods.

Large wading birds

The 30 species in this category (appendix 2) belong to the families of Ardeidae (18), Threskiornithidae (5), Ciconiidae (6), and Scopidae (1). They are large or medium-sized birds that feed on fish, amphibians, and invertebrates, which they capture in shallow waters near banks or in open marshes.

Some species hunt under dense vegetation cover. Small Ardeidae, such as the common little bittern, *Ixobrychus minutus*, and the striated heron, *Butorides striatus*, hunt in reed beds and papyrus beds. The white-crested tiger heron, *Tigriornis leucolopha*, hunts in swamp forests or in flooded forests. Some species also hunt in open dry areas. These include Abdim's stork, *Ciconia abdimii*, the white stork, *C. ciconia*, the western cattle egret, *Bubulcus ibis*, and the black-headed heron, *Ardea melanocephala*. But these species do not nest in the Lower Ogooué deltaic complex, in which they are only visitors. The western cattle egret can be present in large numbers. The species that nest in the deltaic complex are all linked to wetlands. They often nest in mixed colonies on trees or shrubs by the water's edge. However, the white-crested tiger heron and the little bittern are solitary species.

Seasonality

Species that nest locally breed in the dry season when food is at its highest. The rearing of their young is done from July to September.

Distribution

These birds are heavily dependent on environmental conditions and move continuously according to water levels. When waters are high, they are especially fond of the flooded marshes found in the center of the upper deltaic plain, upstream from certain lakes (e.g., Lakes Gomé and Azingo) and in the lake deltas of the Rivière des Pélicans and the Ezanga River (Figure 9.10). When water levels go down, they follow the receding waters, and in the middle of the dry season, they are especially abundant in the marshes of the Mbiné River, around Lake Zilé, in the lake deltas, and all along the main Ogooué river course. Outside of their nesting periods, these birds then disperse across a large area of the Lower Ogooué complex, and probably even far beyond.

Tendencies

There is no indication of overall trends for these species within the region, but the colonies on the islands upstream of Lambaréné appear to have declined significantly in the past 15 years. The population decrease of great egrets, purple herons, *Ardea purpurea*, and black-crowned night herons, *Nycticorax nycticorax*, is evident and ranges from 60% to 70%, though this trend may not be representative of the entire complex.

Threats

(1) These birds do not seem to suffer too much from hunting, at least not in the Lambaréné region, where the colonies are easily accessible and therefore very vulnerable. According to local customs, these colonies may be more or less sacred. In the 19th century, there was a large colony on a sacred island on Lake Onangué, but it

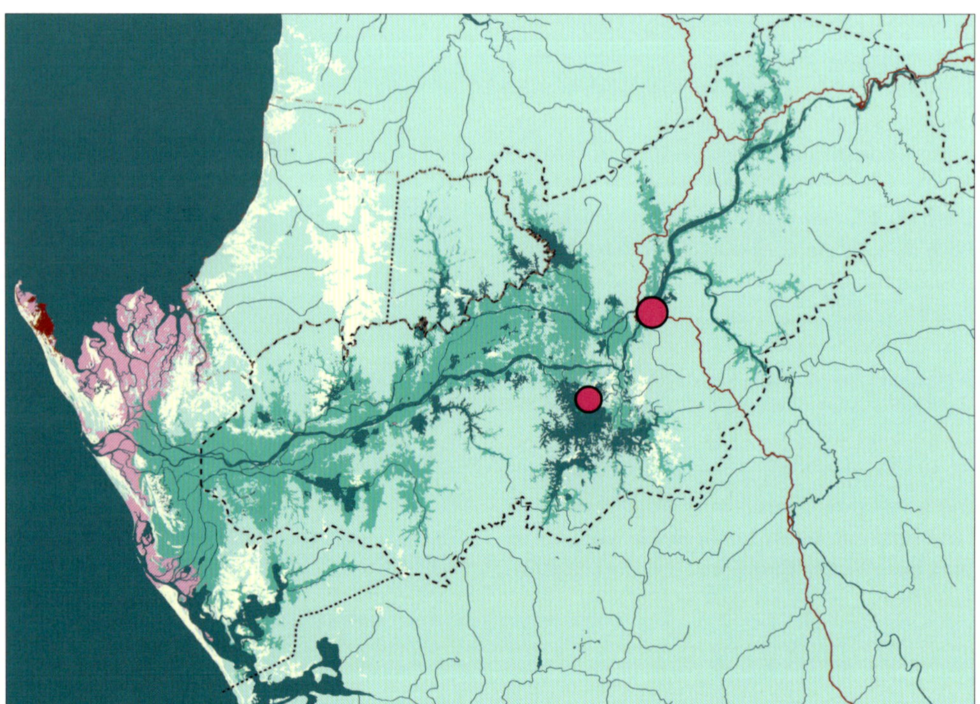

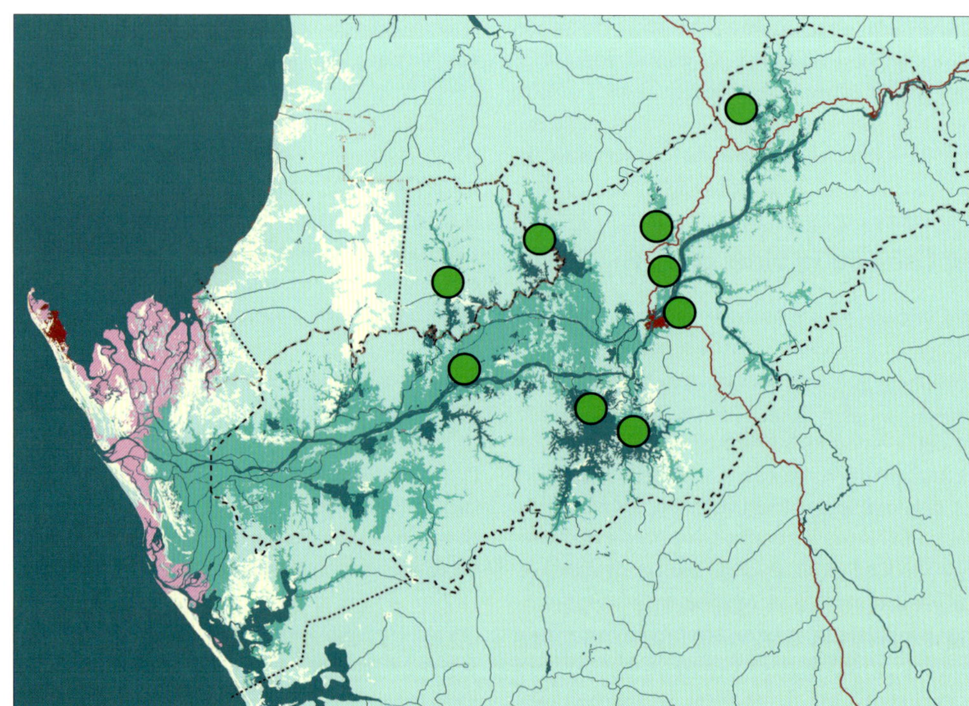

was massacred by French explorers Marche and Compiègne, who sold egret feathers in order to reimburse part of their travel expenses. In the Port-Gentil region, the Nazaré Bay and the surroundings of Lake Gomé, hunting may be more problematic.

(2) Like all fish-eating species, these birds are very sensitive to pollution since they accumulate toxic substances.

(3) Like most species restricted to pioneer, temporary, or seasonal environments (reed beds, papyrus beds, beaches, mudflats, sandbanks), they are very dependent on the proper management of these environments. They may be heavily impacted by hydraulic installations, dams, and draining resulting from anthropic developments—even those with "sustainable" certification.

Ducks and geese

Forested Africa has a low diversity of species and small numbers of Anatidae, unlike the tropical regions or regions with savannas, lakes, and marshes, both north and south of the equator. Ten species have been observed in the Lower Ogooué region, but several of them are only rare or accidental migrants.

Regularly observed species include the African pygmy goose, *Nettapus auritus*, which is fairly common on small, black-water ponds with rich vegetation, and the white-faced whistling duck, *Dendrocygna viduata*, present in small numbers throughout the Lower Ogooué region, including in the swamps found in the urban area extending north of Port-Gentil. The white-faced whistling duck travels a lot. The largest observed groups barely reached 40, and this species is much more abundant in the marshes of the Ngové plain in Loango National Park. The species most adapted to forest environments, Hartlaub's duck, *Pteronetta hartlaubi*, is also present in the region, though not as much as it is on the rivers of the Nkomi Lagoon (e.g., Mpivié) or in Loango National Park.

Three ducks are accidental or rare Palearctic migrants (page 280).

Palearctic waders

In the Lower Ogooué region, this group of birds includes 24 species belonging to the families of Pluvialidae (2), Charadriidae (1), Haematopodidae (1), Recurvirostridae (1), and Scolopacidae (19). These birds feed mainly on intertidal mudflats and beaches, where they often live in multispecific associations. Some species also like open marshes with shallow water and the banks of rivers and lakes. They nest mainly in the North Palearctic and, outside of nesting periods, migrate toward wetlands, especially the coastal wetlands of Eurasia and Africa.

The Pacific golden plover, *Pluvialis fulva* (Pluvialidae), prefers short grasslands near water and sandy banks. The common snipe, *Gallinago gallinago* (Scolopacidae), preferes low grassy marshes and flooded plains. The great snipe, *Gallinago media*, likes wet or dry grassy plains.

The most regular species among the Pluvialidae family is the gray plover, *Pluvialis squatarola;* among the Charadriidae family, it is the common ringed plover, *Charadrius hiaticula*; among the Scolopacidae, they are the Eurasian whimbrel, *Numenius phaeopus*, the Eurasian curlew, *N. arquata*, the bar-tailed godwit, *Limosa lapponica*, the red knot, *Calidris canutus*, the curlew sandpiper, *C. ferrugineus*, the sanderling, *C. alba*, the little stint, *C. minuta*, the marsh sandpiper, *Tringa stagnatilis*, the common greenshank, *T. nebularia*, the common sandpiper, *Actitis hypoleucos*, and the ruddy turnstone, *Arenaria interpres*.

Distribution

Within the Ogooué Delta, most species mainly visit the mudflats of Nazaré Bay (Figure 9.12). At high tide, they take refuge at the tip of Apomandé, on small sandbanks north of Port-Gentil or on the Ozouri sandbar, 40 or 45 km from their feeding grounds. In January to February 1992, the wintering population of Cape Lopez Bay was estimated at 5,845 individuals, all species confused (Schepers & Marteijn, 1993). At that time, Nazaré Bay was therefore, for waders, the third-richest site in Gabon, after the mudflats of the current Akanda and Pongara national parks, in the northern part of the country. The biggest groups number 1,000 individuals or more, mainly of gray plovers, Eurasian whimbrels, and curlew sandpipers.

In March 2013, during the period of passage, about 160 birds, mainly red knots, were observed north of Port-Gentil, but in February 2014 only 15 gray plovers were observed at the tip of Apomandé. An incomplete counting from January 2017 might indicate a few hundred individuals in the bay of Nazaré (Patrice Christy, pers. obs.).

Other species tend to make use of freshwater environments. These include the wood sandpiper, *Tringa glareola*, and the common sandpiper that can be found throughout the complex. The ruff, *Calidris pugnax*, common snipe, and great snipe are restricted to Mandji and Ozouri islands. The common greenshank is also regularly

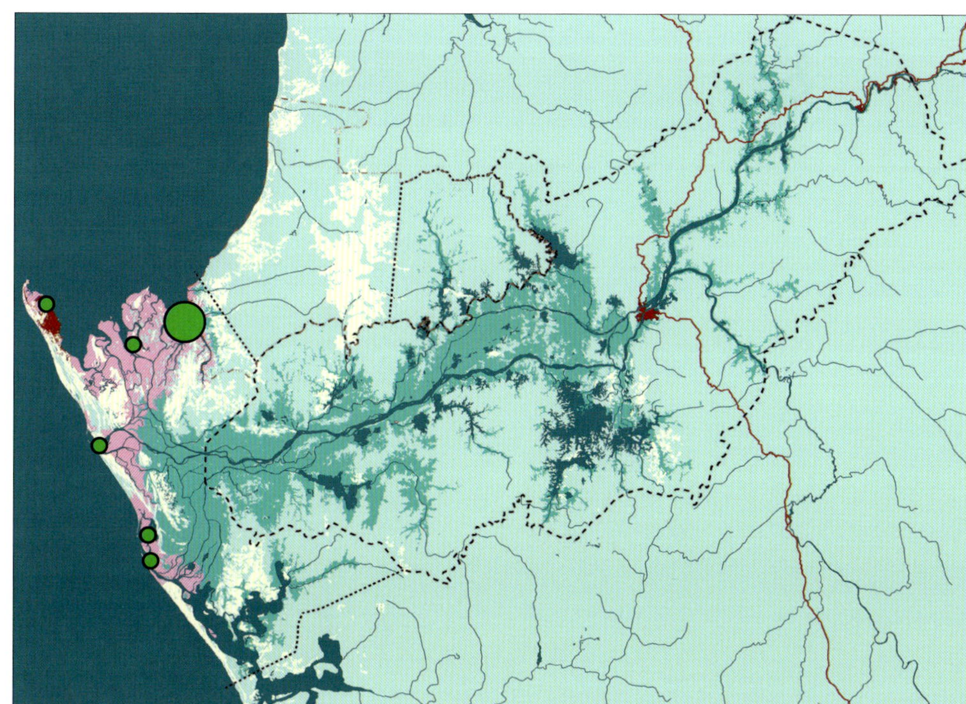

9.12. Areas where waders concentrate.

observed inland. The common sandpiper can be observed everywhere in the Lower Ogooué complex along the banks, and its population is certainly considerable; it can be estimated at 1,000-1,500 birds.

Apart from regular species, occasional migrants have also been observed, in particular American species: the buff-breasted sandpiper, *Calidris subruficollis,* and Wilson's phalarope, *Phalaropus tricolor*. Mandji Island, which juts out into the ocean, therefore seems well placed to attract transoceanic migrants.

Tendencies

The near absence of waders in the Bay of Nazaré as of February 2013 and the low numbers observed by Christy in January 2017 remain unexplained. It is possible that the tides were extremely high during these visits, and that these birds may have moved to another refuge. Though this does not explain the large numbers of terns and skimmers at the tip of Apomandé. On the other hand, it seems that the surface area of mudflats has greatly decreased. According to fishermen in the bay of Nazaré, these mudflats are now only exposed during white water tides. During dead water tides, they now remain submerged. This would mean that large numbers of waders can no longer stay in the Bay of Nazaré. As a result, large groups would only be seen during migration periods. This issue remains to be studied.

Threats

Overall, Palearctic waders are threatened by global warming, which is alarmingly destroying their nesting environments in the Far North, by crucial wetlands drying up along their migratory journey, and because they are hunted in many places they visit throughout the year.

In Gabon, these birds are threatened by:

(1) The rising waters of the Atlantic, linked to global warming, which will gradually reduce the area of mudflats and ultimately make them disappear.

(2) Their refuges getting frequently disturbed by hunters and fishermen at high tide, they force these birds to fly more frequently than necessary. They then consume too much energy and end up abandoning the site. This is why, in many countries, these high tide refuges—often very small areas—are fenced off and strictly monitored.

(3) Chemical pollution can also affect them by destroying mudflats.

Rails and moorhens

Rails and moorhens belong to the families Rallidae (8), Sarothruridae (1), and Heliornithidae (1). Most of these are Afrotropical species, some of which are sedentary and others are migratory species. All of them live more or less hidden within the swamp vegetation or riparian forests. These species are all difficult to spot, and some are more often heard than seen. One species inhabits the undergrowth of dryland forests, namely the Nkulengu rail, *Himantornis haematopus*. The African crake, *Crecopsis egregia*, inhabits open savannas.

Seasonality

Migratory species arrive in the region in October or November, when rains have caused the waters to rise, and leave the region at the start of the dry season. Their movements are, however, poorly known.

Distribution

All species are likely to be found throughout the Lower Ogooué complex, but open marsh species are more common in the inland deltas and swampy areas of the Mandji and Ozouri islands.

Tendencies

No tendency was detected.

Threats

These birds are practically never hunted, but depend on the preservation and soundness of their wetlands.

Raptors

Four species of raptors are present in the Lower Ogooué wetlands: one from the Pandionidae family and three from the Accipitridae family (Appendix 1).

Seasonality

The African fish eagle, *Icthyaetus vocifer*, and the palm-nut vulture, *Gypohierax angolensis*, are sedentary. The western osprey, *Pandion haliaetus,* is a Palearctic migrant, but some few individuals, probably immature, remain in the region year-round. The western marsh harrier, *Circus aeruginosus,* is also a Palearctic migrant.

Distribution

Two species are not exclusively tied to wetlands. The palm-nut vulture is found everywhere in Gabon, but reaches its highest densities in wetlands. It seems most abundant around the Nkomi Lagoon, in the southern delta, and around the Great Lakes. The African fish eagle is present in very low density throughout the complex.

Tendencies

No tendencies are noticeable.

Threats

The African fish eagle and osprey are undoubtedly under pressure from hunting, which makes these birds extremely shy. This pressure must be relatively high; 11 ospreys banded in Sweden and Finland were found killed by hunters in Gabon between 1961 and 1991 (Schepers & Marteyn, 1993). Since the percentage of banded birds was very low, the number of birds slaughtered in the Lower Ogooué deltaic complex during this period would have been somewhere from 110 to 220. Compared to the populations which subsisted at that time in their European countries of origin, this figure is huge. Even today, ospreys summering in the Lower Ogooué region are still very shy and fly away at long distances. The African fish eagle populations of the Lower Ogooué can be estimated at a few dozen individuals, which represents a very low density, much lower than the potential density of this wetland complex. The extreme distrust of this species is also an indication that it is under significant hunting pressure.

Mangrove specialists

In the mangroves of Central Africa, there are two species of passerines restricted to this environment: the swamp boubou, *Laniarius bicolor* (Malaconotidae), and the yellow-breasted apalis, *Apalis flavida* (Cisticolidae). A third species, the mangrove sunbird, *Anthreptes gabonicus*, also frequents mangroves but can be observed in neighboring environments. They are sedentary birds. The boubou and the apalis are present in the delta's mangroves, and the boubou even likes niaouli forests, *Melaleuca quinquenervia*. No tendencies are noticeable, but these birds depend on the preservation of mangroves.

Kingfishers

Kingfishers belong to the Alcedinidae family. They feed on small fish, which they catch on the surface of the water, and nest on vertical riverbanks in the dry season (June to September). The pied kingfisher, *Ceryle rudis*, occupies a particular niche: like terns, it hunts over large open water areas (rivers, lakes) away from the banks and locates its prey by hovering. They are mostly sedentary birds that are found throughout the Lower Ogooué complex. No tendencies are noticeable, but these species, which feed mainly on small fish, are sensitive to pesticides and other chemical pollutants.

Passerines of the open swamps

Tallgrass beds (*Echinochloa*, *Vossia*), Polygonaceae beds and papyrus are inhabited by species of insectivorous passerines from the Acrocephalidae (2) and Cisticolidae (1) families.

Seasonality

The Nile cisticola, *Cisticola marginatus* (Cisticolidae), and the greater swamp warbler, *Acrocephalus rufescens* (Acrocephalidae), are sedentary. The sedge warbler, *Calamodus schoenobaenus*, and the great reed warbler, *Acrocephalus arundinaceus* (Acrocephalidae), are migrants of Palearctic origin.

Distribution

These birds are probably found throughout the complex based on the presence of their prefered habitats (page 296).

Tendencies

No tendencies are noticeable.

Threats

Local species depend on the preservation of periodically flooded grasslands. Palearctic species are especially threatened by the disappearance of their habitats in nesting regions and in transit regions.

The delta's avifauna from a spatial perspective

The relative importance of areas

The littoral and lower deltaic plains of the Ogooué Delta do not change much according to the seasons but instead according to the tides. Their avifauna therefore varies mainly according to migrations (intra-African or Palearctic-African) which condition the presence or absence of certain long-distance migrants. This component of the region's avifauna is very much influenced by phenomena taking place outside of Gabon.

Environments of the upper alluvial plain and the Lower Ogooué valley vary mainly according to the seasons, which determine the water level and therefore the appearance of sandbanks or the flooding of marshes.

Some of the species which appear or disappear seasonally are also Palearctic-African migrants, but the Charadriiformes which come to nest in large numbers on sandbanks or the herons, pelicans, cormorants, and anhingas which nest on the river's tree-covered islets are all Afrotropical species.

These birds are grouped at both ends of the ecosystem: the delta's sandbars to the west; the eastern part of the alluvial plain, the lake deltas, and the river upstream from Lambaréné to the east (Figure 9.13). The presence and abundance of these species are therefore largely conditioned by the state of the delta's environments.

This bipolarity of the deltaic complex is partly linked to the distribution of their prefered environments: with the exception of the sandbars, the sandbanks which are exposed during the dry season, for example, are absent from the lower deltaic plain. This distribution does not explain, however, why the colonies of large wading birds, pelicans, cormorants, and anhinga more or less show this same bipolarity. Or why, among other things, the herons that nest near Lambaréné will feed upstream of the city, especially at Lake Zilé and in the marshes of the Mbiné.

In the absence of more precise data, this problem cannot be answered. It appears however that as in another ecosystem which has been previously studied, these birds' colonies settle in areas of higher productivity. In the eastern Ogooué Delta, this high productivity is probably linked to the fact that these parts of the ecosystem are where nutrients are injected from throughout the river basin. In the west, the high productivity is probably linked to the tides and the mixing of marine and continental waters.

Some of the large wading birds probably only make seasonal movements within the delta. During the dry season, when the water levels are low, they are concentrated along the rivers and lakeshores where they become very visible. On the other hand, in the rainy

Ramsar criteria of international importance

The Ramsar Convention on Wetlands is based on nine criteria to define areas of international importance. Two criteria in particular relate to aquatic birds:

Criterion 5: A wetland should be considered internationally important if it regularly has 20,000 or more waterbirds.

Criterion 6: A wetland should be considered internationally important if it regularly has 1% of the individuals in a population of one species or subspecies of waterbird.

However, given the size of the Ogooué Delta, it is difficult to estimate the total number of aquatic birds present, but it is certain that it very often significantly exceeds the threshold of 20,000 individuals.

Moreover, during the nesting season it shelters more than 1% of the overall nesting population of the following four species:

- The gray pratincole, *Glareola cinerea* (page 290);
- The African skimmer, *Rynchops flavirostris* (pages 290 and 291);
- the *mechowi* race of the white-fronted plover, *Ochthodromus marginatus* (page 287);
- the "Ogooué" tern, of uncertain taxonomic status (page 293), for which the Lower Ogooué region could shelter more than 90% of the world population.

During the boreal winter, the waters bordering the lower deltaic plain also house approximately 10% of the global population of the royal tern, *Thalasseus maximus* (page 292).

Palearctic-African shorebird migration

Of the 39 wader species, 23 are regular migrants, all of Palearctic origin. Among the 16 rare or occasional species, 12 are also of Palearctic origin, and four come from North America. Most Palearctic species come from the taiga and tundra that extend from Scandinavia to Eastern Siberia, where they nest during the brief boreal summer. These birds migrate during the remaining eight to nine months.
A few species mostly frequent freshwater wetlands (the wood sandpiper, the common sandpiper, the green sandpiper, ruffs).
The others are relatively well suited to the exploitation of intertidal mudflats, which extend across parts of the Earth's continents.
Depending on the migration routes they take, the waders of northern Eurasia can be divided among three main populations:

1. The western population migrates toward the coasts of the Baltic Sea, the North Sea, and the Atlantic Ocean, reaching Mauritania and Sierra Leone in the south.

2. The eastern population migrates to East Africa and Southern Africa; some birds cross the continent obliquely to reach Botswana and Namibia.

3. Between these two major populations, there is a third population that passes through eastern Europe, the Middle East, and the Eastern Mediterranean to reach western Central Africa and the coasts of the Gulf of Guinea; some birds of this population continue their journey to Namibia and South Africa.

The majority of Palearctic shorebirds arriving in Gabon therefore follow this transcontinental route. The presence of Asian species or forms confirms from where these birds originate, in particular the oriental form of the Eurasian curlew, *Numenius arquata orientalis* (Figure 9.34), the greater sand plover, *Eupoda leschenaultii*, the lesser sand plover, *Eupoda mongola*, the Pacific golden plover, *Pluvialis fulva*, the Caspian plover, *Eupoda asiatica*, and the Terek sandpiper, *Xenus cinereus*.

This long-distance migration allows these species to exploit temporary or temporarily accessible resources. This is made possible by the fact that these shorebirds can travel several hundreds or thousands of kilometers in one leg thanks to their particular metabolism, provided, however, that at each stage they can find a favorable and undisturbed site where they can rest and build the necessary reserves.
Today, the perpetuation of this Palearctic-African migration is seriously threatened by the draining, drying, and cultivation of intertidal environments, and also hunting—which makes the remaining mudflats unusable for birds, as well as the disappearance of remaining mudflats due to rising ocean levels. The middle route, the one that passes through Gabon, is particularly threatened.

season, especially in April and May when the water level is at its maximum, they disperse to the flooded areas, particularly lake deltas and plains located north of the lakes of the Wonga-Wongué Reserve (Azingo to Gomé) where they escape observation. These movements are not only linked to the presence or absence of nutrient resources, but also to the accessibility of these resources. Aquatic birds are therefore subject to the same constraints as fishermen in the region, whose activities only become truly profitable when the water levels are low.

The Ogooué Delta in the context of the coastal basin

When observed over a one-year period, the avifauna of the lower deltaic plain is fairly comparable to that of Akanda National Park or the Komo Estuary (Pongara National Park), whereas the mainland avifauna resembles that of the Ngové Marshes (Loango National Park), the Bongo River, especially Lake Kivoro, or the Lower Nyanga.

All of these sites are, however, largely complementary. Birds that nest on the sandbanks of the Lower Ogooué July to August thus stay in Akanda National Park, Loango National Park, or in the Nyanga Estuary during the rainy season.

There are, however, differences between various species. The "Ogooué" tern mostly remains in the Great Lakes region and the lake deltas. Skimmers and great egrets mainly move to Akanda National Park, but also to the sites along the country's southwestern coast. However, during the rainy season, white-fronted terns can also be observed in Cape Lopez Bay in Akanda National Park. The white-fronted plover is concentrated mainly in Akanda National Park. As for the gray pratincole, outside of its nesting period, it can be observed in the Nyanga Estuary and especially in Nyonié (between the Wonga-Wongué Reserve and the Pongara National Park), but it is only found there in large numbers in March, April, and May. As to where this species resides outside of this period, perhaps it partially migrates outside Gabon.

Therefore, despite its size and complexity, the Lower Ogooué ecosystem does not function as an independent entity in terms of aquatic avifauna. With the exception of the mangroves and mudflats of the lower deltaic plain, however, it does not harbor significant populations of Palearctic migrants, at least according to existing observations.

The importance of the Ogooué Delta for conservation

Despite the gaps in available observations, the Ogooué Delta is important for bird conservation, especially aquatic birds, both nationally and internationally. Indeed, it represents a very important nesting site for several species with varying threat levels, and it shelters significant "wintering" populations.

Breeding populations

Wide-ranging species

Most of the avifauna in the complex is composed of species with wide or very wide distributions, and for many species, the breeding populations present in the complex represent only a small part of their overall populations. However, for three species, the Lower Ogooué complex is home to populations of international importance (box, page 275).

The African skimmer. The Gabonese population of this species of Laridae, endemic to the African continent, is estimated at 1,800–2,100 birds (page 290). Therefore, it represents 15–27% of the species' West African population–estimated at 7,000–13,000 individuals, or 8–13% of the species' overall population–estimated at 15,000–25,000 individuals (IUCN, 2017). Furthermore, at least 95% of this Gabonese population nests in the Lower Ogooué complex. However, according to the IUCN (2017), this bird is considered "Near Threatened" (IUCN, 2017). Its numbers have declined significantly in southern Africa and it has disappeared from many rivers as a result of hydraulic development such as dams and pipelines that are removing the sandbanks which serve the species' nesting sites. In addition, this species has suffered a lot from disturbances and the looting of its colonies. With the huge increase in human populations, undisturbed sandbanks are becoming increasingly rare.

The gray pratincole. The Gabonese population of this species, which nests almost exclusively in the Lower Ogooué complex, includes around 1,000 birds (page 290), which represents 3-9% of the total population of the species' populations, which are endemic to West Africa and Central Africa. According to the IUCN (2017), this species does not seem to be threatened yet, although it has already disappeared from several rivers in West Africa following hydraulic development. It has almost disappeared from Ghana after a dam was built on the Volta River (IUCN, 2017).

The Caspian tern. The Gabonese population of this cosmopolitan species, which has colonies separated by great distances, is around 500-600 individuals (page 292). This represents around 4% of the population which visited West Africa in the 1990s and probably 8-10% of the breeding population in West Africa (Schepers & Marteijn, 1993). The colonies found on the sandbars of the Ogooué Delta are therefore clearly of international importance. They seem stable at the moment, but they depend on the configuration and, above all, on the tranquility of their prefered sandbanks.

Species with restricted distributions

Some species or taxa have a more heavily restricted distribution, and their Lower Ogooué populations represents a very large percentage of their overall population.

The "Ogooué" tern. Regardless of its taxonomic status (species or subspecies), the Gabonese population of this tern, estimated at around 400-500 individuals, is of great international importance. It probably represents 90-95% of its total population. This bird seems to have existed all around the Gulf of Guinea and all the way up to Senegal, but the Gabonese population is the last viable population. This justifies conservation measures for its nesting sites.

The white-fronted plover. This species has a wide distribution across the African coast, but it is the *mechowi* subspecies which is present in Gabon; a subspecies which inhabits West Central Africa, from Nigeria to the Congo Estuary. Its total population is estimated at 3,500-4,000 individuals (Delany *et al.*, 2009). The nesting population of the Ogooué Delta, estimated at 220-300 individuals (page 287), therefore represents 7-10% of the total population of this taxon. However, this bird nests mainly on the sandbars of the marine delta.

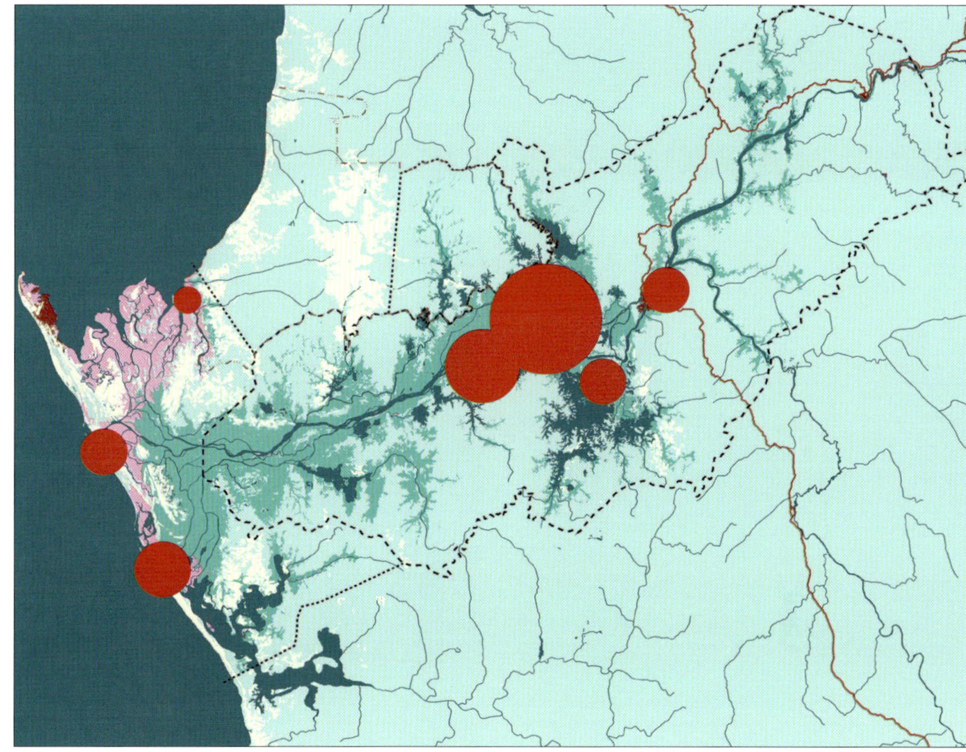

9.13. The parts of the delta most important for birds.

Wintering populations

Waders

Internationally, the largest population is—or was—the Palearctic waders of Nazaré Bay. The 5,845 birds observed by Schepers & Marteijn in 1992 is not huge, but regarding these birds, it is important to preserve as many potential stopping points as possible along their flyway (box, page 276).

The Atlantic route and the eastern route certainly face problems, but not as much as the eastern Mediterranean route (Delany *et al.*, 2009). The wetlands that form it have been drained one after the other, often as a result of ambitious development programs financed through international cooperation, and the rare remaining sites are most often rendered unusable by the assiduous presence of hunters. Furthermore, the mudflats of the west coast of Equatorial Africa are not very extensive and therefore each site is important.

This does not, however, explain why wader populations seem to have decreased significantly in the Bay of Nazaré. There are three possible reasons for this:

(1) refuges being disturbed by fishermen or hunters at high tide;

(2) a significant reduction in mudflats caused by rising waters in the Atlantic Ocean or a change in currents in the Bay of Nazaré;

(3) a sharp reduction in the overall population of these birds linked to climate change, the disappearance of important habitats along their migration route from the Black Sea and the eastern Mediterranean, or the hunting in these same regions, which remains very active.

There is currently no information in favor of one hypothesis over another, and only an in-depth study can answer this question. The mudflats of the Bay of Nazaré are however difficult to approach, and this study will not be easy.

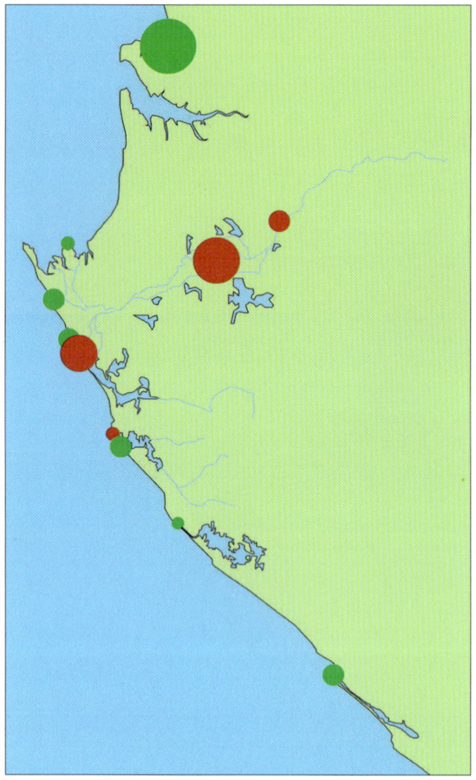

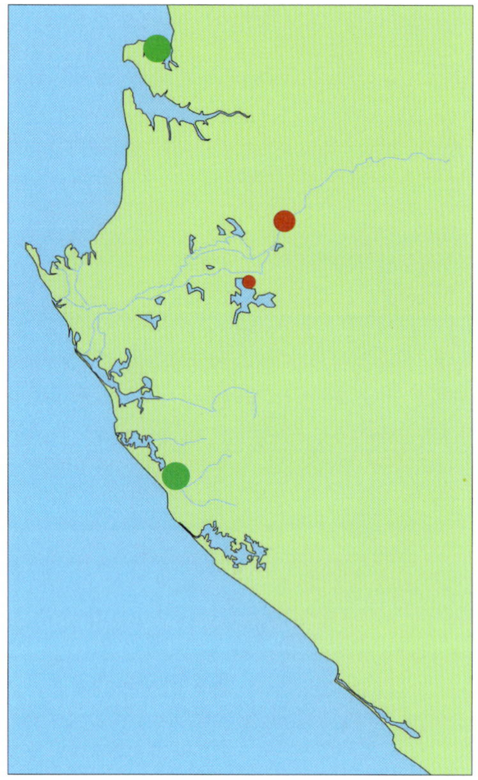

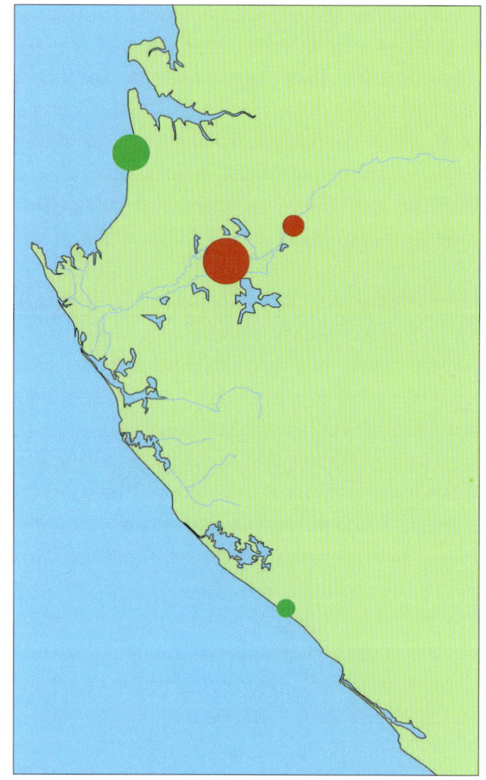

9.14. Schematic representation of the nesting areas of the skimmer (●) and its stopover areas outside the nesting period (●).

9.15. Schematic representation of the nesting areas of the great egret (●) and areas where they are most numerous outside the nesting period (●).

9.16. Schematic representation of the nesting zones of the gray glareole (●) and areas where they are most numerous outside the nesting period (●).

Raptors

The Ogooué deltaic complex must have been very important for the western osprey's wintering. The wintering population of this species also deserves some consideration. It represents only a fraction of the global population of this almost cosmopolitan species, but it certainly represents a significant portion of the Finnish and Swedish population of which it is part (Schepers & Marteijn, 1993). The birds of this species, slaughtered between 1961 and 1991, were killed at a time when the species was at risk of disappearing in Europe due to hunting and pesticides. Gabonese hunters therefore clearly contributed to the near disappearance of the species in Europe at that time.

Passerines

The few great reed warbler observations seem to indicate that large populations of this species could winter in the Ogooué Delta. However, it is currently in sharp decline in most of Western Europe.

The state of the avifauna

Despite its enormous extent, its very low human population, and the almost intact state of its landscapes, the Lower Ogooué wetland complex generally only supports a very low density of aquatic birds. Certain species, which could be much more abundant, are in fact quite scarce. This is the case of the African fish eagle.

Several species have significantly smaller populations today than they did 20 years ago. Heron colonies have significantly decreased, and some species appear to be on their way to extinction. This is the case of the saddle-billed stork, the largest species of stork, which may already be extinct in this region. The marabou stork, *Leptoptilus crumeniferus*, has already disappeared, and probably has not been seen in the region since the 1950s, regardless of the fact that it was reported by Du Chaillu in the second half of the 19th century.

It's very interesting to read Du Chaillu's stories. The landscapes do not seem to have changed much over 150 years, but the fauna seems to have very visibly regressed, as much with regard to large birds as hippopotamuses and crocodiles. Some of his descriptions may not have been correct, and some identifications may have been wrong, but there is no reason to doubt the overall picture painted by Du Chaillu. The catalog of birds brought back by Du Chaillu and drawn up in 1859 by J. Cassin also confirms the identifications and the presence of most of these species. He specifically indicates that Hartlaub's duck was "very common" at the time, whereas it is rather rare today.

Since the landscapes are almost intact, this sharp decrease in avian populations could be due to the impoverishment of the ecosystem as a result of its resources being overexploited (overfishing) or pollution. The decrease in fish populations is obvious—all the housewives of Lambaréné complain about it—and it is probable that there is some pollution in the waters, but this cannot explain that the scarcity mainly affects "large" species.

Hunting practiced all year round, without any constraints, must be the main reason for this. The extreme caution of pelicans, large raptors, large herons, and storks shows that the hunting pressure is of great importance, and on several occasions, direct or indirect indications of the importance of this activity have been observed.

From Du Chaillu's accounts, it appears that in the 19th century birds were not heavily hunted. Local populations lived mainly from agriculture, fishing, and mammal hunting, but rarely targeted large species.

Hunting for birds became only possible with the introduction of the rifle. It is certain that during the colonial era many Westerners hunted birds (ducks especially). Local populations did not commonly have access to rifles until after independence, starting in the 1960s, and especially with the "massive" importation of 12-caliber rifles from the early 1980s. The impact of the introduction of rifles has not been monitored, and is therefore not directly understood, but indirect observations provide a general idea of what it might be: in East Africa, where rifle hunting has never been authorized—unless duly authorized—the avifauna remains very abundant despite a high, or sometimes very high, human population density.

Large forest species are very vulnerable due to their low reproduction rates. In Gabon, the density of the crowned eagle, *Stephanoaetus coronatus*, is extremely low, which is also partly due to the decrease in primate populations, which constitute its main prey, and the large hornbills of the genera *Bycanistes* and *Ceratogymna* have practically disappeared around villages.

For waterbirds, the problem differs depending on the species. Large, long-lived, slow-breeding species are particularly vulnerable, like large forest birds. Smaller but faster-reproducing species may be less directly susceptible, but are very sensitive to disturbance from regular hunting. In many places of the world, sites in a perfect state of conservation have been seen completely abandoned by waterbirds.

Conservation strategies

Given the size of the deltaic complex and the great variability of the distribution of aquatic birds in both time and space, their conservation can only be based on four strategic axes:

(1) Preserving the integrity of environments throughout the delta. Each development action must therefore be subject to a comprehensive and non-compartmentalized impact study.

(2) Implementing a general policy to reduce hunting, starting with a review of existing laws and, above all, strengthening the application of these laws. Certain slow-breeding species must imperatively be fully protected. These especially include the saddle-billed stork, large raptors, and pelicans. All colonies must be fully protected.

(3) Identifying and creating "strict" reserves for critical nesting areas (sandbanks, colonies, etc.) or foraging areas (mudflats, some marshes, etc.).

(4) Implementing awareness programs for local populations addressed especially at fishermen and children of school age.

Bibliography

BOYD, J., 2017. *Taxonomy in Flux*. <jboyd.net/Taxo/index.html>

BROSSET, A. & C. ERARD, 1986. *Les oiseaux des régions forestières du Nord-est du Gabon*. SNPN, Paris.

CASSIN, J., 1859. Catalogue of Birds collected on the Rivers Camma and Ogobai, Western Africa, by Mr. P.B. Duchaillu in 1859, with Notes and Description of New Species. *Miscellanea ortnithologica, 6. Proceedings of the Academy of Natural Sciences of Philadelphia*.

DELANY, S., D. SCOTT, T. DODMAN & D. STROUT, 2009. *An Atlas of Wader Populations In Africa and Western Eurasia*. Wetl. International. Wageningen.

DU CHAILLU, P.B., 1863. *Voyages et aventures dans l'Afrique équatoriale*. Michel Lévy, Paris. 1996. Réédition Sépia-Centre culturel français, Libreville.

GARCÍA-R, J.C., G.C. GIBB & S.A. TREWICK, 2014. Deep global evolutionary radiation in birds: Diversification and trait evolution in the cosmopolitan bird family Rallidae. *Mol. Phylogenet. Evol.*, 81, 96-108.

GILL, F. & D. DONSKER (éds), 2017. *IOC World Bird List* (v4.2). doi:10.14344/IOC. ML.4.2.

IUCN, 2017. 2014. *IUCN Red List of Threatened Species*. <www.redlist.org>.

JARVIS, E.D., S. MIRARAB, A.J. ABERER, B. LI, P. HOUDE, C. LI, S.Y.W. HO, B.C. FAIRCLOTH, B. NABHOLZ, J.T. HOWARD *et al.*, 2014. Whole-genome analysies resolve early branches in the tree of life of moderrn birds. *Science*, 346: 1320. doi:10.1126/science.1253451.

KINGSLEY, M., 1897. *Travels in West Africa. Congo Français, Corisco and Cameroons*. London: MacMillan.

MENGUÉ MEDOU, C., F. ONDAMBA OMBANDA, C. NDJOKOUNDA, M.-D. MOUNGANGA, E. BAYANI & R. MIKALA, 2008. *Site Ramsar du bas Ogooué Fiche descriptive sur les zones humides Ramsar (FDR), version 2006-2008*. <http://www.ramsar.org/ris/key_ris_index.htm>.

SARGEANT, D.E., 1993. *A birders Guide to Gabon*, West Africa.

SCHEPERS, F.J. & E.C.L. MARTEIJN, 1993. *Coastal Waterbirds in Gabon, winter 1992*. Zeist, WIWO.

SWAINSON, W., 1837. The Natural History of the Birds of Western Africa. Vol. 2. (Livre numérique Google: https://books.google.ga).

VANDE WEGHE, J.P., 2003. Les milieux humides côtiers du Gabon: les rias du nord et le bas Ogooué. *Canopée*, 25: 3-8.

VANDE WEGHE, J.P., 2007. *Les parcs nationaux du Gabon: Loango, Mayumba et le bas Ogooué*. WCS, Libreville.

VANDE WEGHE, J.P., 2011. *Les parcs nationaux du Gabon: Akanda et Pongara*. Seconde édition. WCS et ANPN, Libreville.

VANDE WEGHE, J.P., 2012. *Wonga-Wongué*. ANPN, Libreville.

VANDE WEGHE, J.P., P. CHRISTY, M. DUCROCQ, M. LEE, G.R. VANDE WEGHE & O.S.G. PAUWELS, 2016. *Biodiversité des parcs nationaux et réserves du Gabon : 2. Espèces, écosystèmes et populations*. ANPN, Libreville.

WILKS, C., 1990. *La conservation des écosystèmes forestiers du Gabon*. Programme pour les forêts tropicales. UICN, Gland et Cambridge.

Appendix 1: Additional notes on aquatic species

Since a large number of records of waterbirds in the Ogooué Delta have never been published or were published in reports with very limited distribution, it was deemed important to assemble them here.

Anatidae

White-faced whistling duck, *Dendrocygna viduata*. This is an intra-African migratory species (Figure 9.17), seen in Port-Gentil and in the Nkomi Lagoon region (Sargeant, 1993; Schepers & Marteijn, 1993). It was observed in very small numbers (<10 individuals) in December, January, and August in the lake deltas, where it frequents open water holes in floating meadows. By day, it can also be found resting on beaches, sandbanks, and mudflats. This species feeds mainly at night and returns to its feeding places at dusk.

Knob-billed duck, *Sarkidiornis melanotos*. An intra-African migratory species, occasionally observed at Port-Gentil and in the Gamba region (Sargeant, 1993).

African pygmy goose, *Nettapus auritus*. This species is known from Port-Gentil (Sargeant, 1993; Schepers & Marteijn, 1993). A couple was seen on a small pond on the edge of the Ogooué River near Lambaréné on December 31, 2001 (Figure 9.19). This species is especially fond of black water, often in inaccessible sites, and may be less rare in the region than the few observations suggest. The ponds of the alluvial plain should especially suit this species, as well as certain lakes of the Mbiné River.

Hartlaub's duck, *Pteronetta hartlaubii*. This resident and nesting species, endemic to the Guineo-Congolean region, inhabits forest lakes, streams, and ponds (Figure 9.18). It is found throughout Gabon, but avoids salty and brackish environments, although it can occasionally be seen flying over the sea or estuaries. It exists throughout the coastal plain, but mostly frequents forest ponds and rivers, rather than large lakes and the river's open waters. Given its discreet behavior, this bird is perhaps more abundant than the few observations suggest. It is very common on the Mpivié River, a small tributary of the Nkomi Lagoon. Its nests can contain nearly 10 eggs. Usually living in pairs or families, this species can gather, in the dry season, on favorable terrain such as Lake Kivoro or the Ngové marshes in Loango National Park (outside of the Ogooué basin), where groups of several dozen have been observed. Another special aspect of this species is its familiarity when it is not hunted. It is well known to villagers because it comes and eats in river basins where women deposit cassava for retting. At a petroleum exploitation site, north of Port-Gentil, a couple of these ducks came to feed on a pond frequented by domestic ducks, a few meters from the camp buildings.

Northern pintail, *Anas acuta*. This Palearctic species was observed in Port-Gentil: two males and one female were seen on November 29, 1987, and two females on January 3, 1988.

Common pochard, *Aythya ferina*. This accidental Palearctic species was observed on December 8 and 9, 1987, in a marsh on Mandji Island. These are the only data for Gabon.

Garnaney, *Anas querquedula*. This Palearctic species was the least irregular when Peter Alexander-Marrack studied birds on Mandji Island, where it was observed in December and January in small groups of up to 15 individuals.

Phoenicopteridae

Lesser flamingo, *Phoeniconaias minor*. This species was observed in Port-Gentil by Alexander-Marrack in April and December 1987 and January 1988, with a maximum of 11 individuals. Between April and early July 2013, dozens of flamingos, mostly belonging to this species, were observed on a beach north of Port-Gentil (Du Plessis, pers. comm.) and on the mudflats of the Nazaré Bay. At the same time, more than 2,000 individuals were observed in Pongara and Akanda national parks.

[Greater flamingo, *Phoenicopterus roseus*. In April 2013, several dozens of these large flamingos arrived in Akanda National Park. This was the first time the species had been observed in Gabon. It is almost certain that these birds had passed along the Ogooué Delta.]

Pelecanidae

Great white pelican, *Pelecanus onocrotalus*. This Palearctic and Afrotropical species is a transient migrant in Gabon; it does not nest in the country but can be seen throughout the year. Its first sightings date only from 1984 within the current borders of Akanda National Park, where it can now be observed each year, both in the dry and rainy seasons. It feeds mainly in the vast flooded plains, when the water recedes, and moves continuously in search of favorable environments. In Gabon, small groups can stay for a few weeks or months, but large groups of more than 100 individuals do not linger. On August 8, 2014, a group of about 220 birds rested on a sandbank of the Ogooué River near Achouka, 80 km downstream from Lambaréné (Figure 9.21). The origin of this migratory population is unknown, but rare colonies are known to exist in Northern Cameroon and Chad. It is also possible that the birds seen in Gabon come from the southern part of the continent.

Pink-backed pelican, *Pelecanus rufescens*. This is a sedentary Afrotropical species that can fish in deep waters. Like the great white pelican, it can also appear in large numbers when the floodplains dry up, fishing often side by side. The species is present on the entire coast of the country, from the mangroves of Akanda and Pongara national parks in the north, to the border with the Republic of Congo in the south, and also in the mangroves of Mayumba National Park. It can be found several tens of kilometers from the coast in the southern part of the country, especially on Lake Kivoro in the dry season.

Pelicans, which are excellent long-distance flyers, often travel long distances from their colonies to fish at the most favorable sites. As a result, it is likely that individuals in the Lower Ogooué go fishing in the plains of Ngové in Loango National Park and in Lake Kivoro on the edge of the Moukala-Doudou National Park, when the waters recede in June or early July.

The species can be observed from the mangroves of Nazaré Bay and those of Cape Lopez to Lambaréné, but it does not seem to go further upstream along the Ogooué, since it prefers the calm waters of lakes and large rivers to the rapid waters of smaller rivers.

While great white pelicans nest on the ground in reed beds or on rocky peaks in the Sahelian zones of Cameroon and Chad, pink-backed pelicans characteristically form colonies atop large kapok trees (*Ceiba pentandra*) trees, which have been occupied year after year for decades (Figure 9.20). This behavior gave its name to the Agouma River of the Ogooué's upper deltaic plain, which is commonly called the Rivière des Pélicans (French for Pelican River). In the Lower Ogooué complex, a well-known colony is in fact found along the edge of this river, in the lake delta, but other colonies, which could not be located with precision, exist near Lake Gomé and in Nazaré Bay. Birds in breeding plumage

Page 280:
9.17. A pair of white-faced whistling ducks in a marsh on the Ngové Plain in Loango National Park (Photo: Jean P. Vande weghe).

9.18. A pair of Hartlaub's ducks perched on a fallen tree in the Mpivié River, a small tributary of the Nkomi Lagoon (Photo: Jean P. Vande weghe).

9.19. A male African pygmy goose on a water lily pond (Photo: Jean P. Vande weghe).

Page 281:
9.20. Pink-backed pelicans of the colony from the edge of the Rivière des Pélicans (Photo: Jean P. Vande weghe).

9.21. Great white pelicans resting on a sandbank in the Ogooué main channel in August 2014 (Photo: Jean P. Vande weghe).

9.22. A reed cormorant on the banks of the Rivière des Pélicans (Photo: Jean P. Vande weghe).

were observed there in April 2014 with nest-building materials. Given that the Rivière des Pélicans colony comprises approximately 100 individuals, a rough estimate would place the total population for the complex at around 300 birds. Elsewhere in the country, colonies are known in Akanda National Park (150–200 birds around 2010), Perroquet Island in the Komo Estuary (around 220 birds in 2012), and Loango National Park (around 100–150 birds in 2006). The total Gabonese population of this species could therefore be around 770–870 birds. According to Schepers & Marteijn (1993), the total population for the coastal region of Gabon was 1,115–1,435 in 1992. Inland (Rivière des Pélicans, Lake Gomé) it must have even been in the order of 1,500–1,600 birds. These figures could indicate a reduction in the number of individuals of this species. The Lower Ogooué basin is, however, very large, and there are certainly colonies located in uninhabited areas or sites that are difficult to access during the dry season.

Active colonies have been observed from May to November, which suggests that they have an extended nesting period, although the dry season is obviously the most favorable for breeding juvenile individuals.

Phalacrocoracidae

Reed cormorant, *Microcarbo africanus*. This is a resident and nesting species in Gabon. Sargeant (1993) mentioned only a few observations from Port-Gentil. However, the species can be observed throughout the year in small numbers in lake deltas and on lakes bordering the alluvial plain, in both the north and south. It frequents open waters, but most often remains near reed beds or shrub vegetation (Figure 9.22). A small nesting colony is present in the lake delta of Lake Onangué (10–30 pairs in July 1999 and 2001, January 2002, and August 2006; several dozen in August 2013 and August 2014). Some birds also nest with pink-backed pelicans and anhingas along the Rivière des Pélicans. However, the majority nest with purple herons (*Ardea purpurea*) in the large grasslands of the upper deltaic plain. In August 2013 and 2014, a few couples also nested alongside Ardeidae on the island near Lambaréné. The species appeared to be significantly more abundant at that time than in was the early 2000s. The total population is difficult to estimate, but it is probably in the order of several hundred individuals.

Cape cormorant, *Phalacrocorax capensis*. This cormorant nests on rocky islets of the South African coast and feeds in coastal waters. It is included in lists of threatened species. In Gabon, it is a rare or accidental migrant that has only been observed along the beaches of Cape Lopez Bay, near the petroleum terminal. This species was discovered in Gabon by Peter Alexander-Marrack on July 17, 1986, and later in the dry season of 1988 and 1989, between June 11 and September 4, with a maximum of seven individuals. There have been no sightings in recent years, but this is probably due more to an absence of observers.

[**White-breasted cormorant, *Phalacrocorax lucidus*.** This species was supposedly collected on the Ogooué–without further details–by Marche and Compiègne in the 19th century. This occurrence remains possible since isolated individuals of this species have been observed on rare occasions in the Libreville region.]

Anhingidae

African darter, *Anhinga rufa*. This is an Afrotropical species that resides and nests in Gabon and can be observed throughout the year throughout the country's wetlands. This bird fishes in the open waters of rivers and lakes, most often closer to the banks. It is present throughout the Lower Ogooué complex, except in saltwater areas. It nests most often in the crown of large kapok (*Ceiba pentandra*) trees, sometimes alongside pink-backed pelicans or herons. It nests on large kapok trees with pink-backed pelicans along the Rivière des Pélicans (40–50 pairs in July and August 1999, July 2001, January 2002, August 2006, and August 2014). It also nests with herons on the island of Lambaréné (5–10 pairs in August 2013). These birds are sedentary. The largest colony, comprising around 200 pairs, is located on the left bank of the main course of the Ogooué, 80 km downstream from Lambaréné, shortly after Achouka and before the river flows into Lake Anengué (Figure 9.23). Other smaller colonies occur upstream from Lambaréné, in the delta of Lake Evaro and not far from the mouth of the Ogooué in the central delta. The central delta colony is established on small trees emerging from the waters of the river, and the nests are only a few meters off the ground. It is certain that some colonies have escaped observation because they are in areas that are difficult to access. The total counted population in the Lower Ogooué region is 550 individuals, but it is estimated to reach at least 700–900 individuals. According to Schepers & Marteijn (1993), the population of the coastal zone of Gabon comprised 215–435 birds in early 1992, not counting those of the Lower Ogooué. This species also nests in other parts of the country, in particular in Bakoumba in Haut-Ogooué, along the Haut-Ivindo in Minkébé National Park, and in Ngové Lagoon in Loango National Park. Contrary to pelicans, anhingas are distributed throughout the country because of their fishing behavior, which makes them rather solitary birds, both on average-sized lakes and ponds and on fish ponds and narrow rivers. However, they seem to avoid mangrove rivers when the water is charged with too much alluvium.

Heliornithidae

African finfoot, *Podica senegalensis*. This is a resident and nesting Afrotropical species, observed regularly throughout the Lower Ogooué region, except in mangroves. It lives mainly in riparian forests found above the banks of rivers, where it gathers during the dry season. In the rainy season, it probably disperses throughout flooded forests. A barely hatched chick was seen in late December at Lake Evaro in 2001.

Sarothruridae

White-spotted flufftail, *Sarothrura pulchra*. This is a resident and nesting Afrotropical species, probably present throughout all the humid forests of Gabon. It is, however, very discreet and only reveals its position by its calls. This species has been heard several times in December and January in the forests bordering the Ogooué, both upstream and downstream from Lambaréné, and along certain channels of the upper deltaic plain.

Gray-throated rail, *Canirallus oculeus*. This resident and nesting Afrotropical species has been observed in the Port-Gentil region (Alexander-Marrack, pers. comm.). It inhabits dense, relatively marshy forests. This species was included in the family Rallidae, but the work of Garcia-R. *et al.* (2014) has shown that it is close to the genus *Sarothrura*.

9.23. *Part of the large colony of darters present along the Ogooué (Photo: Jean P. Vande weghe).*

Rallidae

African rail, *Rallus caerulescens*. This is an intra-African migrant, rare and very localized in West-Africa and Central-West Africa. It is occasionally observed in Gabon (Vande weghe, pers. obs.); on December 30, 2001, at dusk, an individual was heard inside a floating herbaceous mat of the delta of the Rivière des Pélicans. This may have been an occasional migratory traveller, but it is not impossible that the species is more frequent in the region; it is indeed very difficult to observe and probably remains silent during migration

9.24. *A male African finfoot at the edge of the Rivière des Pélicans, swimming under overhanging roots (Photo: Jean P. Vande weghe).*

Yellow-beaked rail, *Limnocorax flavirostris*. This Afrotropical species is resident and nesting. It is quite common in the wetlands of Mandji Island, mainly in small swamps that extend between the dune cords (Sargeant, 1993). It is also common in the interior deltas of the Ogooué (Evaro and Ezanga), where it has been observed during all months of the year.

Lesser moorhen, *Paragallinulla angulata*. This migratory Afrotropical species is fairly abundant in the partially floating herbaceous mats of the Evaro and Ezanga interior deltas when water levels are high, from November to April. Some individuals remain during the dry season, but most of the population appears to be migratory. A nest was found near Makokou on the Ivindo in February (Brosset & Erard, 1986).

Allen's gallinule, *Porphyrio alleni*. This is an Afrotropical, intra-African migratory species that nests in Gabon (Christy & Vande weghe, 1999). It has been observed in Port-Gentil from the end of October

9.25. *A hadada ibis in the Lake Onangué delta (Photo: Jean P. Vande weghe).*

9.26. *A woolly-necked stork in humid savanna (Photo: Jean P. Vande weghe).*

to June, where it nests in December and January (Alexander-Marrack), and was observed in December 2001 in a large floating herbaceous mat in Lake Evaro.

Scopidae

Hamerkop, *Scopus umbretta*. The species is resident, nesting, endemic to sub-Saharan Africa, and can be observed in small numbers throughout the Lower Ogooué wetlands. It generally hunts in shallow water near brush or tree cover, as well as in flooded grasslands. Its population size is difficult to estimate, especially since it prefers the interior parts of mangroves, where it cannot be observed. A few unoccupied nests have been found scattered throughout the wetland complex, including along the rivers of the lower deltaic plain.

Threskiornitidae

African sacred ibis, *Threskiornis aethiopicus*. This species is common in most of sub-Saharan Africa, but in Gabon it is limited to the coastal region, where it inhabits the intertidal zone. In January and February 1992, 40–60 birds were observed in Cape Lopez Bay (Schepers & Marteijn, 1993). On August 3, 2014, a dozen individuals were observed on a mudflat at the mouth of the Kondjo River.

Hadada ibis, *Bostrychia hagedash*. This species is resident and breeds in small numbers, and is present throughout the Lower Ogooué region along rivers and in flooded forests (Figure 9.25). Elsewhere in Africa, this bird is often not very shy and can even visit urban gardens, but in the Lower Ogooué region it is very cautious and generally flies far away when approached—an indication of significant hunting pressure.

Olive ibis, *Bostrychia olivacea*. This forest species is found in the Gamba region (Sargeant, 1993), in Omboué (Christy, pers. obs.), and around the Bongo River (Vande weghe, pers. obs.). It typically is hidden and often goes unnoticed except when flying in groups, calling out at dawn and before dusk. In the Ogooué deltaic complex, it is known to be found in the eastern part of the Ramsar site in the lower river valley.

Spot-breasted ibis, *Bostrychia rara*. This forest species, mentioned as present in the Gamba region (Sargeant, 1993) and also observed in the *Lecomtedoxa* swamp forests of the Mpivié River, a tributary of the Nkomi Lagoon, is known to occur at Lake Gomé and in the lower river valley above Lambaréné.

African Spoonbill, *Platalea alba*. This species has been observed in Port-Gentil (Alexander-Marrack, pers. comm.), in Lambaréné in 1981 and 1988 (Christy, pers. obs.) and in the Nkomi Lagoon region.

Ciconiidae

Yellow-billed stork, *Mycteria ibis*. This Afrotropical species was observed in the region of Port-Gentil and the Nkomi Lagoon (Schepers & Marteijn, 1993; Sargeant, 1993), as well as on Lake Onangué on August 8, 1988 (Christy, pers. obs.).

Abdim's stork, *Ciconia abdimii*. This Afrotropical species is a transequatorial migrant. It passes through the entire coastal plain of Gabon from September to November and in March or April, generally in small numbers (a maximum of a few tens of individuals) but sometimes in bands of 100–200. It rests on the sandbanks along the Ogooué.

White stork, *Ciconia ciconia*. This migratory species of Palearctic origin has occasionally been recorded in Gabon, notably near Port-Gentil in July and August 1987 and in July 1989 (Alexander-Marrack, pers. comm.), and in the Gamba region on December 2, 2005 (Vande weghe, 2007).

Woolly-necked stork, *Ciconia episcopus*. This resident and nesting Afrotropical species has been observed throughout the coastal region of Gabon (Figure 9.26). It can be seen all year round in the Lower Ogooué complex, although less abundantly than in the Wonga-Wongué Reserve. It frequents both freshwater environments (flooded savannahs, grassy beds, and banks) as well as marine environments (mudflats and mangrove rivers). Significant seasonal movements have been observed.

Saddle-billed stork, *Ephippiorhynchus senegalensis*. This Afrotropical species has been observed near Port-Gentil and further south in the Great Lagoons (Schepers & Marteijn, 1993; Sargeant, 1993). An isolated bird has since been spotted, in April 2006, near Lake Anengué. From available observations, it appears that it was less rare in the 1980s and 1990s. Since it is large and slow flying, this species is unfortunately an easy target for hunters, who are omnipresent throughout the Lower Ogooué complex. It may now have disappeared. The large marshes of the upper deltaic plain and those of the Inguessi Plain are the most suitable environments for this species, which does not yet appear to be threatened at a global scale but shows a clear decrease throughout its area of distribution.

Ardeidae

White-crested heron, *Tigriornis leucolopha*. This species, endemic to Guineo-Congolean forests, is resident and nesting in Gabon. It has been observed in many places throughout the coastal basin (Sargeant, 1993; Vande weghe, 2003). It favors swamp forests and riparian forests, and seems relatively common in the flooded forests of the Nkomi Lagoon. Within the Lower Ogooué complex, it has occasionally been seen on Mandji Island, in Ozouri, in Lake Gomé, along the Mpolounié River, and in a riparian forest near the confluence of the Ogooué and the Ngounié. Its population could be relatively large, but there is nothing available on which to base a quantitative estimate.

Little bittern, *Ixobrychus minutus*. This is a Palearctic and Afrotropical species. A male of the nominal Palearctic form was seen on December 31, 2000, in a floating herbaceous mat at Lake Evaro, and several birds of this form were also seen between November and May in the ponds of Port-Gentil. Birds of the Afrotropical entity, form *payesii*, were seen in February 1992 near Port-Gentil (Schepers & Marteijn, 1993) and on several occasions in the alluvial plain, the Lake Evaro region, in Lambaréné, and in Port-Gentil (Sargeant, 1993; Schepers & Marteijn, 1993; pers. obs.). This bird frequents floating grass mats, reed beds, and papyrus beds. It is very discreet, solitary, and generally goes unnoticed, except when disturbed. It is therefore difficult to judge its abundance, but it is probably much more frequent in the interior delta and the large grassy marshes of the alluvial plain than the low number of observations suggest.

Black-crowned night heron, *Nycticorax nycticorax*. This Palearctic and Afrotropical species is resident and migratory in Gabon. It has been observed almost annually since 1999, from July to September, upstream of Lambaréné, where 25-50 individuals nest on an island in the Ogooué (Figure 9.27). Some birds also prefer interior deltas. However, since it is most active around twilight, this species is difficult to follow and can easily go unnoticed. Sargeant (1993) mentions having seen the species in Port-Gentil, where Alexander-Marrack considers it to be fairly common from November to May. Schepers & Marteijn (1993) did not, however, observe this species in the coastal region during their 1992 census. The individuals of the Lower Ogooué complex appear to be sedentary, and there is no indication that Palearctic birds have reached Gabon.

White-backed night heron, *Gorsachius leuconotos*. This Afrotropical species also favors twilight and is very difficult to observe. Sightings in the Lower Ogooué complex involve three birds seen and heard at Lambaréné in September 2006, birds seen at night at Lake Gomé, and an individual observed on the track from Port-Gentil to Ozouri in December 1988.

Squacco heron, *Ardeola ralloides*. This species includes Afrotropical and Palearctic populations. In the Ogooué Delta, it has been seen year-round, but especially between October and April (Sargeant, 1993), which suggests the presence of Palearctic birds during that period. These birds are fond of open marshes, flooded savannas, papyrus beds, and floating grass mats, although never of saltwater areas. In July and August 2000 and 2001, several birds with breeding plumage were observed along the Ogooué in the Lambaréné region. Perhaps there were then a few nesting couples on an island in the river among the great egrets and purple herons. No birds were seen in July and August 2014.

Western cattle egret, *Bubulcus ibis*. This species includes a resident population and a (probably transequatorial) migratory population that passes through Gabon in March and April and in November. It has been observed in Lambaréné and Port-Gentil, but it is much less abundant along the Lower Ogooué than in the Libreville region. Schepers & Marteijn (1993) estimated the population in the complex's coastal region at 1,250-2,000 birds. Including the rest of the Lower Ogooué would probably add at least 1,000-2,000 individuals to this count. In August 2013, a few pairs nested among other Ardeidae on the island northeast of Lambaréné.

Striated heron, *Butorides striatus*. This Afrotropical species is resident and nesting in the Lower Ogooué complex. It is frequently observed throughout the area's wetlands, including on the mudflats of Cape Lopez Bay, but always alone and most often hidden in the dense vegetation along the shores or in papyrus beds. A bird was building a nest in July 2001 along the Ogooué, 2 km upstream from Lambaréné. A couple were feeding at a nest on the island northeast of Lambaréné in August 2013. Large juvenile individuals were observed on the same island in August 2014. The population size in the complex must be large, but no quantified estimate can be given.

Great egret, *Casmerodius albus*. This species is resident and nesting in Gabon (Figure 9.28). During the nesting period (July to September) it is abundant in the interior deltas of Lakes Evaro and Ezanga, in the alluvial plain, and upstream from Lambaréné, where it frequents the banks of large channels, sandbanks, and especially large beds of *Echinochloa* and *Vossia*. A colony can be found on an island a few kilometers east of Lambaréné (60-100 nests in July 2001; about as many in August 2006, but only 15-20 in August 2013 and 2014). These nesting birds feed mainly around Lake Zilé and in the Mbiné marshes. Another colony occurs in the Rivière des Pélicans river delta, and it is likely that colonies also exist in inaccessible places on the upper deltaic plain. Outside of the nesting period, when the waters of the Ogooué are high, this species seems to move toward the intertidal areas of Cape Lopez Bay, where Schepers & Marteijn (1993) estimated the "wintering" population at 150-250 individuals. Many birds also appear to reach the mudflats of Mondah Bay and the Komo Estuary, as well as the swampy plains of the Ngove in Loango National Park. The total population of the coastal region of Gabon was estimated at 1,050-1,350 birds in January to March 1992.

Intermediate egret, *Mesophoyx intermedius*. Nesting of this migratory Afrotropical species has never been observed in Gabon. In August 2013, two couples in breeding plumage were present in the colony of herons northeast of Lambaréné. Overall, some isolated individuals can be observed in the lake deltas. This species had previously been observed in small numbers by Schepers & Marteijn (1993), Alexander-Marrack, and Sargeant (1993) in the region

9.27. A black-crowned night heron in the colony just northeast of Lambaréné on a small river island (Photo: Jean P. Vande weghe).

9.28. *A great egret with a chick in its nest from the colony just northeast of Lambaréné (Photo: Jean P. Vande weghe).*

9.29. *A purple heron in the Rivière des Pélicans (Photo: Jean P. Vande weghe).*

of Port-Gentil, the lagoons region, and in the vicinity of Lambaréné. It remains to be sighted in Akanda National Park.

Little egret, *Egretta garzetta*. This Afrotropical and Palearctic species is resident and nesting in Gabon. The presence of Palearctic individuals, indistinguishable from local ones, has not, however, been proven. This is the second most abundant species of herons after the cattle egret, and it is found in all coastal wetlands. It was occasionally observed nesting in the mangroves of Akanda National Park in 2003, but in the Lower Ogooué complex, the only data available are of a bird carrying a twig in its beak on August 18, 1986. In the dry season, when water levels are low, this species can be observed throughout the Ogooué, where it frequents riverbanks with more or less dense vegetation, as well as sandbanks. During seasons when the water levels are high, the highest concentrations of this species are observed in the Port-Gentil region, where it frequents mudflats and beaches, as well as mangrove rivers. According to Schepers & Marteijn (1993), the total population in the coastal part of the complex was 850–1,200 individuals, which represented 30–40% of the total population of the Gabonese coast, estimated at 2,350–3,100 in January and February 1992. At least 1,000 birds should be added to this to account for the other parts of the Lower Ogooué region, which suggests that the Lower Ogooué complex has 20–65% of this species' population in Gabon.

Black heron, *Egretta ardesiaca*. This Afrotropical species has been seen in the vicinity of Port-Gentil (four in mangroves on August 18, 1986, two on September 6, 1986, and one on October 12, 1986) and near Lambaréné (one on a sandbank upstream of Lake Zilé on August 10, 1988).

Western reef heron, *Egretta gularis*. This strictly coastal Afrotropical species is common in parts of West Africa and has been regularly observed in Akanda and Pongara national parks, even on the beaches from Libreville. It was mentioned by Sargeant (1993) near Port-Gentil, where it favors beaches and mudflats. Schepers & Marteijn (1993) estimated the Gabonese population at 5–15 individuals in February 1992, which was surely an underestimate. More than 10 individuals can be observed on the mudflats of Akanda at any given time.

Gray heron, *Ardea cinerea*. This Afrotropical and Palearctic species is resident and nesting in Gabon, but mostly restricted to intertidal coastal environments. A colony of a few dozen birds exists in Akanda National Park. In the Ogooué Delta, it is frequently observed on mudflats and in mangroves, yet rarely seen in the freshwater wetlands found upstream. In February 1992, the population in Cape Lopez Bay was estimated at 100–175 individuals, which represented 10–20% of the population of the Gabonese coast totaling an estimated 750–1,025 birds (Schepers & Marteijn, 1993).

Goliath heron, *Ardea goliath*. This Afrotropical species is probably sedentary and resident in Gabon; it has been observed nesting in several parts of the country. In the Lower Ogooué complex, isolated individuals have been spotted at the edge of all open water areas (lakes, lagoons, rivers), both in saltwater and freshwater areas. In late August 2013, a bird was present on a sandbank upstream from Lambaréné, and it is possible that two to three pairs nested in the eastern part of the complex (especially in the alluvial plain and the lake deltas). In January and February 1992, the total population of this species across the entire coastal area was around 10–20 birds (Schepers & Marteijn, 1993), to which can be added at least 10 individuals from the alluvial plain and the lake deltas of the Ogooué.

Purple heron, *Ardea purpurea*. This Afrotropical and Palearctic species (Figure 9.29) is resident and nesting in Gabon, where colonies have been observed in several places in the coastal basin, notably in the marshes of Lower Nyanga and in Loango National Park. In the Lower Ogooué complex, it is present year-round. It rarely hunts in the open, but rather in flooded herbaceous vegetation, reed beds, floating grass mats, or even flooded thickets of *Alchornea*, and therefore often goes unnoticed. It appears almost never to frequent saltwater environments, and the highest densities have been observed in lake deltas, the upper deltaic plain, and upstream from Lambaréné. A colony was found on an island in the Ogooué east of Lambaréné (100–200 individuals in July 2001; 40–60 in August 2013 and August 2014). These birds feed mainly in the area of Lake Zilé and in the large grassy marshes of the Mbiné. Other colonies occur in the Rivière des Pélicans river delta, the Ezanga river delta, and the lakes to the north of the alluvial plain, notably Lake Gomé. Some colonies in the central alluvial plain have most likely gone unnoticed. The total population for the entire deltaic complex must include at least 1,500–2,500 birds, which probably represents more than 75% of this population in Gabon. During periods when the waters are high, these birds disperse throughout the complex. Their numbers could then be augmented by birds of Palearctic origin (Schepers & Marteijn, 1993), but this has never been proven.

Black-headed heron, *Ardea melanocephala*. This Afrotropical species is known in Gabon from the southwestern coast (Mayumba, Basse Nyanga, Loango). Further north, it was observed on Mandji Island, where a maximum of six birds were seen (Schepers & Marteijn, 1993).

Burhinidae

Water thick-knee, *Burhinus vermiculatus*. This Afrotropical bird lives in pairs or small groups of three to five. It is found all along the coast of Gabon, on beaches, sandbanks, and even in areas with bare ground at the edges of forests (such as old lumberyards). It is found throughout the Lower Ogooué complex, from Port-Gentil to Lambaréné, where it nests in the dry season. It has also been observed in the southern delta along the Mpolounie River and on the shoal of the Nkomi Lagoon. Given that this species is most active at twilight, it often goes unnoticed, and it is very difficult to estimate its population size.

Pluvialidae

American golden plover, *Pluvialis dominica*. This is an occasional migrant from North America, observed in Port-Gentil by Alexander-Marrack (one individual on December 26, 1987, and January 3, 1988).

Pacific golden plover, *Pluvialis fulva*. This species migrates from the Eastern Palearctic and Western Nearctic realms (from the tundras of the eastern Urals to the coasts of Alaska), and regularly winters in very small numbers in the region of Port-Gentil, Setté-Cama, and the mouth of the Nyanga River, from October to April (Sargeant, 1993; Schepers & Marteijn, 1993). The wintering population in Gabon is estimated at 25–50 individuals (Schepers & Marteijn, 1993). It frequents various environments, including relatively sandy mudflats and beaches, as well as dunes at high tide. However, the sites most favorable for this species are the Animba and Inguéssi plains, which are largely inaccessible. It is therefore possible that the wintering population is larger than estimated.

Gray plover, *Pluvialis squatarola*. This is a migrating species of Palearctic origin (Taimyr Peninsula), which can be seen from August to April, especially in Mondah Bay and in the western part of the Komo Estuary, but also on the mudflats of Cape Lopez Bay, which is said to host 10% of the individuals present in Gabon (Figure 9.30).

Charadriidae

Common ringed plover, *Charadrius hiaticula*. This migrating species of Palearctic origin can be observed in small numbers in the area of Port-Gentil and on the mudflats of Cape Lopez Bay (Sargeant, 1993; Schepers & Marteijn, 1993).

Kittlitz's plover, *Ochthodromus pecuarius* (ex *Charadrius pecuarius*). This species is resident on Mandji Island, where it frequents the short grasses found near water (Sargeant, 1993; Schepers & Marteijn, 1993). It nests there from April to October.

White-fronted plover, *Ochthodromus marginatus* (ex *Charadrius marginatus*). This Afrotropical bird is present, as the form *mechowi*, in the area (Figure 9.31), nesting in isolated pairs on large expanses of bare sand. It covers its eggs with hot sand in order to incubate them. It is found in small numbers throughout most of the complex, but the highest concentration of nesters has been found on shoals. Given their number and dispersal, it has not been possible to georeference each individual, but approximately one individual was recorded for every 100–150 m of shore. The populations from the shoals of Ogooué can thus be estimated at 80–120 individuals, including 100–160 from the shoal of the Nkomi Lagoon. This brings the total population of the complex to 220–300 birds. During the period when water levels are highest, this species leaves the interior of the complex and disperses across the beaches. The highest concentration is found in Akanda National Park, where it can reach 300 individuals.

Forbes's plover, *Afroxyechus forbesi* (ex ***Charadrius forbesi***). This Afrotropical species, which has a wide distribution in sub-Saharan Africa, can be found, in Gabon, within the coastal basin and the savannas of Lopé National Park. In the Lower Ogooué complex, it has been observed in very small numbers on Mandji Island (Schepers & Marteijn, 1993). The species favors short meadows and areas with bare soil near water. Breeding takes place in the dry season, from July to September.

9.30. A gray plover at Apomandé Point in the Bay of Nazaré (Photo: Jean P. Vande weghe).

9.31. A white-fronted plover on a sandbank in the Ogooué (Photo: Jean P. Vande weghe).

White-crowned lapwing, *Vanellus albiceps*. This Afrotropical species is endemic to the Guineo-Congolean forest region and its periphery (Figure 9.35). It lives in pairs all along the Ogooué, where it favors sandbanks and small muddy beaches. It is mainly found in the eastern part of the complex (Figure 9.33). The species also nests further upstream, along the Middle Ogooué (Lopé National Park), and around the lakes of the Wonga-Wongué Reserve (Vande weghe, 2012). The highest concentrations of this species have been observed in Lake Kivoro in the Ndogo Lagoon basin, where the breeding population was estimated at 100–200 individuals in September 2006. In the Lower Ogooué complex its population is estimated to comprise at least 100 birds. When water levels are high, they remain in the region, although they take refuge on the remaining sandbanks and wet lawns near banks.

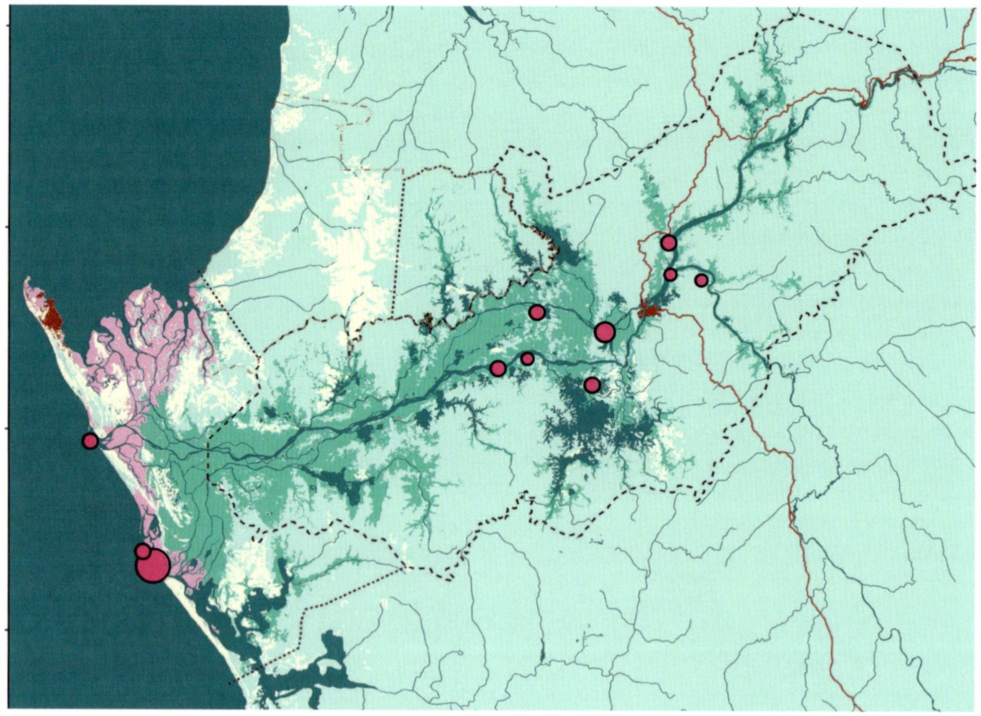

9.32. *Distribution of the white-fronted plover during the nesting period.*

9.33. *Distribution of the white-crowned lapwing during the nesting period*

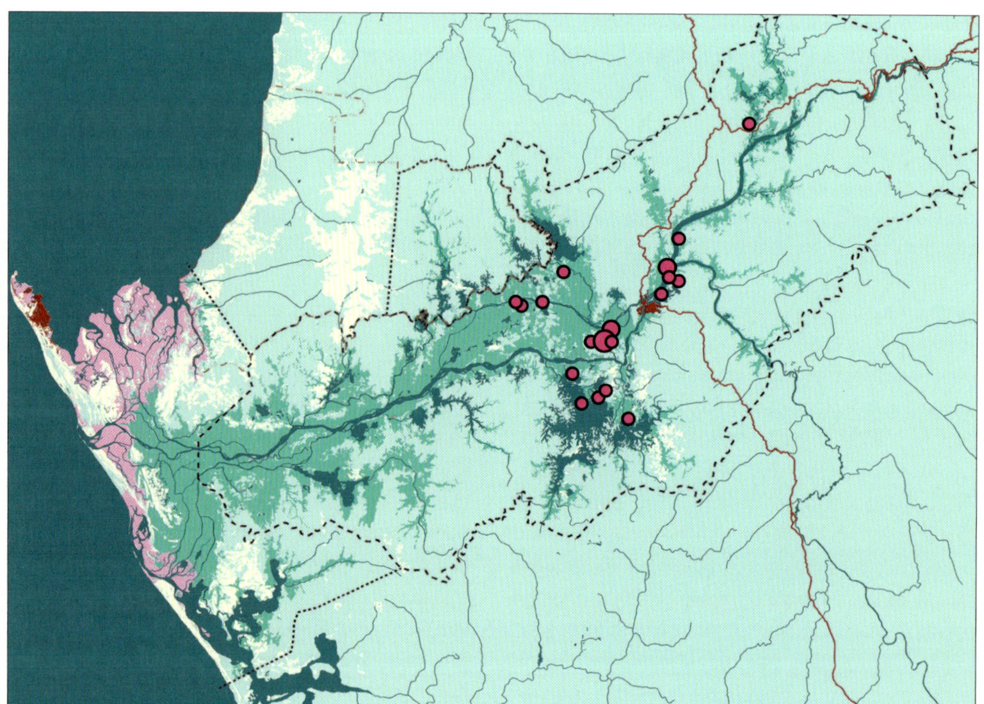

Haematopodidae

Eurasian oystercatcher, *Haematopus ostralegus*. This migratory species of Palearctic origin has been observed in Port-Gentil, in February 1976, December 1986, and October 1988 (Sargeant, 1993). As for the rest of Gabon, this species has been seen in Akanda National Park (Vande weghe, 2011) and at Cape Estérias.

Scolopacidae

Great snipe, *Gallinago media*. This migratory species of Palearctic origin is known to occur in the regions of Port-Gentil and Ozouri, between October and April (Alexander-Marrack). It frequents flooded grasslands and meadows with short grasses.

Common snipe, *Gallinago gallinago*. This migratory bird of Palearctic origin is found in the Port-Gentil region between November and May (Sargeant, 1993). It inhabits the flooded meadows and low marsh vegetation located between dune ridges, especially north of Omboué.

Bar-tailed godwit, *Limosa lapponica*. This migrant of Palearctic origin can be seen in Gabon from late July to early May; a few dozen birds are also observable in June and early July. The species is abundant in Akanda National Park, where the wintering population was around 4,400–4,800 birds in January 1992 (Schepers & Marteijn, 1993). It is less frequent further south (Vande weghe, 2003), but is present around Port-Gentil (Sargeant, 1993): 175–225 individuals were observed on the mudflats of Cape Lopez Bay in February 1992 (Schepers & Marteijn, 1993).

Eurasian whimbrel, *Numenius phaeopus*. This migratory species of Palearctic origin can be observed from the end of July to the end of May, as well as a few summer visitors in June and July. It is abundant in Akanda National Park and the Komo Estuary, as well as in Nazaré Bay (Sargeant, 1993; Schepers & Marteijn, 1993). In June 2016, more than 200 individuals from Nazaré Bay gathered at the end of the day on a sandbank of the Ogooué shoal.

Eurasian curlew, *Numenius arquata*. This migratory species of Palearctic origin is regularly found in Akanda National Park, but is rare in the vicinity of Port-Gentil (Sargeant, 1993). In January and February 1992, 15–30 individuals were observed in Cape Lopez Bay (Schepers & Marteijn, 1993). In February 2014, an isolated bird was seen on a small mudflat a few kilometers north of Port-Gentil (Figure 9.34).

Common redshank, *Tringa totanus*. This migratory species of Palearctic origin is occasionally found in Port-Gentil (Sargeant, 1993).

Marsh sandpiper, *Tringa stagnatilis*. This migratory species of Palearctic origin can be seen from October to April, mainly in Cape Lopez Bay, where 75–100 individuals were noticed in February 1992 (Schepers & Marteijn, 1993).

Common greenshank, *Tringa nebularia*. This migratory species of Palearctic origin can be observed mainly from late July to early May. In the Lower Ogooué complex, the species is especially present in Cape Lopez Bay (Sargeant, 1993), where 300–425 individuals were counted in February 1992 (Schepers & Marteijn, 1993). Isolated birds have, however, been observed throughout the complex, including in the interior deltas and upstream from Lambaréné. Occasional solitary birds are observed in June or July (pers. obs.).

Wood sandpiper, *Tringa glareola*. This Palearctic migrant can be seen in small numbers from September to April in all the wetlands of the coastal basin. It is quite common in the Lower Ogooué complex, especially in the interior deltas. Its numbers are difficult to estimate because these birds often frequent isolated ponds within large, dense plant formations.

Common sandpiper, *Actitis hypoleucos*. This Palearctic migrant can be observed from July to May, with some sightings in June. It is common throughout the Lower Ogooué region, along the riverbanks and mangrove rivers, and in mudflats. According to Schepers & Marteijn (1993), the wintering population along the Gabonese coast is around 1,300–2,200, to which more than 500–1,000 individuals must be added for the alluvial plain and the interior deltas of the Ogooué.

Ruddy turnstone, *Arenaria interpres*. This migratory species of Palearctic origin is found in the vicinity of Port-Gentil. Many individuals can be seen passing over the Libreville region from September to November, as well as in the Port-Gentil region in September, with a maximum of 42 birds observed on September 3, 1988 (Alexander-Marrack).

Red knot, *Calidris canutus*. This migratory species of Palearctic origin is present in Cape Lopez Bay from September to April (Sargeant, 1993; Schepers & Marteijn, 1993; Vande weghe, pers. obs.). About 60 birds were observed on a small refuge at high tide in April 2013.

Sanderling, *Calidris alba*. This Palearctic migrant is seen from August to April, most frequently on sandy beaches in the region of Port-Gentil (Sargeant, 1993; pers. obs.). The wintering population in the Lower Ogooué region was estimated at 220–240 individuals in February 1992 (Schepers & Marteijn, 1993).

Little stint, *Calidris minuta*. This Palearctic migrant can be observed in small numbers in Lopez Bay from August to April (Sargeant, 1993; pers. obs.) The wintering population there comprised 200–300 individuals in February 1992 (Schepers & Marteijn, 1993). In April 2013, 40 birds were noted north of Port-Gentil (pers. obs.).

Dunlin, *Calidris alpina*. This migratory species of Palearctic origin has occasionally been observed in Gabon: two individuals were seen on January 30, 1992, at the "Barre de l'Arabe," north of the Nkomi Lagoon (Schepers & Marteijn, 1993). This is the only sighting from Gabon and one of the few south of the equator in sub-Saharan Africa.

Curlew sandpiper, *Calidris ferruginea*. This migratory species of Palearctic origin can be observed from September to April in Mondah Bay, the Komo Estuary, and Nazaré Bay, which is said to harbor 10% of its Gabonese population (Schepers & Marteijn, 1993).

Buff-breasted sandpiper, *Calidris subruficollis*. This migratory species of Nearctic origin was recorded in the sand dunes north of Port-Gentil on March 26, 1988, by Alexander-Marrack.

Ruff, *Calidris pugnax*. This Palearctic migrant can sometimes be observed from August to May in the vicinity of Port-Gentil, with a maximum recorded of 13 birds on November 28th, 1987 (Sargeant, 1993; Schepers & Marteijn, 1993).

9.34. A Eurasian curlew photographed just north of Port-Gentil in April 2013. This photo was taken from far away, but it clearly shows that the feathers of the underside of its wings are whitish and without dark spots. This is a diagnostic trait for the suschkini *and* orientalis *forms of this species, which breed in Asia, from the Urals to Manchuria (Photo: Jean P. Vande weghe).*

9.35. A couple of white-crowned lapwings on a sandbank upstream of Lambaréné (Photo: Jean P. Vande weghe).

Wilson's phalarope, *Phalaropus tricolor*. This migratory species of Nearctic origin was observed north of Port-Gentil from October 10 to December 6, 2009 (G. Passary, pers. comm.)

Jacanidae

African jacana, *Actophilornis africana*. This resident and breeding species is known from all freshwater environments in the coastal region of Gabon; it is well distributed throughout the interior deltas of the Ogooué. It is also frequently found in the Ogooué sandbanks when water levels are low and they are covered with herbaceous vegetation.

9.36. A gray pratincole on an Ogouoé sandbank (Photo: Jean P. Vande weghe).

9.37. Skimmers on a sandbank at the mouth of the Rivière des Pélicans, with a young "Ogooué" tern in the foreground (Photo: Jean P. Vande weghe).

9.38. Distribution of the gray pratincole in August 2014, during the nesting period.

Glareolidae

Gray pratincole, *Glareola cinerea*. This Afrotropical species nests in large numbers in July, August, and September on the large sandbanks of the Ogooué, both upstream from Lambaréné and throughout the upper deltaic plain (Figure 9.36). During its nesting period it is absent from the lower deltaic plain and sandbars (Figure 9.38). After nesting, it scatters to the coast, where it has been observed from Nyonié to the mouth of the Nyanga. There, it frequents beaches, dunes being colonized by vegetation, and grassy savannas with a steppe-like appearance. In the Gamba and Great Lagoons region, this species is common from October to March (Sargeant, 1993). It can then reach concentrations of 150-200 individuals (Sargeant, 1993; Schepers & Marteijn, 1993). The species was also spotted several times on the beach at Nyonié, especially from March to May, in very large gatherings: 850 birds in early May 2005 and 350-400 in April 2013. The breeding population in Gabon must comprise at least 500 couples.

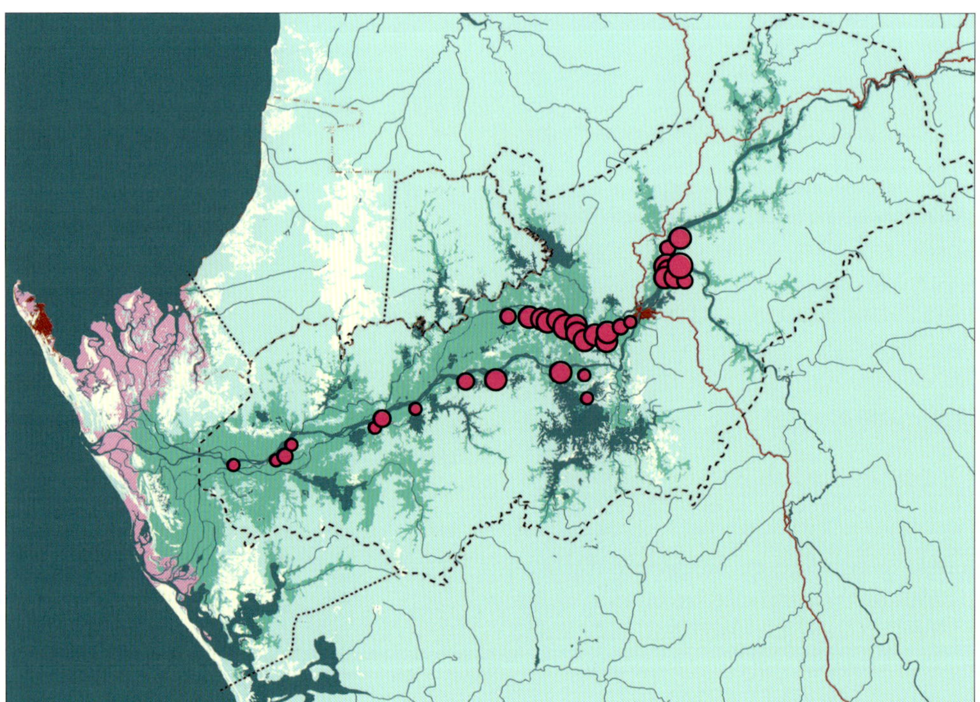

Rock pratincole, *Glareola nuchalis*. This Afrotropical bird lives on rocks emerging from rivers. On August 4, 2014, an isolated bird was observed on a sandbank of the alluvial plain. The majority of the Gabonese population lives east of Lambaréné, outside of the coastal basin.

Black-winged pratincole, *Glareola nordmanni*. This rare migratory species of Palearctic origin has been observed in the regions of Port-Gentil and Gamba in October, November, December, and April, on meadows with short herbaceous vegetation (Sargeant, 1993). Its last sighting dates back to October 25, 2008.

Laridae

African Skimmer, *Rynchops flavirostris*. This species feeds on small fish taken at the surface of water, especially at dusk and at night. It breeds on sandbanks all over the delta (Figure 9.39), but the highest numbers are found on the shoals (30%) and in the upper deltaic plain (43%). In July and August 2014 the total population was estimated at 688 birds, but was probably closer to 800-900. These birds form loose colonies of 5-15 nests spaced 10-20 m apart. A two-week-old chick was observed at the end of July 2014, which places the start of the laying period mid-June. Eggs are still produced around mid-August.

When the waters rise, these birds leave the region and gather in Akanda National Park (800-1,200 birds), Loango National Park (350-500 individuals maximum), Cape Lopez Bay (180 birds in February 2014), the mouth of the Nyanga river (140 birds), and the mouth of the Banio Lagoon (50-100 individuals). The total Gabonese population is therefore between 1,600 and 2,100 individuals, of which only 50% seem to nest. In January and February 1992, it was estimated at 900-1,000 (Schepers & Marteijn, 1993); either a part of the population escaped observations, which seems unlikely, some individuals are migratory, or the Gabonese population must have increased over about 20 years, which also seems implausible.

Kelp gull, *Larus dominicanus*. This migratory bird from southern Africa has been observed mainly from April to September on the beach at Port-Gentil (Sargeant, 1993). However, there are no recent sightings.

Lesser black-backed gull, *Larus fuscus*. This migratory bird of Palearctic origin is observed from September to April on the beach in Port-Gentil, with a recorded maximum of 160 individuals in the Akori Lagoon on February 18, 1990 (Sargeant, 1993). However, it has not been observed since 2012, and the beach at Cape Lopez, where this species was observed in the 1980s, has been transformed into a canoe park by West African fishermen.

[**Heuglin's gull, *Larus heuglini*.** Sargeant (1993) reports a sighting of a "herring gull," *Larus argentatus*, on February 25, 1990, at Akori in the Port-Gentil region, without giving a precise description of the bird. This Palearctic species does not, however, occur further south than the Mediterranean and "herring gull" sightings in tropical Africa all relate to the Heuglin's gull, *Larus heuglini*, a species close to the lesser black-backed gull, but which has a much lighter coat and wing-cover. Like the lesser black-backed gull, it is a long-distance migrant that winters mainly in the Persian Gulf, the Red Sea, and East Africa. It is regularly observed on Lake Edouard in Uganda and on Lake Tanganyika in Burundi. It would therefore be logical for this Gabonese sighting to also relate to this species.]

Caspian tern, *Hydroprogne caspia*. A couple belonging to this cosmopolitan species nested on a sandbank upstream from Lambaréné in July and August 1999, but the eight individuals seen in this sector and in the alluvial plain in July and August 2014 were not breeding. Several were in transition plumage. Colonies of this species are located on shoals (Figures 9.41 and 9.43): the Ogooué shoals totaled 160 individuals, and the Nkomi Lagoon shoals had around 300 individuals. Taking into account the fact that a certain number of birds must have been away fishing in the sea, the total population of this species in the Lower Ogooué complex could be estimated at 500–600 individuals. A small population of 10–20 birds also nest on the sandbar of the Ngové Lagoon in Loango National Park, at least in some years. The population of the Lower Ogooué region therefore represents at least 95% of the total Gabonese population. The size of this population does not seem to have changed significantly: the Ogooué sandbar had the same number of Caspian terns in 1990 during the surveys conducted for the IUCN report (Wilks, 1990) as in August 2014. It is difficult to know what these birds do outside the nesting period, but the species can be observed throughout the year at Cape Lopez and in Cape Lopez Bay. The "wintering" population is not very large: Schepers & Marteijn (1993) counted only 35–50 for all of coastal Gabon in February 1992. Palearctic birds visit the Gabonese coast during the boreal winter; a bird banded in

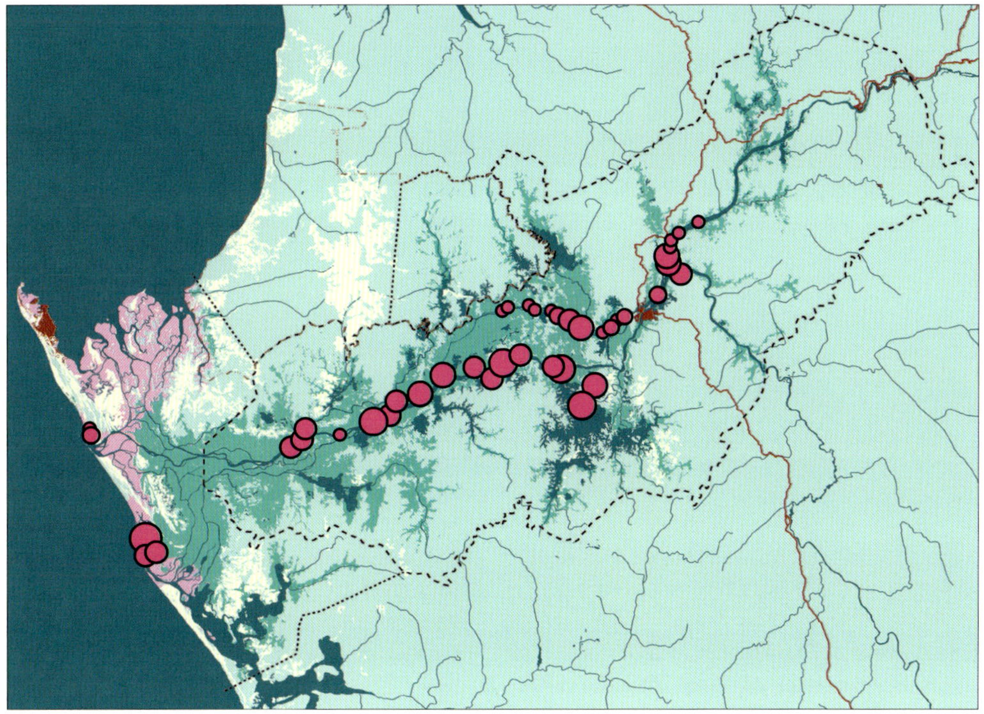

9.39. Distribution of the African skimmer during the nesting period.

9.40. A group of seven skimmers over the estuarine waters of the Ozouri region (Photo: Jean P. Vande weghe).

9.41. *Caspian terns near their colony on the "Barre des Portugais" (Photo: Jean P. Vande weghe).*

9.42. *Royal terns in breeding plumage in April 2013 on a beach just north of Port-Gentil. In the foreground there is a little stint (Photo: Jean P. Vande weghe).*

June 1976 in Finland was found in Libreville in May 1977 (Schepers & Marteijn, 1993).

Royal tern, *Thalasseus maximus*. This migratory species of West African origin (Senegal, Mauritania) can be observed throughout the year along the coast. According to Schepers & Marteijn (1993), the "wintering" population in Gabon was estimated at 7,500-9,500 individuals in January and February 1992, 4,000-4,500 of which (about 50%) were observed in Cape Lopez Bay. This population represents 10-15% of the total African population. These birds are fond of coastal waters and range into river mangroves for several kilometers beyond the coast (Figure 9.42).

Sandwich tern, *Thalasseus sandvicensis*. This is a migratory species of Palearctic origin found on beaches and in coastal waters from the end of July to May, and especially from October to March. In Port-Gentil, a maximum of 153 individuals was observed by Alexander-Marrack in October 1986, and the "wintering" population of Cape Lopez Bay was estimated at 50-100 individuals in January and February 1992 (Schepers & Marteijn, 1993). Banded birds found in Gabon came from the United Kingdom (4 individuals), Denmark (2), Estonia (2), Germany (1), the Netherlands (2), and Sweden (1).

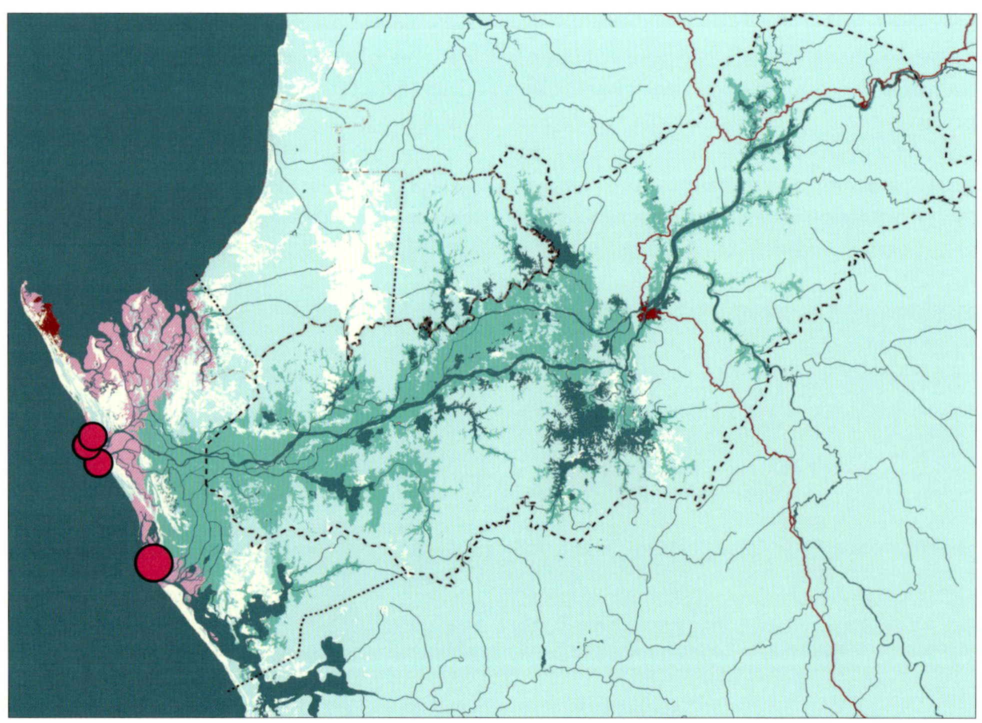

9.43. *Distribution of Caspian tern colonies.*

Common tern, *Sterna hirundo*. This Afrotropical and Palearctic species is represented in Gabon by a migratory population from Europe and a very small locally breeding population (Figure 9.6). Migratory individuals pass in very large numbers and remain confined to coastal waters and beaches. Several hundred spend the months of May, June, and July on the edge of the delta, especially near the Ozouri shoals, the "Barre de l'Arabe," and "Barre des Portugais" (Figure 9.4). The Palearctic population passes off the coast of Gabon in large numbers in September-October and March-April. On October 8, 2004, 2,000-3,000 individuals, including Arctic terns, *Sterna paradisaea*, were observed migrating off Nyonié. This migration seems very fast, and most of these birds stop in Gabon only to spend a night on a beach. Gatherings of several thousand can be observed on the sandbars of the Ogooué Delta, the Ngové Lagoon, and the Banio Lagoon, as well as at the mouth of the Nyanga. A few hundred birds, probably immature, remain in June and July. The banded birds found in Gabon came from the United Kingdom (1 individual), Denmark (3), Estonia (6), Finland (21), Germany (2), Lithuania (2), the Netherlands (1), Norway (1), and Sweden (12).

A small number of birds reach breeding plumage (Figure 9.6), and some nest on the banks of the Ogooué. They are larger, have red beaks that are clearly stronger than local terns, and have whiter underparts. These stand apart and have never been observed in or near "Ogooué" tern colonies, but they have been seen in the Lower Ogooué valley and in the upper deltaic plain. Some breeding pairs have been observed upstream of Lambaréné and on the banks of the Ngounié.

"Ogooué" tern, *Sterna* sp.

The majority of medium-sized terns nesting in the Ogooué Delta belong to a population of undetermined taxonomic status. They are distinguished by their smaller size, smaller and generally black bill, even during the nesting season, their underside of their body which is much darker than that of the Eurasian common tern, and the shorter black feathers on the top of the head (Figures 9.2–9.5). This taxon, although clearly distinguishable in the field, is currently not recognized (Gill & Donsker, 2017), even though it was described under the name *Sterna senegalensis* in 1837 by W. Swainson. Previously it probably nested from Mauritania to Angola, but its populations have been eliminated almost everywhere. Some couples (<10) may survive in Senegal, but the Gabonese population seems to be the only one that is still viable. This taxon could be sufficiently different from Eurasian common tern to constitute an "independent evolutionary unit," or even a species in its own right–subendemic to Gabon in this case.

An unknown but significant part of the Ogooué population remains in the inland wetlands year-round and therefore does not come into contact with Palearctic migrants. The two forms only meet on sandbars, and the few observations seem to indicate that they remain separate, even during the nesting period. Only a genetic study could clarify the real status of this unknown form.

This local population nests in colonies of several dozen pairs (20–40) on the sandbars of the marine delta, generally near the Caspian terns, and, more scattered, on the sandbanks further upstream in the delta. Two couples were observed at Lake Nguéné in July 2004, north of Ebel Abanga.

Nesting in a scattered manner on the Ogooué sandbanks is probably a response to predation. Some couples can easily go unnoticed. The highest concentrations of breeding birds were observed upstream of Lambaréné and on the sandbars around 1999 and 2000. Taking into account unexplored areas and individuals that went unnoticed, the total nesting population for the complex can be estimated at 480–550 individuals.

These terns remain during the period when the waters of the Great Lakes are high, especially in the delta of the Rivière des Pélicans and along the main course of the Ogooué. They are therefore associated with white-winged terns, *Chlidonias leucoptera*, of Palearctic origin, which also winter in substantial numbers on the large lakes of the delta (Figure 9.44). These terns hunt above floating grass mats in the open waters of lake deltas and also further from the shores, on Lakes Onangué, Ezanga, and Oguémoué. "Ogooué" terns therefore do not mix with the Palearctic migrants,

9.44. Two "Ogooué" terns and a white-winged tern (in the middle) in December 2004 in the Rivière des Pélicans delta (Photo: Jean P. Vande weghe).

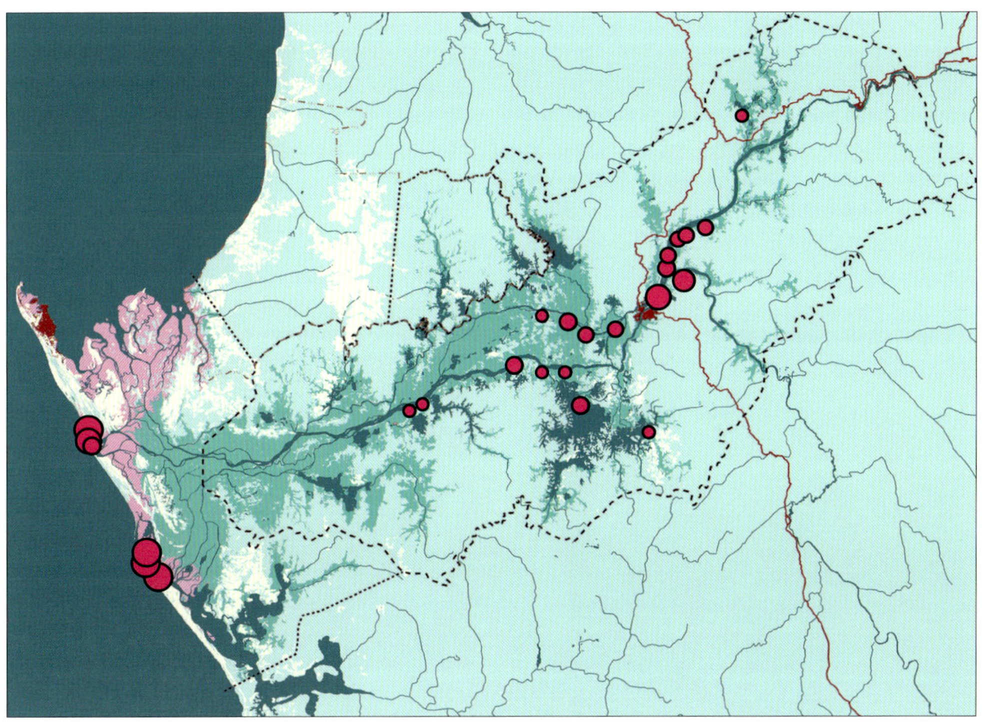

9.45. Distribution of common tern and the "Ogooué" tern during the nesting period.

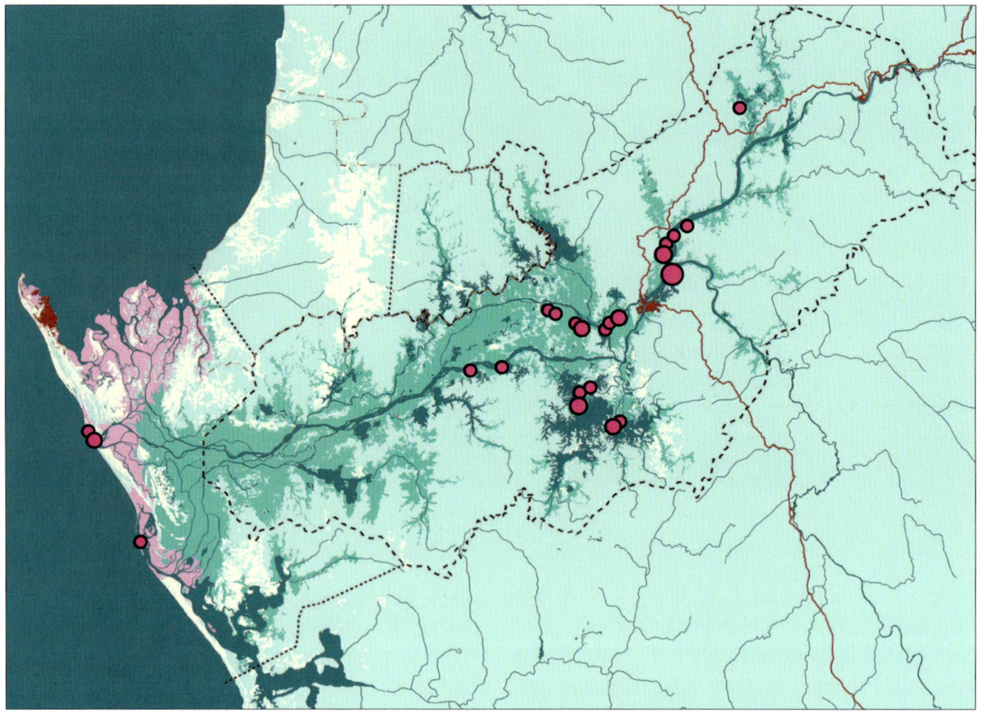

9.46. Distribution of the little tern during the nesting period, in August 2015.

who remain along the Gabonese coasts in September-October and March-April.

As for the population tendencies of this nesting population, it should be noted that the numbers observed in August 2014 from Lambaréné were one-half to one-third of those observed in the same region around 1999 and 2000. Based on numbers from December 2001, the total population of this species in the Ogooué Delta was nevertheless estimated at 400 individuals, which was not significantly different from the currently estimated total.

Arctic tern, *Sterna paradisaea*. This migratory species of Palearctic origin can be observed in October and November in Port-Gentil (Sargeant, 1993). While en route to the Indian Ocean, the Australian coast, and the Antarctic Ocean, these birds only pass Gabon offshore and do not linger. Banded birds found in Gabon came from the United Kingdom (1 individual), Estonia (1), the Faroë islands (1), Finland (4), Germany (1), and Sweden (3).

Little tern, *Sternula albifrons*. This Afrotropical and Palearctic species (Figure 9.47) nests on sandbanks throughout the complex, and it is even more widespread than the "Ogooué" tern. Since it is a small species, more individuals may have gone unnoticed, but the population likely does not exceed 150-200 individuals (Figure 9.46). When water levels are high, this species leaves the Ogooué and moves toward the bays of Cape Lopez and Nazaré (160 birds were observed at Apomandé Point in February 2014) and toward Mondah Bay in Akanda National Park, where 100-200 birds can be observed on a regular basis. This species is, however, rarely observed south of Port-Gentil. Its total wintering population of the Gabonese coast may thus be greater than the population in the Lower Ogooué complex. This probably includes indistinguishable birds of Palearctic origin.

Damara tern, *Sternula balaenarum*. This migratory species is native to southern Africa and can be observed in small numbers from April to November in Port-Gentil (Sargeant, 1993). The maximum number of birds counted at Cape Lopez was 927 individuals on October 26, 1986 (Alexander-Marrack). In August 2014, several dozen birds were seen on the "Barre des Portugais" (Figure 9.48).

White-winged tern, *Chlidonias leucoptera*. This migratory species of Palearctic origin can be observed in December and January on the lakes of the Lower Ogooué region, mainly in areas of open water between floating grass mats. They are found in groups of up to 50 birds, alongside "Ogooué" terns. This species is rare in Port-Gentil, with only four sightings between 1987 and 1989 (Alexander-Marrack). The population "wintering" on the coast of Gabon was estimated at 85-175 individuals in 1992 (Schepers & Marteijn, 1993). To this, at least 100-200 birds should be added to account for the rest of the Lower Ogooué complex.

Black tern, *Chlidonias niger*. This species of Palearctic origin can be observed throughout the year in Port-Gentil, but most abundantly from October to April (Sargeant, 1993). The maximum number of individuals counted at Port-Gentil was 1,600 on October 5, 1986. A summering bird was sighted in August 2014 on the "Barre des Portugais." The species has not been observed in the interior of the Lower Ogooué Delta.

Accipitridae

African fish eagle, *Ichthyophaga vocifer*. This Afrotropical species is resident and nesting in Gabon. It can be observed along the Ogooué, both upstream and downstream from Lambaréné, near Lakes Evaro, Ezanga, and Onangué, as well as near Lakes Azingo and Gomé, and also along rivers of the marine delta. This species is known to occur in the Port-Gentil region (Sargeant, 1993). Its population density is extremely low.

9.47. A couple of little terns in the Rivière des Pélicans delta (Photo: Jean P. Vande weghe).

Palm-nut vulture, *Gypohierax angolensis*. This resident and nesting species is abundant throughout the Lower Ogooué region and the Nkomi Lagoon.

Western marsh harrier, *Circus aeruginosus*. This migratory species of Palearctic origin winters in the interior Ogooué deltas and in the open areas of the alluvial plain, where it hunts over floating grass mats, papyrus groves, reed beds, and various *Alchornea* thickets. These birds pass by in October-November and February-March. The species was previously noted in Port-Gentil (Sargeant, 1993).

Pandionidae

Western osprey, *Pandion haliaetus*. This Palearctic species can be observed between October and April along the various channels of the Ogooué, lakes, and the coast (Sargeant, 1993; Schepers & Marteijn, 1993). In July and August 2014, several individuals were observed in the Great Lakes region and on the Ogooué. They were summering individuals and were probably immature. The wintering population in Cape Lopez Bay was estimated at 10-15 birds in February 1992 (Schepers & Marteijn, 1993). The total wintering population across the Lower Ogooué complex could reach 60-100 individuals. Banded birds were found to come from Finland (10 individuals) and Sweden (1).

Alcedinidae

Shining-blue kingfisher, *Alcedo quadribrachys*. This resident and nesting Afrotropical species is common along all watercourses bordered by shrubby or tree formations.

Malachite kingfisher, *Corythornis cristatus*. This resident and nesting Afrotropical species is common along open waterways, especially on the lakes of the Lower Ogooué.

Blue-breasted Kingfisher, *Halcyon malimbica*. This resident and nesting Afrotropical species is common in riparian and flooded forests, both in the eastern part of the wetland complex and in the southern delta, as well as in mangroves.

Giant kingfisher, *Megaceryle maxima*. This resident and nesting Afrotropical species is common all along the Ogooué, in forests and more open environments.

Pied kingfisher, *Ceryle rudis*. This resident and nesting Afrotropical species is common along all the major rivers of the Lower Ogooué region. It nests during the dry season on steep banks (Figure 9.49).

Meropidae

Rosy bee-eater, *Merops malimbicus*. This Afrotropical species is endemic to Lower Guinea and nests mainly in the white sands of the old dunes of the coastal zone, but it seems that small colonies of a few dozen pairs also occur in the alluvial plain and in the inland deltas, where they nest in August.

European bee-eater, *Merops apiaster*. This Palearctic species is not generally associated with aquatic environments, but wintering birds are often encountered near the water. In February 2014, a large concentration was observed around islands in Nazaré Bay.

Blue-cheeked bee-eater, *Merops persicus*. This Palearctic species is more closely associated with aquatic environments than the European bee-eater. A small wintering population was found at Lake Gomé in December 2011 and further downstream in December 2008 and November 2009.

Malaconotidae

Swamp boubou, *Laniarius bicolor*. This Afrotropical species is restricted to mangroves, at least in Gabon. It can be heard throughout the mangroves of the Ogooué Delta, and it even frequents the niaouli forests of Mandji Island. Schepers & Marteijn (1993) mention it from the "Barre des Portugais," Cape Lopez, and Three Rivers, north of Port-Gentil.

9.48. Terns on the beach of the "Barre des Portugais" in August 2014. In the foreground, six Damara terns, and in the background, common terns in "winter" plumage (Photo: Jean P. Vande weghe).

Hirundinidae

Gray-rumped swallow, *Pseudhirundo griseopyga*. This bird nests on all the sandbanks of the Lower Ogooué complex but not on the sandbars (Figure 9.50). It does so in loose colonies on the top of these sandbanks, usually where the surface is marked by a small, steep, 5-10 cm threshold in which it digs an inclined gallery. Its total

9.49. A pied kingfisher in front of its nest on the banks of the Rivière des Pélicans (Photo: Jean P. Vande weghe).

9.50. A gray-rumped swallow (Photo: Jean P. Vande weghe).

population can be estimated at 2,500–3,000 individuals. This species also nests elsewhere in the white sand savannas of the coastal basin (e.g., in Nyonié, in the Wonga-Wongué Reserve, and in Pongara and Loango national parks), but its Ogooué population is by far the largest. This population, however, appears to have declined significantly over the past 10–15 years. In August 2004, more than 2,000 birds nested on large sandbanks located between Lambaréné and the river's confluence with the Ngounié. At the end of August 1999, i.e., just after nesting, gatherings of 5,000–10,000 individuals could be observed above these same sandbanks at the end of the day. When these birds all flew together, they resembled a cloud of mosquitoes rising above the river. A decrease in population size could be linked to the fact that school children collect their eggs and young from the sandbanks when on holiday. This phenomenon was often observed between 1999 and 2006.

White-throated blue swallow, *Hirundo nigrita*. This species nests along rivers in forested areas. It is common on small tributaries, but also inhabits the banks of the Ogooué's main river course, especially near Lambaréné.

9.51. An orange weaver in the upper deltaic plain in July 2006 (Photo: Jean P. Vande weghe).

Acrocephalidae

Sedge warbler, *Calamodus schoenobaenus*. This Palearctic migrant frequents the high reed beds found in the alluvial plain and also upstream of Lambaréné. It is uncommon.

Great reed warbler, *Acrocephalus arundinaceus*. This Palearctic migrant, which is larger than the previous species, lives mainly in *Alchornea* thickets. It is a very discreet bird, more often heard than seen. It has been noted from December to April in several parts of the alluvial plain, and especially in the inland deltas. It is impossible to estimate the number of individuals wintering in the region, but it may be quite large.

Cisticolidae

Nile cisticola, *Cisticola marginatus* (ex *Cisticola galactotes*). This sedentary Afrotropical species is present in all herbaceous marshes (reed beds with *Echinochloa* and *Vossia*, stands of *Polygonum*, etc.).

Yellow-throated apalis, *Apalis flavida*. This species has a vast distribution in tropical Africa, but in Gabon it is mainly inhabiting mangroves. In the Lower Ogooué region it was recorded in the northern delta, near Port-Gentil, and at the outlet of the Nkomi Lagoon.

Muscicapidae

White-browed forest flycatcher, *Fraseria cinerascens*. This Afrotropical species lives in galleries found along rivers, where it frequents the lower strata of the forest and feeds mainly on the ground. It is common in the eastern part of the alluvial plain and also upstream from Lambaréné in floodplain forests.

Cassin's flycatcher, *Muscicapa cassini*. This bird inhabits forest rivers. It is common in the Lower Ogooué valley, especially on the river's small tributaries.

Ploceidae

Orange weaver, *Ploceus aurantius*. This bird nests in large colonies along the rivers of the alluvial plain. Its population must comprise tens of thousands of individuals (Figure 9.51).

Village Weaver, *Ploceus cucullatus*. This weaver, common throughout Gabon, is not limited to areas surrounding wetlands, but it nevertheless nests in large colonies along the rivers of the alluvial plain and also upstream from Lambaréné, often near other weavers.

Slender-billed weaver, *Ploceus pelzelni*. This species nests along streams. It often builds its nest among papyrus, and has adapted to the urban environment of Port-Gentil.

Appendix 2: List of birds of the Ogooué Delta

IUCN (level of threats): CR = Critically Endangered; EN = Endangered; VU = Vulnerable; NT = Near Threatened
Biogeography: Pal = Palearctic; (Pal) = Palearctic and/or Afrotropical; N = Nearctic
Status: R = Resident (observable all year); N = Breeder; M = Migratory (absent at certain times of the year); Irr = Irregular or very rare; Occ = Occasional.
The status indicated is that in the Lower Ogooué complex and may therefore be different from the status at the national level.
Ecology: F = Forest; N = Nonforest; A = Aquatic; M = Marine*
Distribution: A = Lower Ogooué valley; B = Upper deltaic plain; C = Lower deltaic plain; D= Coastal sands and lagoon.

	IUCN	Biogeo.	Status	Ecology	A	B	C	D
GALLIFORMES								
NUMIDIDAE								
Black Guinea-fowl , *Agelastes niger*			RN	F	■	■		
Plumed Guinea-fowl, *Guttera plumifera*			RN	F	■	■		
PHASIANIDAE								
Gallinae								
Latham's Francolin, *Peliperdrix lathami*			RN	F	■	■		
Tetraogallinae								
Harlequin Quail, *Coturnix delegorguei*			M	N				
Blue Quail, *Excalfactoria adansonii*			M	N				
Scaly Francolin, *Pternistis squamatus*			RN	F	■	■	■	■
ANSERIFORMES								
ANATIDAE								
Dendrocygninae								
Fulvous Whistling Duck, *Dendrocygna bicolor*			Occ	A				
White-faced Whistling Duck, *Dendrocygna viduata*			RN	A		■		■
Plectropterinae								
Spur-winged Goose, *Plectropterus gambensis*			Occ	A				■
Anserinae								
African Pygmy Goose, *Nettapus auritus*			R	A		■		■
Anatinae								
Egyptian Goose, *Alopochen aegyptiaca*			Occ	A				■
Hartlaub's Duck, *Pteronetta hartlaubii*			RN	A	■	■		■
Knob-billed Duck, *Sarkidiornis melanotos*			Occ	A				■
Northern Pintail, *Anas acuta*		Pal	Occ	A				■
Garganey, *Anas querquedula*		Pal	M	A				■
Common Pochard, *Aythya ferina*		Pal	Occ	A				■
PODICIPEDIFORMES								
PODICIPEDIDAE								
Little Grebe, *Tachybaptus ruficollis*			RN	A	■	■	■	■
PHOENICOPTERIFORMES								
PHOENICOPTERIDAE								
Lesser Flamingo, *Phoeniconaias minor*	NT		Irr	A			■	■
COLUMBIFORMES								
COLUMBIDAE								
Columbinae								
Wzestern Bronze-naped Pigeon, *Columba iriditorques*			RN	F	■	■		■

	IUCN	Biogeo.	Status	Ecology	A	B	C	D
Afep Pigeon, *Columba unicincta*			RN	F	■	■		
Red-eyed Dove, *Streptopelia semitorquata*			RN	FN	■	■	■	■
Ring-necked Dove, *Streptopelia capicola*			Irr	N				■
Laughing Dove, *Spilopelia senegalensis*			RN	N				■
Raphinae								
African Green Pigeon, *Treron calvus*			RN	FN	■	■	■	■
Namaqua Dove, *Oena capensis*			Irr	N				■
Blue-headed Wood Dove, *Turtur brehmeri*			RN	F	■		■	■
Tambourine Dove, *Turtur tympanistria*			RN	F	■	■	■	■
Blue-spotted Wood Dove, *Turtur afer*			RN	F	■	■	■	■
Emerald-spotted Wood Dove, *Turtur chalcospilos*			RN	N				■
CUCULIFORMES								
CUCULIDAE								
Centropodinae								
Gabon Coucal, *Centropus anselli*			RN	F	■	■		■
Black Coucal, *Centropus grillii*			MN	N				■
Senegal Coucal, *Centropus senegalensis*			RN	N	■	■	■	■
Blue-headed Coucal, *Centropus monachus*			RN	N	■	■	■	■
Cuculinae								
Yellowbill, *Ceuthmochares aereus*			RN	F	■	■	■	■
Jacobin Cuckoo, *Clamator jacobinus*			M	N		■		■
Levaillant's Cuckoo, *Clamator levaillantii*			M	N	■			■
Great Spotted Cuckoo, *Clamator glandarius*		(Pal)	Irr	N				■
Thick-billed Cuckoo, *Pachycoccyx audeberti*			R	F		■		
African Emerald Cuckoo, *Chrysococcyx cupreus*			RN	F	■	■		■
Yellow-throated Cuckoo, *Chrysococcyx flavigularis*			RN	F	■	■		
Klaas's Cuckoo, *Chrysococcyx klaas*			RN	FN	■	■		■
Didric Cuckoo, *Chrysococcyx caprius*			RN	N	■	■	■	■
Red-chested Cuckoo, *Cuculus solitarius*			RN	FN	■	■	■	■
Black Cuckoo, *Cuculus clamosus*			RN	F	■	■	■	■
Common Cuckoo, *Cuculus canorus*		Pal	M	N				■
African Cuckoo, *Cuculus gularis*			M	N				■
MUSOPHAGIFORMES								
MUSOPHAGIDAE								
Corythaeolinae								
Great Blue Turaco, *Corythaeola cristata*			RN	F	■	■		
Musophaginae								
Green Turaco, *Tauraco persa*			RN	F	■	■	■	■
Yellow-billed Turaco, *Tauraco macrorhynchus*			RN	F	■	■	■	■

* "Marine" refers to species that, outside of their nesting period, live on the open sea and are completely independent of the coasts, which they most often avoid.

	IUCN	Biogeo.	Status	Ecology	A	B	C	D
OTIDIFORMES								
OTIDIDAE								
Black-belied Bustard, *Lissotis melanogaster*			RN	N		■		
CAPRIMULGIFORMES								
CAPRIMULGIDAE								
Bates's Nightjar, *Caprimulgus batesi*			RN	F	■	■		
Swamp Nightjar, *Caprimulgus natalensis*			RN	N		■		■
Long-tailed Nightjar, *Caprimulgus climacurus*			M	N		■		■
Square-tailed Nightjar, *Caprimulgus fossii*			RN	N	■	■	■	■
Plain Nightjar, *Caprimulgus inornatus*			Occ	N				■
European Nightjar, *Caprimulgus europaeus*		Pal	M	N		■		
Pennant-winged Nightjar, *Macrodipteryx vexillarius*			M	N				■
APODIFORMES								
APODIDAE								
Sabine's Spinetail, *Rhaphidura sabini*			R	F	■	■		
Black Spinetail, *Telacanthura melanopygia*			R	F	■			
Mottled Spinetail, *Telacanthura ussheri*			R	NF	■			
Cassin's Spinetail, *Neafrapus cassini*			R	F	■	■		■
Böhm's Spinetail, *Neafrapus boehmi*			Occ	N				■
Palm Swift, *Cypsiurus parvus*			R	N	■	■	■	■
Common Swift, *Apus apus*		Pal	M	N	■	■	■	■
Bates's Swift, *Apus batesi*			R	N	■			
Little Swift, *Apus affinis*			R	N	■	■	■	■
GRUIFORMES								
SAROTHRURIDAE								
White-spotted Flufftail, *Sarothrura pulchra*			RN	A	■	■		
Red-chested Flufftail, *Sarothrura rufa*			R	A		■		
Grey-throated Rail, *Canirallus oculeus**			RN	A				■
HELIORNITHIDAE								
African Finfoot, *Podica senegalensis*			RN	A	■	■	■	■
RALLIDAE								
Nkulengu Rail, *Himantornis haematopus*			RN	F	■	■		
African Crake, *Crex egregia*			MN	N		■		■
African Water Rail, *Rallus caerulescens*			M	A		■		
Black Crake, *Amaurornis flavirostra*			RN	A	■	■	■	■
Allen's Gallinule, *Porphyrio alleni*			RN	A		■		■
Lesser Moorhen, *Paragallinula angulata*			RN	A		■		■
CHARADRIIFORMES								
TURNICIDAE								
Little Buttonquail, *Turnix sylvaticus*			MN	N		■		
Black-rumped Buttonquail, *Turnix nanus*			M	N		■		■
BURHINIDAE								
Water Thick-knee, *Burhinus vermiculatus*			RN	N	■	■	■	■
Spotted Thick-knee, *Burhinus capensis*			Occ	N				■

* This rail was thought to belong to the family Rallidae (Vande weghe *et al.*, 2016), but recent studies have shown that it belongs to Sarothruridae (Boyd, 2017).

	IUCN	Biogeo.	Status	Ecology	A	B	C	D
CHARADRIIDAE								
Vanellinae								
White-headed Lapwing, *Vanellus albiceps*			RN	A	■	■	■	■
Senegal Lapwing, *Vanellus lugubris*			M	N		■	■	■
Charadriinae								
Little Ringed Plover, *Tinornis dubius*		Pal	M	A				■
Common Ringed Plover, *Charadrius hiaticula*		Pal	M	A			■	■
Forbes's Plover, *Afroxyechus forbesi*			RN	N				■
Anarhynchinae								
White-fronted Plover, *Ochthodromus marginatus*			RN	A	■	■	■	■
Kittlitz's Plover, *Ochthodromus pecuarius*			RN	N				■
Caspian Plover, *Eupoda asiatica*		Pal	M	N				■
PLUVIALIDAE								
American Golden Plover, *Pluvialis dominica*		N	Occ	N				■
Pacific Golden Plover, *Pluvialis fulva*		Pal	M	N/A			■	■
Grey Plover, *Pluvialis squatarola*		Pal	M	A			■	■
RECURVIROSTRIDAE								
Common Stilt, *Himantopus himantopus*		?	M	A				■
Pied Avocet, *Recurvirostra avosetta*		?	M	A		■		
HAEMATOPODIDAE								
Eurasian Oystercatcher, *Haematopus ostralegus*		Pal	M	A				■
ROSTRATULIDAE								
Greater Painted-snipe, *Rostratula benghalensis*			M	A				■
JACANIDAE								
African Jacana, *Actophilornis africanus*			RN	A	■	■	■	■
SCOLOPACIDAE								
Numeniinae								
Upland Sandpiper, *Bartramia longicauda*		N	Occ	N				■
Whimbrel, *Numenius phaeopus*		Pal	M	A			■	■
Eurasian Curlew, *Numenius arquata*	NT	Pal	M	A			■	■
Limosinae								
Black-tailed Godwit, *Limosa limosa*		Pal	Occ	A				■
Bar-tailed Godwit, *Limosa lapponica*	NT	Pal	M	A			■	■
Arenariinae								
Ruddy Turnstone, *Arenaria interpres*		Pal	M	A			■	■
Red Knot, *Calidris canutus*	NT	Pal	M	A			■	■
Ruddy Turnstone, *Calidris alba*		Pal	M	A			■	■
Little Stint, *Calidris minuta*		Pal	M	A			■	■
Pectoral Sandpiper, *Calidris melanotos*		Pal	Occ	A				■
Curlew Sandpiper, *Calidris ferrugineus*	NT	Pal	M	A			■	■
Dunlin, *Calidris alpina*		Pal	Occ	A				■
Buff-breasted Sandpiper, *Calidris subruficollis*		N	Occ	A				■
Ruff, *Calidris pugnax*		Pal	M	A				■
Scolopacinae								
Common Snipe, *Gallinago gallinago*		Pal	M	A				■
Great Snipe, *Gallinago media*	NT	Pal	M	A/N				■
Tringiinae								
Common Redshank, *Tringa totanus*		Pal	M	A	■		■	■
Marsh Sandpiper, *Tringa stagnatilis*		Pal	M	A			■	■
Common Greenshank, *Tringa nebularia*		Pal	M	A	■	■	■	■

	IUCN	Biogeo.	Status	Ecology	A	B	C	D
Green Sandpiper, *Tringa ochropus*		Pal	M	A				■
Wood Sandpiper, *Tringa glareola*		Pal	M	A	■	■	■	■
Common Sandpiper, *Actitis hypoleucos*		Pal	M	A	■	■	■	■
Wilson's Phalarope, *Phalaropus tricolor*		Nt	Occ	A				■
GLAREOLIDAE								
Temminck's Courser, *Cursorius temminckii*			RN	N				■
Collared Pratincole, *Glareola pratincola*			M	N				■
Black-winged Pratincole, *Glareola nordmanni*		Pal	Irr	N				■
Rock Pratincole, *Glareola nuchalis*			RN	A	■	■		
Grey Pratincole, *Glareola cinerea*			RN	A	■	■	■	■
STERCORARIIDAE								
Arctic Skua, *Stercorarius parasiticus*		Pal	Occ	Mar				■
LARIDAE								
Rynchopinae								
African Skimmer, *Rynchops flavirostris*	NT		RN	A	■	■	■	■
Anoinae								
Black Noddi, *Anous minutus*			Occ	Mar				■
Brown Noddi, *Anous stolidus*			Occ	Mar				■
Sterninae								
Bridled Tern, *Onychoprion anaethetus*			Occ	Mar				■
Little Tern, *Sternula albifrons*		(Pal)	RN	A	■	■	■	■
Damara Tern, *Sternula balaenarum*	VU		M	A			■	■
Caspian Tern, *Hydroprogne caspia*		(Pal)	RN	A	■	■	■	■
Black Tern, *Chlidonias niger*		Pal	M	A			■	■
White-winged Tern, *Chlidonias leucopterus*		Pal	M	A	■			■
Royal Tern, *Thalasseus maximus*			M	A			■	■
Sandwich Tern, *Thalasseus sandvicensis*		Pal	M	A			■	■
Common Tern, *Sterna hirundo*		(Pal)	RN	A	■	■	■	■
Arctic Tern, *Sterna paradisaea*		Pal	M	Mar				■
Larinae								
Sabine's Gull, *Xema sabini*		Nt	M	Mar				■
Black-headed Gull, *Chroicocephalus ridibundus*		Pal	Occ	A				■
Little Gull, *Hydrocoloeus minutus*		Pal	Occ	Mar				■
Western Gull, *Larus dominicanus*			M	A				■
Lesser Black-backed Gull, *Larus fuscus*		Pal	M	A			■	■
PHAETHONTIFORMES								
PHAETHONTIDAE								
Ascencion Frigatebird, *Phaethon aethereus*			Occ	Mar			□	□
PROCELLARIIFORMES								
PROCELLARIIDAE								
Sooty Shearwater, *Puffinus griseus*			M	Mar				□
OCEANITIDAE								
Wilson's Petrel, *Oceanites oceanicus*			M	Mar				□
SULIFORMES								
SULIDAE								
Cape Gannet, *Morus capensis*	VU		M	Mar				□
Brown Booby, *Sula leucogaster*			M	Mar				□

	IUCN	Biogeo.	Status	Ecology	A	B	C	D
ANHINGIDAE								
African Darter, *Anhinga rufa*			RN	A	■	■	■	■
PHALACROCORACIDAE								
Reed Cormorant, *Microcarbo africanus*			RN	A	■	■	■	■
Cape Cormorant, *Phalacrocorax capensis*			Occ	A				■
CICONIIFORMES								
CICONIIDAE								
Marabou Stork, *Leptoptilos crumeniferus*			Occ	AN	■			■
Abdim's Stork, *Ciconia abdimii*			M	AN	■	■		■
Woolly-necked Stork, *Ciconia episcopus*			RN	AN	■	■	■	■
White Stork, *Ciconia ciconia*		Pal	M	N				■
Saddle-billed Stork, *Ephippiorhynchus senegalensis*			?	A		■	■	■
Yellow-billed Stork, *Mycteria ibis*			M	A		■	■	■
ARDEIFORMES								
ARDEIDAE								
Tigriornithinae								
White-crested Tiger Heron, *Tigriornis leucolopha*			RN	AF	■	■		■
Botaurinae								
Great Bittern, *Botaurus stellaris*		Pal	Occ	A				■
Little Bittern, *Ixobrychus minutus*		(Pal)	RN	A	■	■	■	■
Dwarf Bittern, *Ixobrychus sturmii*			M	A	■			■
Nycticoracinae								
White-backed Night Heron, *Gorsachius leuconotus*			R	A	■	■		■
Black-crowned Night Heron, *Nycticorax nycticorax*			RN	A	■	■		■
Ardeinae								
Cattle Egret, *Bubulcus ibis*			RN	AN	■	■	■	■
Intermediate Egret, *Mesophoyx intermedia*			RN	A	■	■	■	■
Great Egret, *Casmerodius albus*			RN	A	■	■	■	■
Purple Heron, *Ardea purpurea*			RN	A	■	■	■	■
Gray Heron, *Ardea cinerea*			R	A	■	■	■	■
Black-headed Heron, *Ardea melanocephala*			M	AN				■
Goliath Heron, *Ardea goliath*			R	A	■	■	■	■
Green-backed Heron, *Butorides striata*			RN	A	■	■	■	■
Squacco Heron, *Ardeola ralloides*			M	A	■	■	■	■
Black Heron, *Egretta ardesiaca*			Irr	A		■		■
Western Reef Egret, *Egretta gularis*			R	A			■	■
Little Egret, *Egretta garzetta*			R	A	■	■	■	■
PLATALEIFORMES								
THRESKIORNITHIDAE								
Threskiornithinae								
Hadada Ibis, *Bostrychia hagedash*			RN	A	■	■	■	■
Olive Ibis, *Bostrychia olivacea*			R	AF	■			
Spot-breasted Ibis, *Bostrychia rara*			R	AF	■	■		
Sacred Ibis, *Threskiornis aethiopicus*			M	A			■	■
African Spoonbill, *Platalea alba*			M	A		■	■	■

	IUCN	Biogeo.	Status	Ecology	A	B	C	D
PELECANIFORMES								
SCOPIDAE								
Hamerkop, *Scopus umbretta*			RN	A	■	■	■	■
PELECANIDAE								
Great White Pelican, *Pelecanus onocrotalus*			M	A		■		■
Pink-backed Pelican, *Pelecanus rufescens*			RN	A	■	■	■	■
ACCIPITRIFORMES								
PANDIONIDAE								
Western Osprey, *Pandion haliaetus*		Pal	R	A	■	■	■	■
ACCIPITRIDAE								
Elaninae								
Black-shouldered Kite, *Elanus caeruleus*			M	N				■
Gypaetinae								
African Harrier Hawk, *Polyboroides typus*			RN	NF	■	■	■	■
Palm-nut Vulture, *Gypohierax angolensis*			RN	AF	■	■	■	■
Perninae								
European Honey Buzzard, *Pernis apivorus*		Pal	M	F	■	■		■
African Cuckoo Hawk, *Aviceda cuculoides*			R	F	■	■		■
Gypinae								
White-headed Vulture, *Trigonoceps occipitalis*	CR		Occ	N				■
Circaetinae								
Short-toed Snake Eagle, *Circaetus gallicus*		Pal	Occ	N				■
Black-breasted Snake Eagle, *Circaetus pectoralis*			M	N				■
Brown Snake Eagle, *Circaetus cinereus*			Occ	N				■
Congo Serpent Eagle, *Circaetus spectabilis*			RN	F	■	■		■
Harpiinae								
Bat Hawk, *Macheiramphus alcinus*			RN	NF	■	■		■
Aquilinae								
Crowned Eagle, *Stephanoaetus coronatus*	NT		RN	F	■	■		■
Long-crested Eagle, *Lophaetus occipitalis*			RN	N		■		■
Wahlberg's Eagle, *Hieraaetus wahlbergi*			Occ	N				■
Ayres's Hawk Eagle, *Hieraaetus ayresii*			M	N	■			■
Booted Eagle, *Hieraaetus pennatus*		Pal	M	N		■		■
Tawny Eagle, *Aquila rapax*			M	N				■
Cassin's Hawk Eagle, *Aquila africana*			RN	F	■	■		
Buteoninae								
Yellow-billed Kite, *Milvus aegyptius*			M	N	■	■	■	■
African Fish Eagle, *Icthyophaga vocifer*			RN	A	■	■	■	■
Common Buzzard, *Buteo buteo*		Pal	M	N			■	■
Red-necked Buzzard, *Buteo auguralis*			RN	N		■		■
Accipitrinae								
Lizard Buzzard, *Kaupifalco monogrammicus*			R	N	■	■		■
Gabar Goshawk, *Micronisus gabar*			Occ	N				■
Long-tailed Hawk, *Urotriorchis macrourus*			RN	F	■	■		
Red-thighed Sparrowhawk, *Tachyspiza erythropus*			RN	F	■	■		
Chestnut-flanked Sparrowhawk, *Aerospiza castanilius*			R	F	■			
Red-chested Goshawk, *Aerospiza toussenelii*			RN	F	■			■
Black Sparrowhawk, *Astur melanoleucus*			RN	f	■	■		■
Eurasian Marsh Harrier, *Circus aeruginosus*		Pal	M	N		■	■	■

	IUCN	Biogeo.	Status	Ecology	A	B	C	D
STRIGIFORMES								
TYTONIDAE								
Barn Owl, *Tyto alba*			RN	N				■
STRIGIDAE								
Surninae								
Red-chested Owlet, *Glaucidium tephronotum*			R	F	■			
Striginae								
Fraser's Eagle Owl, *Bubo poensis*			R	F	■			
Pel's Fishing Owl, *Scotopelia peli*			RN	AF	■	■		
Vermiculated Fishing Owl, *Scotopelia bouvieri*			RN	AF	■	■		
African Wood Owl, *Strix woodfordii*			RN	F	■	■		■
COLIIFORMES								
COLIIDAE								
Speckled Mousebird, *Colius striatus*			RN	N	■			
TROGONIFORMES								
TROGONIDAE								
Bare-cheeked Trogon, *Apaloderma aequatoriale*			RN	F	■			
Narina's Trogon, *Apaloderma narina*			RN	NF	■	■		
BUCEROTIFORMES								
UPUPIDAE								
Hoopoe, *Upupa epops*		(Pal)	M	N		■		■
BUCEROTIDAE								
White-crested Hornbill, *Tropicranus albocristatus*			RN	F	■	■		■
Black Dwarf Hornbill, *Horizocerus hartlaubi*			RN	F	■			
Red-billed Dwarf Hornbill, *Lophoceros camurus*			RN	F	■	■		■
African Pied Hornbill, *Lophoceros fasciatus*			RN	F	■	■	■	■
Piping Hornbill, *Bycanistes fistulator*			RN	F	■	■	■	■
White-thighed Hornbill, *Bycanistes albotibialis*			RN	F	■	■		
Black-casqued Hornbill, *Ceratogymna atrata*			RN	F	■	■		■
CORACIIFORMES								
CORACIIDAE								
European Roller, *Coracias garrulus*		Pal	M					■
Blue-throated Roller, *Eurystomus gularis*			RN	F	■	■	■	
Broad-billed Roller, *Eurystomus glaucurus*			M	N	■	■		■
ALCEDINIDAE								
Alcedininae								
African Dwarf Kingfisher, *Ispidina lecontei*			RN	F	■	■		
African Pygmy Kingfisher, *Ispidina picta*			RN	N	■	■	■	■
White-bellied Kingfisher, *Corythornis leucogaster*			RN	AF	■	■		
Malachite Kingfisher, *Corythornis cristatus*			RN	A	■	■	■	■
Shining-blue Kingfisher, *Alcedo quadribrachys*			RN	AF	■	■	■	■
Daceloninae					■			
Chocolate-backed Kingfisher, *Halcyon badia*			RN	F	■	■		
Grey-headed Kingfisher, *Halcyon leucocephala*			M	N		■		■
Blue-breasted Kingfisher, *Halcyon malimbica*			RN	F	■	■	■	■
Woodland Kingfisher, *Halcyon senegalensis*			RN	F	■	■	■	■

	IUCN	Biogeo.	Status	Ecology	A	B	C	D
Cerylinae								
Giant Kingfisher, *Megaceryle maxima*			RN	A	■	■	■	■
Pied Kingfisher, *Ceryle rudis*			RN	A	■	■	■	■
MEROPIDAE								
Black-headed Bee-eater, *Merops breweri*			RN	F		■		■
Black Bee-eater, *Merops gularis*			RN	F	■	■	■	■
Blue-breasted Bee-eater, *Merops variegatus*			RN	N	■	■	■	■
White-fronted Bee-eater, *Merops bullockoides*			RN	N		■		■
White-throated Bee-eater, *Merops albicollis*			M	N		■		■
Blue-cheeked Bee-eater, *Merops persicus*		Pal	M	N		■	■	
European Bee-eater, *Merops apiaster*		Pal	M	N			■	■
Rosy Bee-eater, *Merops malimbicus*			MN	N	■	■	■	■
PICIFORMES								
LYBIIDAE								
Grey-throated Barbet, *Gymnobucco bonapartei*			RN	F	■	■		
Naked-faced Barbet, *Gymnobucco calvus*			RN	F	■	■		
Speckled Tinkerbird, *Pogoniulus scolopaceus*			RN	F	■	■	■	■
Red-rumped Tinkerbird, *Pogoniulus atroflavus*			RN	F	■	■		
Yellow-throated Tinkerbird, *Pogoniulus subsulphureus*			RN	F	■	■	■	■
Yellow-rumped Tinkerbird, *Pogoniulus bilineatus*			RN	F	■	■	■	■
Yellow-spotted Barbet, *Buccanodon duchaillui*			RN	F	■	■		■
Streaky-throated Barbet, *Tricholaema hirsuta*			RN	F	■	■	■	■
Yellow-billed Barbet, *Trachylaemus purpuratus*			RN	F	■	■		
INDICATORIDAE								
Cassin's Honeybird, *Prodotiscus insignis*			RN	F	■	■		
Lyre-tailed Honeyguide, *Melichneutes robustus*			RN	F	■	■		
Spotted Honeyguide, *Indicator maculatus*			RN	F	■			
Thick-billed Honeyguide, *Indicator conirostris*			RN	F	■	■		■
Least Honeyguide, *Indicator exilis*			RN	F	■	■		■
Willcocks's Honeyguide, *Indicator willcocksi*			RN	F	■	■		■
PICIDAE								
Picumninae								
African Piculet, *Sasia africana*			RN	F	■			
Picinae								
Green-backed Woodpecker, *Campethera cailliautii*			RN	F	■	■		
Buff-spotted Woodpecker, *Campethera nivosa*			RN	F	■	■		
Brown-eared Woodpecker, *Campethera caroli*			RN	F	■	■	■	■
Gabon Woodpecker, *Dendropicos gabonensis*			RN	F	■	■		■
Cardinal Woodpecker, *Dendropicos fuscescens*			RN	NF	■	■		■
Golden-crested Woodpecker, *Dendropicos xantholophus*			RN	F	■	■		■
FALCONIFORMES								
FALCONIDAE								
Falconinae								
Common Kestrel, *Falco tinnunculus*		(Pal)	M	N				■
Gray Kestrel, *Falco ardosiaceus*			M	N		■		
Red-footed Falcon, *Falco vespertinus*		Pal	M	N				■
Lanner Falcon, *Falco biarmicus*			M	N				■
Peregrine Falcon, *Falco peregrinus*		(Pal)	M	N				■

	IUCN	Biogeo.	Status	Ecology	A	B	C	D
PSITTACIFORMES								
PSITTACIDAE								
Psittacinae								
Grey Parrot, *Psittacus erithacus*	EN		RN	F	■	■	■	■
Red-fronted Parrot, *Poicephalus gulielmi*			RN	F	■	■		■
PASSERIFORMES								
SMITHORNITHIDAE								
Rufous-sided Broadbill, *Smithornis rufolateralis*			RN	F	■	■		
PITTIDAE								
African Pitta, *Pitta angolensis*			RN	F	■			
ORIOLIDAE								
Black-winged Oriole, *Oriolus nigripennis*			RN	F	■	■		
Western Black-headed Oriole, *Oriolus brachyrhynchus*			RN	F	■	■		
Eurasian Golden Oriole, *Oriolus oriolus*		Pal	M	N				■
CAMPEPHAGIDAE								
Purple-throated Cuckoo-shrike, *Campephaga quiscalina*			RN	F	■	■		
Blue Cuckoo-shrike, *Cyanograucalus azureus*			RN	F	■	■		
PLATYSTEIRIDAE								
Chestnut Wattle-eye, *Dyaphorophyia castanea*			RN	F	■	■		■
White-spotted Wattle-eye, *Dyaphorophyia tonsa*			RN	F	■	■		
Common Wattle-eye, *Platysteira cyanea*			RN	F	■		■	■
VANGIDAE								
Prionopinae								
Rufous-bellied Helmet-shrike, *Prionops rufiventris*			RN	F	■	■		
Tephrodornithinae								
Shrike Flycatcher, *Megabyas flammulatus*			RN	F	■	■		
Black-and-white Flycatcher, *Bias musicus*			RN	N	■	■		
MALACONOTIDAE								
Fiery-breasted Bush-shrike, *Malaconotus cruentus*			RN	F	■			
Sabine's Puffback, *Dryoscopus sabini*			RN	F	■	■		
Northern Puffback, *Dryoscopus gambensis*			M	N				■
Black-shouldered Puffback, *Dryoscopus senegalensis*			RN	F	■	■		
Sooty Boubou, *Laniarius leucorhynchus*			RN	F	■	■		
Swamp Boubou, *Laniarius bicolor*			RN	F		■	■	■
DICRURIDAE								
Square-tailed Drongo, *Edolius ludwigii*			RN	F	■			
Shining Drongo, *Edolius atripennis*			RN	F	■	■		
Velvet-mantled Drongo, *Edolius coracinus*			RN	F	■	■		
MONARCHIDAE								
Blue-headed Crested Flycatcher, *Trochocercus nitens*			RN	F	■	■		
African Paradise Fly, *Terpsiphone viridis*			RN	NF	■	■		
Rufous-vented Paradise Flycatcher, *Terpsiphone rufocinerea*			RN	F	■	■	■	■
Red-bellied Paradise Flycatcher, *Terpsiphone rufiventer*			RN	F	■	■		■
LANIIDAE								
Lesser Grey Shrike, *Lanius minor*		Pal	M	N				■
Isabelline Shrike, *Lanius isabellinus*		Pal	Occ	N				■
Red-backed Shrike, *Lanius collurio*		Pal	M	N				■
Woodchat Shrike, *Lanius senator*		Pal	Occ	N				■

	IUCN	Biogeo.	Status	Ecology	A	B	C	D
CORVIDAE								
Pied Crow, *Corvus albus*			RN	N	■	■	■	■
HYLIOTIDAE								
Violet-backed Hyliota, *Hyliota violacea*			RN	F	■	■		
STENOSTIRIDAE								
African Blue Flycatcher, *Elminia longicauda*			M	F				■
Dusky Crested Flycatcher, *Elminia nigromitrata*			RN	F		■		
REMIZIDAE								
Forest Penduline Tit, *Anthoscopus flavifrons*			RN	F	■			
ALAUDIDAE								
Flappet Lark, *Mirafra rufocinnamomea*			RN	N		■		■
Red-capped Lark, *Calandrella cinerea*			M	N				■
NICATORIDAE								
Western Nicator, *Nicator chloris*			RN	F	■	■	■	■
MACROSPHENIDAE								
Yellow Longbill, *Macrosphenus flavicans*			RN	F	■	■		
Grey Longbill, *Macrosphenus concolor*			RN	F	■	■		■
Green Crombec, *Sylvietta virens*			RN	F	■	■	■	■
Lemon-bellied Crombec, *Sylvietta denti*			RN	F	■	■		
HIRUNDINIDAE								
Pseudochelidoninae								
African River Martin, *Pseudochelidon eurystomina*			MN	N	■	■	■	■
Hirundininae								
Square-tailed Saw-wing, *Psalidoprocne nitens*			RN	F	■	■		
Black Saw-wing, *Psalidoprocne pristoptera*			RN	N	■	■		
Grey-rumped Swallow, *Pseudhirundo griseopyga*			?N	N	■	■	■	■
Banded Martin, *Riparia cincta*			?	N		■		■
White-throated Blue Swallow, *Hirundo nigrita*			RN	FA	■	■	■	
Red-chested Swallow, *Hirundo lucida*			RN	N	■	■		
Barn Swallow, *Hirundo rustica*		Pal	M	N	■	■	■	■
Common House Martin, *Delichon urbicum*		Pal	M	N				■
Rufous-chested Swallow, *Cecropis semirufa*			RN	N	■	■		■
Lesser Striped Swallow, *Cecropis abyssinica*			RN	N	■	■		■
Red-throated Cliff Swallow, *Petrochelidon rufigula*			RN	N		■		
ACROCEPHALIDAE								
Icterine Warbler, *Hippolais icterina*		Pal	M	N				■
Sedge Warbler, *Calamodus schoenobaenus*		Pal	M	A		■		■
Great Reed Warbler, *Acrocephalus arundinaceus*		Pal	M	A		■		■
Greater Swamp Warbler, *Acrocephalus rufescens*			RN	A		■		■
LOCUSTELLIDAE								
Broad-tailed Warbler, *Schoenicola platyurus*			RN	N		■		
PYCNONOTIDAE								
Pycnonotinae								
Common Bulbul, *Pycnonotus barbatus*			RN	N	■	■	■	■
Crinigerinae								
Golden Greenbul, *Calyptocichla serinus*			RN	F	■	■		
Slender-billed Greenbul, *Stelgidillas gracilirostris*			RN	F	■	■		
Eastern Bearded Greenbul, *Criniger chloronotus*			RN	F	■			
Eastern Bearded Greenbul, *Criniger calurus*			RN	F	■			
White-bearded Greenbul, *Criniger ndussumensis*			RN	F	■			

	IUCN	Biogeo.	Status	Ecology	A	B	C	D
Little Greenbul, *Eurillas virens*			RN	F	■	■		
Cameroon Sombre Greenbul, *Eurillas curvirostris*			RN	F	■			
Little Grey Greenbul, *Eurillas gracilis*			RN	F	■	■		■
Ansorge's Greenbul, *Eurillas ansorgei*			RN	F	■	■		
Yellow-whiskered Greenbul, *Eurillas latirostris*			RN	F	■	■		
Icterine Greenbul, *Phyllastrephus icterinus*			RN	F	■	■		■
Red-tailed Bristlebill, *Bleda syndactylus*			RN	F	■	■		
Lesser Bristlebill, *Bleda notatus*			RN	F	■	■	■	■
Yellow-throated Leaflove, *Atimastillas flavicollis*			RN	NF	■	■	■	■
Spotted Greenbul, *Ixonotus guttatus*			RN	F	■	■		
Swamp Palm Bulbul, *Thescelocichla leucopleura*			RN	F	■	■	■	■
Honeyguide Greenbul, *Baeopogon indicator*			RN	F	■	■		
Sjöstedt's Honeyguide Greenbul, *Baeopogon clamans*			RN	F	■	■		
Simple Leaflove, *Chlorocichla simplex*			RN	N	■	■		
CISTICOLIDAE								
Eremomelinae								
Rufous-crowned Eremomela, *Eremomela badiceps*			RN	F	■	■		■
Grey-backed Camaroptera, *Camaroptera brachyura*			RN	N	■	■	■	■
Yellow-browed Camaroptera, *Camaroptera superciliaris*			RN	F	■	■		
Olive-green Camaroptera, *Camaroptera chloronota*			RN	F	■	■		■
Yellow-breasted Apalis, *Apalis flavida*			RN	F			■	■
Buff -throated Apalis, *Apalis rufogularis*			RN	F	■	■		
Priniinae								
Banded Prinia, *Prinia bairdii*			RN	F	■	■		
Cisticolinae								
Black-faced Rufous Warbler, *Bathmocercus rufus*			RN	F	■	■		
Chattering Cisticola, *Cisticola anonymus*			RN	N	■	■	■	■
Nile Cisticola, *Cisticola marginatus*			RN	A	■	■	■	■
Croaking Cisticola, *Cisticola natalensis*			RN	N		■		
Short-winged Cisticola, *Cisticola brachypterus*			RN	N		■		
Zitting Cisticola, *Cisticola juncidis*			RN	N		■		■
Pectoral-patch Cisticola, *Cisticola brunnescens*			RN	N		■		■
HYLIIDAE								
Green Hylia, *Hylia prasina*			RN	F	■	■	■	■
CETTIIDAE								
Chestnut-capped Flycatcher, *Erythrocercus mccallii*			RN	F	■	■		
PHYLLOSCOPIDAE								
Willow Warbler, *Phylloscopus trochilus*		Pal	M	N		■		■
Wood Warbler, *Rhadina sibilatrix*		Pal	M	N		■		■
SYLVIIDAE								
Garden Warbler, *Sylvia borin*		Pal	M	N				■
Greater Whitethroat, *Curruca communis*		Pal	M	N				■
PELLORNEIDAE								
Pale-breasted Illadopsis, *Illadopsis rufipennis*			RN	F	■	■		■
Brown Illadopsis, *Illadopsis fulvescens*			RN	F	■	■		■
Blackcap Illadopsis, *Illadopsis cleaveri*			RN	F	■	■		
BUPHAGIDAE								
Yellow-billed Oxpecker, *Buphagus africanus*			RN	N		■		

	IUCN	Biogeo.	Status	Ecology	A	B	C	D
STURNIDAE								
Sturninae								
Forest Chestnut-winged Starling, *Onychognathus fulgidus*			RN	F	■	■		■
Splendid Glossy Starling, *Lamprotornis splendidus*			RN	F	■	■	■	■
Purple-headed Glossy Starling, *Hylopsar purpureiceps*			RN	F	■	■		■
Violet-backed Starling, *Cinnyricinclus leucogaster*			M	N				■
Wattled Starling, *Creatophora cinerea*			M	N				■
TURDIDAE								
Red-tailed Ant Thrush, *Neocossyphus rufus*			RN	F	■	■		■
White-tailed Ant Thrush, *Neocossyphus poensis*			RN	F	■	■		
Rufous Flycatcher Thrush, *Stizorhina fraseri*			RN	F	■	■		■
African Thrush, *Turdus pelios*			RN	N	■	■		■
MUSCICAPIDAE								
Muscicapinae								
Fire-crested Alethe, *Alethe castanea*			RN	F	■	■		■
Sooty Flycatcher, *Bradornis fuliginosus*			RN	F	■	■		
White-browed Forest Flycatcher, *Fraseria cinerascens*			RN	F	■	■	■	■
Fraser's Forest Flycatcher, *Fraseria ocreata*			RN	F	■	■		■
Ashy Flycatcher, *Fraseria caerulescens*			RN	F	■			■
Olivaceous Flycatcher, *Fraseria olivascens*			RN	F	■			
Grey-throated Flycatcher, *Fraseria griseigularis*			RN	F	■	■		
Lead-colored Flycatcher, *Fraseria plumbea*			RN	F	■	■	■	■
Spotted Flycatcher, *Muscicapa striata*		Pal	M	N	■	■	■	■
Yellow-footed Flycatcher, *Muscicapa sethsmithi*			RN	F	■			
Cassin's Flycatcher, *Muscicapa cassini*			RN	AF	■	■		
Cossyphinae								
Brown-chested Alethe, *Pseudalethe poliocephala*			RN	F	■	■		
Red-capped Robin Chat, *Cossypha natalensis*			?					■
Snowy-crowned Robin Chat, *Cossypha niveicapilla*			RN	F	■	■		■
Gabon Robin, *Stiphrornis gabonensis*			RN	F	■	■		
Saxicolinae								
Common Stonechat, *Saxicola torquatus*			RN	N		■		■
Whinchat, *Saxicola rubetra*		Pal	M	N				■
Northern Wheatear, *Oenanthe oenanthe*		Pal	M	N				■
NECTARINIIDAE								
Fraser's Sunbird, *Deleornis fraseri*			RN	F	■	■		
Violet-tailed Sunbird, *Anthreptes aurantius*			RN	F	■	■	■	■
Brown Sunbird, *Anthreptes gabonicus*			RN	F	■	■	■	■
Green Sunbird, *Anthreptes rectirostris*			RN	F	■	■		■
Little Green Sunbird, *Anthreptes seimundi*			RN	F	■	■		■
Collared Sunbird, *Hedydipna collaris*			RN	F	■	■		■
Reichenbach's Sunbird, *Anabathmis reichenbachii*			RN	F	■	■	■	■
Green-headed Sunbird, *Cyanomitra verticalis*			RN	F	■	■		■
Blue-throated Brown Sunbird, *Cyanomitra cyanolaema*			RN	F	■	■	■	■
Olive Sunbird, *Cyanomitra olivacea*			RN	F	■	■	■	■
Green-throated Sunbird, *Chalcomitra rubescens*			RN	NF	■	■		■
Carmelite Sunbird, *Chalcomitra fuliginosa*			RN	NF	■	■	■	■
Olive-bellied Sunbird, *Cinnyris chloropygius*			RN	N	■	■	■	■
Tiny Sunbird, *Cinnyris minullus*			RN	F	■	■		
Johanna's Sunbird, *Cinnyris johannae*			RN	F	■	■		■

	IUCN	Biogeo.	Status	Ecology	A	B	C	D
Superb Sunbird, *Cinnyris superbus*			RN	F	■	■	■	■
Bates's Sunbird, *Cinnyris batesi*			RN	F	■	■		
Copper Sunbird, *Cinnyris cupreus*			RN	N		■		■
PLOCEIDAE								
Red-billed Quelea, *Quelea quelea*			RN	N				■
Red-headed Quelea, *Quelea erythrops*			RN	NF	■	■		■
Yellow-mantled Widowbird, *Euplectes macroura*			RN	N				■
Blue-billed Malimbe, *Malimbus nitens*			RN	F	■	■	■	■
Crested Malimbe, *Malimbus malimbicus*			RN	F	■	■	■	
Cassin's Malimbe, *Malimbus cassini*			RN	F	■	■	■	■
Red-headed Malimbe, *Malimbus rubricollis*			RN	F	■	■		
Slender-billed Weaver, *Ploceus pelzelni*			RN	A	■	■	■	■
Loango Weaver, *Ploceus subpersonatus*	VU		RN	NF				■
Black-necked Weaver, *Ploceus nigricollis*			RN	N	■	■	■	■
Orange Weaver, *Ploceus aurantius*			RN	N	■	■		
Vieillot's Black Weaver, *Ploceus nigerrimus*			RN	N	■	■	■	■
Village Weaver, *Ploceus cucullatus*			RN	N	■	■	■	■
Yellow-mantled Weaver, *Ploceus tricolor*			RN	F	■			
VIDUIDAE								
Pin-tailed Whydah, *Vidua macroura*			RN	N	■	■		■
ESTRILDIDAE								
Lonchurinae								
Bronze Mannikin, *Spermestes cucullatus*			RN	N	■	■	■	■
Black-and-white Mannikin, *Spermestes bicolor*			RN	N	■	■		
Magpie Mannikin, *Spermestes fringilloides*			RN	N	■	■		
Estrildinae								
Black-chinned Quailfinch, *Ortygospiza gabonensis*			RN	N		■		■
Black-bellied Seedcracker, *Pyrenestes ostrinus*			RN	F	■	■	■	■
Western Bluebill, *Spermophaga haematina*			RN	F	■	■		■
Grey-crowned Negrofinch, *Nigrita canicapillus*			RN	F	■	■	■	■
Pale-fronted Negrofinch, *Nigrita luteifrons*			RN	F	■	■		
Chestnut-breasted Negrofinch, *Nigrita bicolor*			RN	F	■	■		
White-breasted Negrofinch, *Nigrita fusconotus*			RN	F	■	■		
Green Twinspot, *Mandingoa nitidula*			RN	F	■	■		
Orange-cheeked Waxbill, *Estrilda melpoda*			RN	N	■	■		■
Common Waxbill, *Estrilda astrild*			RN	N	■	■		■
PASSERIDAE								
Gray Sparrow, *Passer griseus*			RN	N	■	■		■
MOTACILLIDAE								
Yellow Wagtail, *Motacilla flava*		Pal	M	N		■		■
Mountain Wagtail, *Motacilla clara*			RN	F	■			
African Pied Wagtail, *Motacilla aguimp*			RN	N	■	■	■	■
Yellow-throated Longclaw, *Macronyx croceus*			RN	N		■		■
Long-legged Pipit, *Anthus pallidiventris*			RN	N	■	■		■
Tree Pipit, *Anthus trivialis*		Pal	M	N	■			■
Red-throated Pipit, *Anthus cervinus*		Pal	Occ	N				■
FRINGILLIDAE								
Black-faced Canary, *Crithagra capistrata*			RN	N		■		■
Yellow-fronted Canary, *Crithagra mozambica*			M	N				■

10. Mammals

Olivia Scholtz, Heather Arrowood & Jean P. Vande weghe

In many parts of the Tropics as well as elsewhere in the world, it is clearly apparent that wildlife is desperately hanging on to survival in a world being crushed by anthropic pressures. The Lower Ogooué gives a very different impression, at least at first glance, since natural environments are still largely dominant. Since the majority of Gabon's human population benefits from oil revenues (Vande weghe, 2017, page 49), they are concentrated in urban areas, while rural populations live in very low density and still largely subsist off the resources offered by nature through hunting, fishing, and low-intensity farming. However, since the 1930s, these rural populations have been concentrated along the main roads. Huge areas of forest have lost their inhabitants. The low density of human populations and their concentration along roads has favored the conservation of vast, nearly intact forests that shelter a substantial mammalian fauna.

However, this clustering has not taken place in the Lower Ogooué region, where roads were almost nonexistent and the population has mainly settled along waterways. The human population has therefore remained more dispersed and its impact is more insidious and evenly distributed across the landscape. The larger fauna of the Lower Ogooué has suffered as a result, and is still under strong hunting pressure. What appears as intact nature is thus very misleading, as indicated below.

Available data

To date, 56 species of mammals have been listed within the limits of the area studied (see the table in the appendix). Two, however, could not be identified with certainty at the species level. These are the two introduced species: the rat, *Rattus* sp., and the mouse, *Mus* sp. Two other species have been observed just outside the site's limits; both are Delphinidae. The list presented in the appendix is undoubtedly very incomplete given that it does not contain many species commonly found in Gabon and that certain families are completely absent, in particular shrews (Soricidae) and most of the families of bats (Chiroptera).

Hunters in the region mention many species not included in this list, which would grow considerably based on their observations, provided they are verified (Appendix, pages 314 and 315). Caution is essential, since Gabonese hunters, like all hunters around the world, do not necessarily view species in the same way as scientists. In fact, these hunters also make mention of species whose presence in the Lower Ogooué region is in fact highly unlikely, in particular the large porcupine, *Hystrix africaeaustralis* (page 315). Their testimonies can, however, be useful in guiding future research. This is why the table presented here includes their most plausible observations in quotation marks. Among the observations reported by WWF investigators (see Le-Duc Yeno *et al.*, 2015), the white-bellied duiker, *Cephalophus leucogaster*, is very suspect since this species is only known to exist, within Gabon, in the far eastern parts of the country.

Gabon is currently known to have 160 species of nonmarine mammals. This number might increase somewhat if systematic studies of micromammals were carried out in marginal areas that have been poorly explored, but the small amount of intensive research that has been carried out in recent years has not added many species new to the national list.

As for the mammalian fauna of the Lower Ogooué region, it is particularly poorly known because very little census work has been done on mammals in the region. Furthermore, the studies that have been conducted focused on large species and on the dryland forests found at the periphery of wetlands. These forests are perhaps the only environments that lend themselves to typical census techniques, but they are unfortunately not representative for the immense wetland

10.1. Two herds of buffalo in the large marshy plain east of Lake Nyaye (March 2017). It was not until 2017 that the interest of this area was understood; it is very difficult to access but is unique within Gabon. More than 100 buffalo were observed during this short exploratory flight, which only skirted along the southern edge of the plain (Photo: Jean P. Vande weghe).

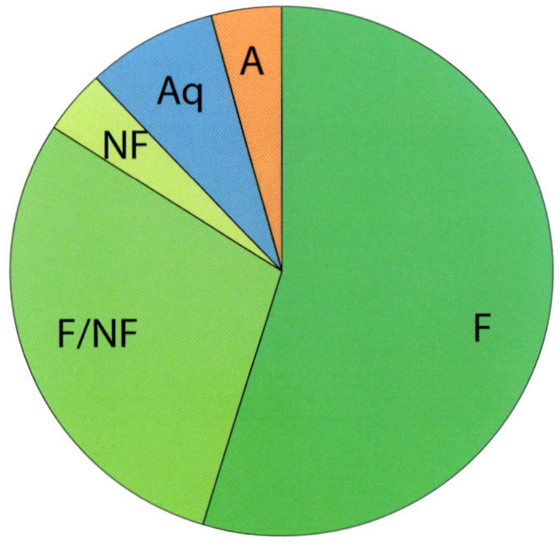

10.2. *Ecological composition of the mammalian fauna: F = forest; F/NF = forest and nonforest; NF = nonforest; Aq = aquatic; A = anthropogenic.*

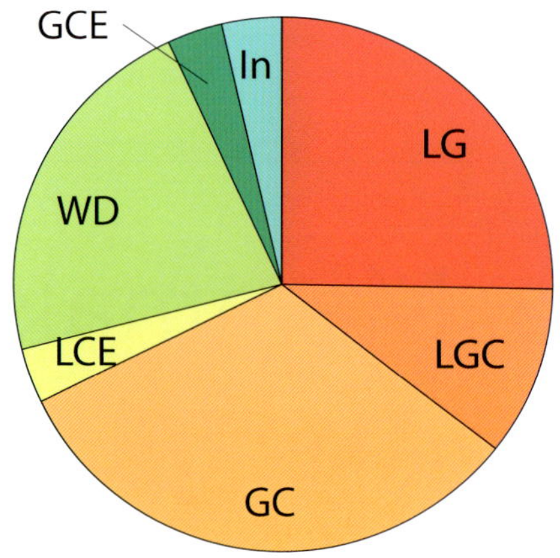

10.3. *Biogeographic composition of the mammalian fauna: LG = endemic to Lower Guinea; LGC = Lower Guinea and the Congolean region; GC = Guineo-Congolean region; LCE = Lower Guinea, Congolean region, and East Africa; WD = wide distribution; GCE = Guineo-Congolean Region and East Africa; In = introduced.*

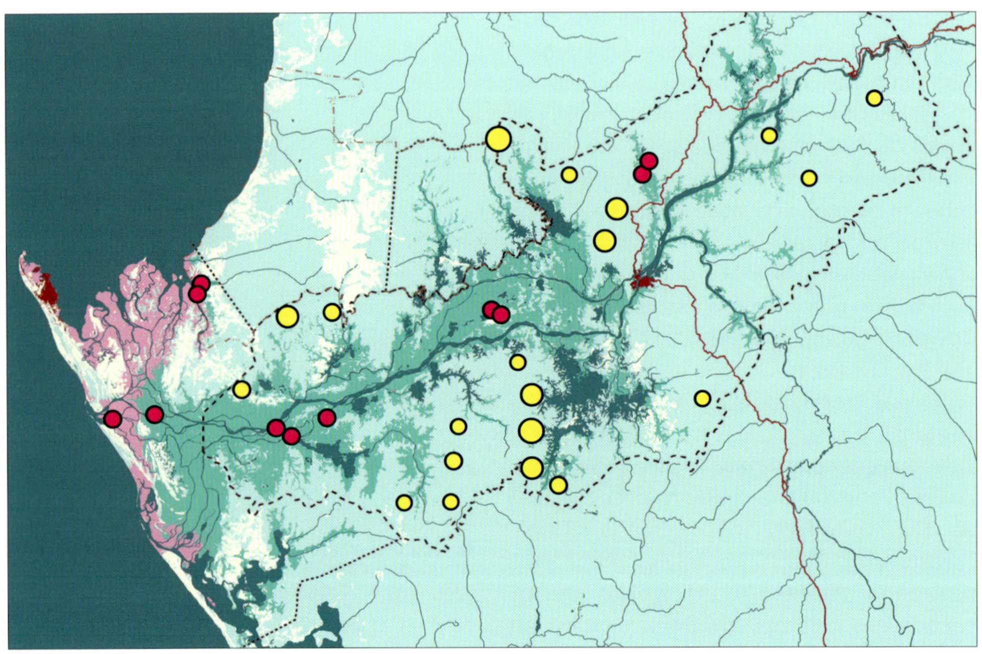

10.4. *Elephant tracks observed since 2012 in the Lower Ogooué complex. In red, the tracks in wetlands, observed on the ground or from the air. In yellow, the observations made in the dryland forests at the periphery of wetlands in 2014 and 2015 (Le-Duc Yeno et al., 2015). Only terrestrial transects were used to calculate a Kilometric Abundance Index (KAI) expressed by the size of the yellow circles*

complex. In fact, the only studies available are those of Latour (2005), which deal only with the Great Lakes region, and Le-Duc Yeno *et al*. (2015), which cover the entire Ramsar site.

Surveys of animals being sold in bushmeat markets, which were organized by the Ministry of Waters and Forests of Gabon (MINEF), WWF, and OELO (box page 134), have also provided many interesting observations. Unfortunately, the origin of animals being sold can rarely be determined with certainty, and it is very difficult to estimate what percentage of the region's existing populations these animals represent. Some species may also have come from afar. A sun-tailed monkey, *Allochrocebus solatus*, was spotted for sale along the road near Bifoun, which shows that some game may come from far away. In the Port-Gentil market, it is certain that some game comes from the Nkomi Lagoon or even from the area surrounding Loango National Park. Game from the Lambaréné market could come from Mouila or the foothills of the Du Chaillu Mountains.

Regarding the mammalian fauna that inhabits the vast swampy expanses of the flooded valley of the Ogooué, upstream from Lambaréné, and from the upper deltaic plain—the heart of the deltaic complex—it was not until March 2017 that significant observations were conducted out on this subject during an aerial survey that targeted verifications needed for vegetation mapping. Finally, regarding the lower deltaic plain and the coastal sandy formations, only fortuitous observations are available, made during botanical or ornithological work.

The fauna

Ecological distribution

Among the 57 species listed, 32 (56%) inhabit forests, 17 (29%) occur in either savannas or forests, 2 (3.5%) inhabit only savannas or other open environments, and 4 (7%) are restricted to aquatic environments (Figure 10.2). The two introduced species are anthropogenic and confined to villages and camps. Since the list is very incomplete, these figures are not particularly significant, and it is quite likely that any species added in the future will be associated with forest environments.

Biogeographic distribution

Among the 53 terrestrial species, 19 (36%) are Guineo-Congolean and 11 (21%) are endemic to Lower Guinea (Figure 10.3). These include seven primates (the mandrill, *Mandrillus sphinx*, the red-capped mangabey, *Cercocebus torquatus,* the Gabon talapoin, *Miopithecus ogoouensis,* the crested mona monkey, *Cercopithecus pogonias*, the moustached guenon, *Cercopithecus cephus,* the black colobus, *Colobus satanas*, and the western gorilla, *Gorilla gorilla*), two duikers (Peter's duiker, *Cephalophus callipygus*, and Ogilby's duiker, *Cephalophus ogilbyi*), and a mongoose (the black-footed mongoose, *Bdeogale nigripes*). Among the two aquatic species, the African manatee, *Trichechus senegalensis*, is restricted to the Atlantic coastal regions of Africa, but the common hippopotamus, *Hippopotamus amphibius*, has, or at least had, a wide distribution in sub-Saharan Africa. In general, the Guineo-Congolean species are dominant, and faunistic affinities are more pronounced with the Congolean region than with West Africa (Figure 10.3).

The species

Elephantidae

The African forest elephant, *Loxodonta cyclotis*, is known throughout the Guineo-Congolean forest region, but its largest surviving populations are in Lower Guinea. In Gabon, this species was very abundant and present in almost the entire country. Despite full legal protection, however, it has declined significantly over the past decade as a result of intensive poaching. In the Lower Ogooué region and its periphery, the species has been known for a long time and has been heavily hunted for ivory in the past (Du Chaillu, 1863, 1868). Nineteenth-century accounts suggest that the Kélé specialized in hunting for ivory, which they sold to ships stopping over in Cape Lopez Bay (Raponda-Walker, 1960).

Currently, a large population lives in the Wonga-Wongué Reserve (Motsaba & Aba'a, 2011; Vande weghe, 2012), just north of the delta complex, where it is, however, still subject to poaching. Terrestrial censuses have indicated that this species is also present in very small numbers in the forests surrounding the Great Lakes (Latour, 2005; Le-Duc Yeno *et al.*, 2015), where it stays away from villages and camps.* In recent years, elephant tracks have been observed in many parts of the deltaic complex, notably in the island of Ozouri, the Ompingo Plain, and the Inguéssi Plain, as well as in the forests located between the Gangwé and Ozima rivers (very close to Wonga-Wongué), the Nkengué marsh, the vast marsh that extends between the Ogooué and the Ouango, and in the marshes of the Mbiné River (Figure 10.4).

Tracks observed by plane in the Nkengué marsh in March 2017 were clearly oriented north-south and appeared to indicate the passage of elephants through the marshes of the Lower Ogooué region, either going from Wonga-Wongué to the forests located south of the river or vice-versa (Figure 10.6). These traces seemed to converge on a section of the Ogooué where there are large sandbanks – the last ones before the lower deltaic plain. It is not easy to interpret these indications, but we can assume that they relate to seasonal movements. Some Wonga-Wongué elephants migrate between the Komo Estuary in the north and the Ogooué in the south (Vande weghe *et al.*, 2016). Certain individuals probably cross the Ogooué on their way even farther south. Given the dangers in this area, these movements are probably made at night.

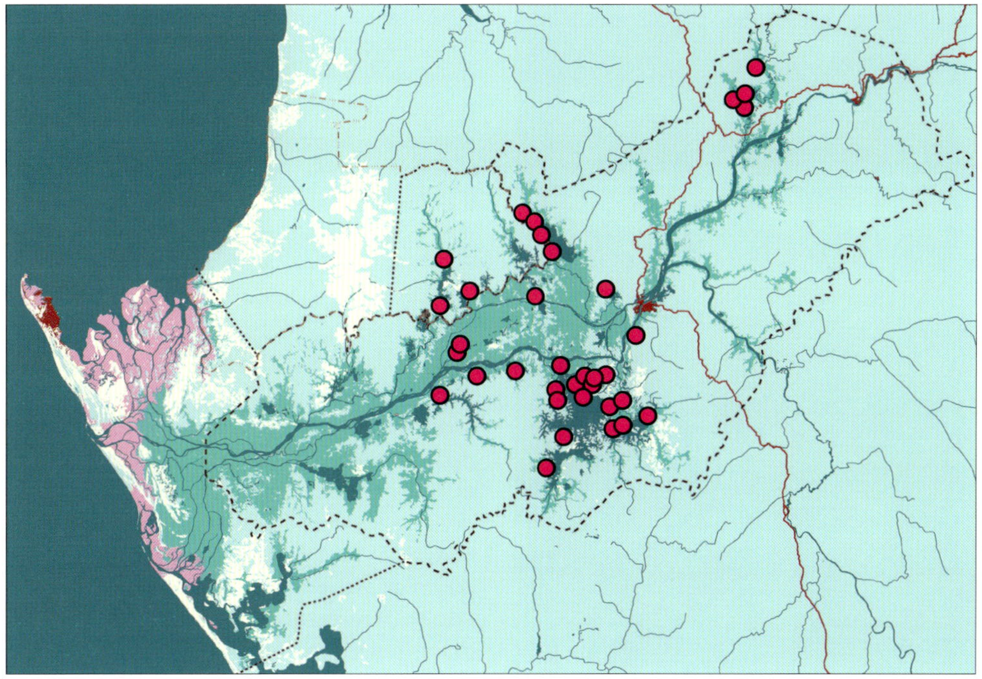

10.5. Manatees observed (dead and alive) in recent years. This species is also found in the lower deltaic plain, but no recent observations have been made in this region.

Trichechidae

This family is represented by a single species, the African manatee. It lives along the Gabonese coast (Keith, 2007), and it is present in the Lower Ogooué complex. It has been mainly observed, and especially hunted, in the eastern part of the deltaic complex, with many observations from the Abanga River and Lake Nguéné, as well as lakes located north of the upper deltaic plain (e.g., Azingo, Gomé), the Azingo River, Ouango River, and Lakes Déguèlié, Onangué, Evaro, Ezanga, and Oguémoué (Figure 10.5).

The only field studies regarding this species are those of Mbina (2001), Louembet (2012), and OELO (2013-2016), but they only concern the eastern part of the Lower Ogooué complex.† The actual distribution of this species across the complex and the overall size of its population thus remain unclear. Given the large number of animals killed with total impunity by hunters and fishermen (Vande weghe & Bidault, 2017, page 133), and despite the species being fully protected, the African manatee population must have suffered a considerable decline and must previously have been very large. When they were hunted with harpoons, or even with rifles, the impact may have been sustainable.‡ During the last few years, the region's inhabitants have,

10.6. Elephant tracks in the Nkengué Plain (March 2017).

* Some elephants still come at night around the villages on the edge of Lake Oguémoué and cause damage to crops. Elephant-human conflicts are therefore quite frequent.

† In his study on the manatee in Gabon, Keith (2007) mentions the Lower Ogooué, but without making any specific observations.

‡ Manatees are very sensitive to hunting due to their very low reproduction rate. Hunters are encouraged by the huge prices paid by wealthy Gabonese for manatee meat offered at important ceremonies. In addition, some parts of the manatee (skull, oils, penis) are highly sought after by Chinese and Malaysian workers in the region.

10.7. A red-capped mangabey on the edge of the Ezanga River. Despite being hunted, this species is not particularly timid and can be observed quite easily.

unfortunately, started hunted this species using "lambeaux," very large mesh nets from which manatees cannot escape.

Lorisidae

Bosman's potto, *Perodicticus potto*, is a widely distributed species. It is uncommon in the Lower Ogooué complex, although it probably often goes unnoticed.

Galagidae

The only listed species is the Prince Demidoff's bushbaby, *Galago demidoff*. It is widely distributed across the mainland forests of the Guineo-Congolean region. Two other species are likely present, especially in the forests of the far east of the Ramsar site (Appendix), but their presence remains to be confirmed.

Cercopithecidae

Among the eight species present in the Lower Ogooué complex, six are endemic to Lower Guinea. Three, namely the mandrill, the moustached guenon, and the crowned monkey, are widely distributed throughout Gabon and are essentially dryland forest species. The moustached guenon nevertheless visits the canopy of flooded or floodable forests. These species do not, however, cross deep rivers and are therefore absent from the central parts of the upper and lower deltaic plains, isolated by wide river courses. The crested mona monkey is represented by the form *nigripes*.

The red-capped mangabey is known to Gabon only from the coastal sedimentary basin, where it mainly inhabits swamp forests, *Raphia* groves, and floodable riparian forests. In the absence of hunting, this species probably also favors large dryland forests. It still appears to be fairly common in the Lower Ogooué deltaic complex, where it is often seen on the edges of rivers and lakes. It also frequents mangroves and coastal forests near beaches. Its diet is very varied, and in the intertidal zone, it often feeds on crabs.

The Gabon talapoin is found throughout most of Gabon, including in the Lower Ogooué deltaic complex, where it is particularly fond of riparian forests. It has been observed 2 km from Lambaréné along the river and is also present on Mandji Island. Nothing is known, however, of its numbers or distribution.

The black colobus was once abundant in the dryland forests of the Lower Ogooué region. However, its population has greatly decreased due to hunting because it is a large animal whose meat is very popular with local populations, and also because it does not flee when facing a hunter, but rather takes refuge at the top of a tree. Before independence, hunters who visited the Wonga-Wongué Reserve were allowed to kill four to eight black colobus monkeys in addition to the other species, for which they paid hunting fee. This colobus must therefore have been particularly abundant at the time. Currently the species must have become very rare. In fact, it has not been seen over the past decade, except at the market in Lambaréné.

Overall, Cercopithecidae are highly sought after by hunters and constitute an important source of bushmeat, and therefore represent an important source of income for hunters, who are generally content with less profitable game (brush-tailed porcupines or cane rats) for their personal consumption (Starkey *et al.*, 2013).

Hominidae

The common chimpanzee, *Pan troglodytes*, and the western gorilla are both present in the Lower Ogooué complex. The first is extremely rare and is only known to occur in large forested areas south of the Ogooué. The second is not abundant but remains somewhat more common. Despite the full protection they are supposed to receive, western gorillas are regularly killed, both because their meat is appreciated and supposed to have magic powers, and because they frighten hunters, who very often simply shoot them out of fear.

Nesomyidae

The only listed species are Emin's pouched rat, *Cricetomys emini,* a species of rat *Rattus* sp. (probably the black rat, *Rattus rattus*), and the house mouse, *Mus* sp. (probably *Mus musculus*). The latter two are common in Port-Gentil and probably also in Lambaréné. These are introduced species that live in association with humans.

Hystricidae

The African brush-tailed porcupine, *Atherurus africanus*, is common and is frequently sold in bushmeat markets (Louembet, 2012). It is in fact the most sold species.* It could perhaps be hunted sustainably (OELO, 2013–2016), although its reproduction rate is low and it could be extirpated by hunting.

According to some hunters, the Cape porcupine, *Hystrix africaeaustralis*, also occurs in the region, but the presence of this species is highly improbable: it does not reach farther north than Mayombe region in Congo.

* This species is sometimes so abundant in market stalls that it has not been included in surveys made by WWF and MINEF (Le-Duc Yeno *et al.*, 2015). However, the most commonly available bushmeat on the market today is slender-snouted crocodile, and the most common mammal is the forest hog (Arrowood, pers. obs.).

Sciuridae

Several species of squirrels have been sighted in the forests of the deltaic complex. According to hunters, they belong to five species. Only the African giant squirrel, *Protoxerus stangeri,* and the green bush squirrel, *Paraxerus poensis*, have been identified with certainty; the others remain to be confirmed. In general, however, squirrels are relatively rare in the region's forests, which can only be the result of hunting. In forests where they are not hunted, their density is much higher.

Thryonomyidae

The great cane rat, *Thryonomys swinderianus*, is often captured by hunters.

Manidae

The order Pholidota is represented only by the family Manidae, three species of which are known in the Lower Ogooué region. Despite being entirely protected, they are hunted for both their meat and their scales.

Suidae

The red river hog, *Potamochoerus porcus,* is present throughout the complex but in small numbers and is strictly nocturnal, as in all regions where it is subjected to strong hunting pressure. The highest densities have been observed west of Lake Onangué, whereas densities are low across the entire Lower Ogooué valley (Le-Duc Yeno *et al*., 2015).

Hippopotamidae

The common hippopotamus was once present in much of Gabon, but has disappeared from most areas. According to Du Chaillu (1863), the species was abundant in the Fernan Vaz Lagoon, and herds of up to 30 individuals could be observed on the low grasslands that separate the lagoon from the ocean. These open savannas served as pastures for them. Du Chaillu also indicated that this species was even more abundant in the Ogooué. Some Nkomi people with rifles hunted hippopotamus for their ivory, but those without guns never bothered them.

Today, this species remains present in the Lower Ogooué complex, but its numbers have decreased significantly. It has disappeared from the Nkomi Lagoon, and its numbers continue to decrease. Around 1999, several dozen animals still lived in the section of the Ogooué between Lambaréné and the confluence with the Ngounié (Vande weghe, pers. obs.; Shirley, 2012). Today, only a few extremely fierce individuals remain in this area. In recent years, hippopotamus tracks have nevertheless been found in many parts of the complex (Figure 10.8). Downstream, the species has been observed as far as where the Djembagombé River runs alongside the Djembagombé savannah. Upstream, it has been seen as far as Makouké. The highest density seems to be in the lake deltas of the Ezanga River and the Rivière des Pélicans, where these animals take refuge in vast, partially floating grass mats with *Vossia* or *Echinochloa*, where hunters cannot reach them. In these two deltas, populations must be in the order of several tens of individuals at least, perhaps about 50. In March 2017, tracks were also observed from the air in the southern part of the Inguéssi Plain, an ideal habitat for this species, and the number of individuals could be very much larger there.

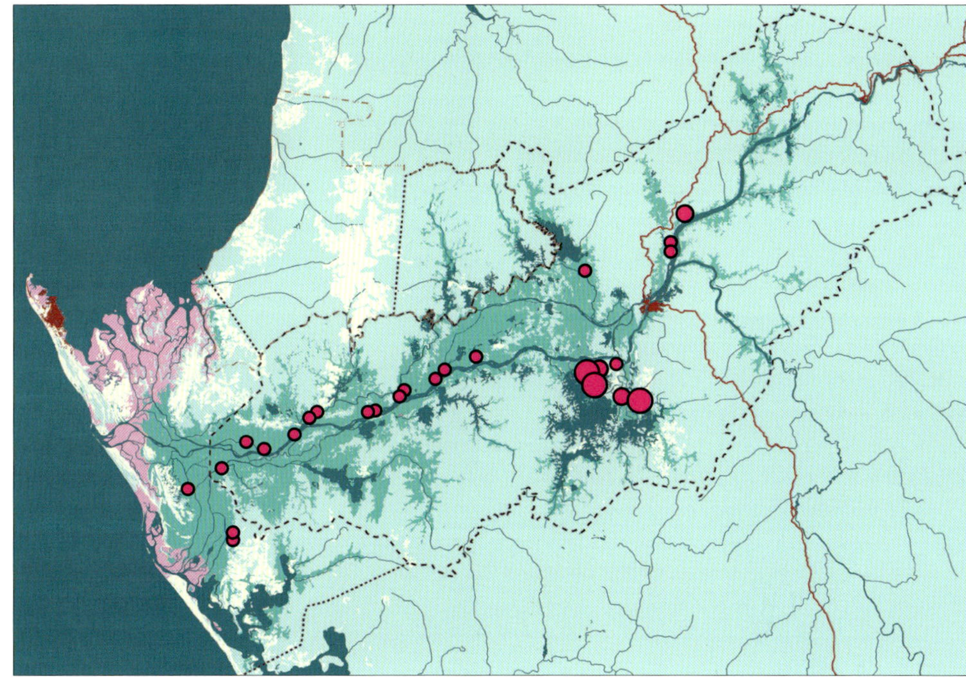

10.8. Observations of hippopotamuses and their tracks over the past five years.
- *1-2 individuals*
- *3-9 individuals*
- *10 or more individuals*

Tragulidae

The water chevrotain, *Hyemoschus aquaticus*, is present in the region, but its population size must be very small. Despite being fully protected, this species is highly sought after by hunters and regularly sold in restaurants in Lambaréné and Port-Gentil.

10.9. A tree pangolin, Phataginus tricuspis, *surprised by a camera-trap (Photo: Laila Bahaa-el-Din).*

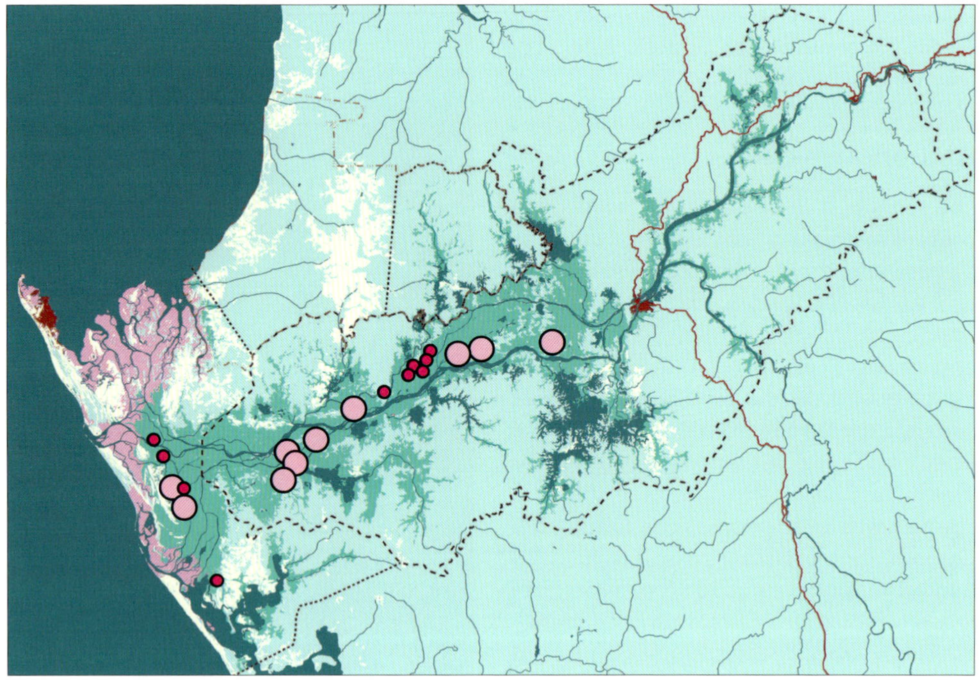

10.10. *Observations of buffalo (●) and buffalo tracks (○) over the last five years in the humid environments of the Lower Ogooué. This species is also found in very low density in the dryland forests of the complex. According to the WWF-MINEF study in 2014-2015, the average Kilometric Abundance Index for all of the dryland forests was 0.06 (Le-Duc Yeno et al., 2015). In 2005, Latour (2005) reported an index of 0.05 for the forest block that encompasses the Great Lakes. In fact, the dwarf buffalo is not a forest animal and is concentrated in open areas with savannas or marshy plains.*

10.11. *Peters' duiker,* Cephalophus callipygus *(Photo: Laila Bahaa-El-Din).*

10.12. *The black-fronted duiker,* Cephalophus nigrifrons *(Photo: Laila Bahaa-El-Din).*

Bovidae

The dwarf buffalo,* *Syncerus nanus*, is present throughout a large part of Gabon, where its populations are concentrated around blocks of savanna. A large population, probably the largest in Gabon, is found on the Wonga-Wongué Plateau, just north of the deltaic complex. Inside the complex, this species is present at very low density throughout most of the Ramsar site (Le-Duc Yeno *et al.*, 2015), and it has been observed in the Inguéssi and Ompingo plains (Vande weghe, pers. obs.). The only significant population inhabits the vast grassy marshes situated between the Ogooué proper and Ouango (Vande weghe, pers. obs.). During an aerial survey on March 18, 2017, at least 80 buffalo, forming five herds (including one with 25 individuals and one with 28), were observed in this area (Figure 10.10 and Figure 10.13). Countless tracks were also noted. The total population in this marsh area could include several hundred individuals, for which it should be possible to organize aerial counts.

The bushbuck, *Tragelaphus scriptus*, is present in the Wonga-Wongué Reserve and in the savannas of Loango National Park. Its presence in the deltaic complex would not be surprising, but remains to be proven.

The sitatunga, *Tragelaphus spekii*, is present in a large part of the Ogooué complex, and its tracks have been found in many places in recent years. This species is even found at the center of the northern delta, and it is one of the animals mostly frequently sold in the bushmeat market of Lambaréné. As in all regions with high hunting pressure, this species is rarely seen during the day.

The region has six species of duiker, *Cephalophus* spp. (Figures 10.11 and 10.12), but few observations have been identified to species. The only accurate observations come from markets. In general, the density of duikers seems to be very low in the dryland forests of the Great Lakes region, where the Kilometric Abundance Index is only 0.33 (Latour, 2005).

* The name "African forest buffalo" is often used but very poorly chosen; dwarf buffalo is preferable. This species occurs in forests, but its populations are centered on areas of open environments. Furthermore, this species also lives in the savannas located north of the forest block, from the CAR to Senegal.

Pteropodidae

Despite the fact that bats provide important ecosystems services (pollination, seed dispersal, insect population control, etc.) and that they may play an important role as a vector for Ebola hemorrhagic fever, they have been poorly studied in Gabon and not studied at all in the Lower Ogooué region–where there has never been an Ebola fever epidemic.

Of the 45 species of bats listed in Gabon (Vande weghe *et al.*, 2016), only five have been found in the Lower Ogooué region. Four belong to the family Pteropodidae, fruit-eating bats. The straw-colored fruit bat, *Eidolon helvum*, and the Egyptian fruit bat, *Rousettus aegyptiacus*, are common in Lambaréné (Arrowood, pers. obs.). The first of these species gathers in the large trees that surround Schweitzer Hospital, and the second one forms groups below the bridge that crosses the northern branch of the Ogooué. The hammer-headed bat, *Hypsignathus monstrosus*, is found in the forests of the Great Lakes region.

Felidae

The leopard, *Panthera pardus*, is present in the deltaic complex (OELO, pers. comm.), but its distribution and population size are unknown. The African golden cat, *Caracal aurata,* is present throughout the forests of Gabon. It is an extremely discreet animal, and it would be surprising if it were absent from the forests of the deltaic complex.

10.13. A group of 28 buffalo in the great central alluvial plain of the upper deltaic plain in March 2017 (Photo: Jean P. Vande weghe).

Viverridae

Four species have been listed. All of them, apart from the Central African oyan, *Poiana richardsonii,* are widely distributed. The presence of this species, like that of the servaline genet, *Genetta servalina,* was revealed by market surveys (OELO) and was reported by Bahaa-El-Din (2013).

Nandinidae

The African palm civet, *Nandinia binotata*, is probably omnipresent throughout the area, although in small numbers. This species is heavily hunted.

Herpestidae

Three species have been listed. The marsh mongoose, *Atilax paludinosus,* is probably the most common. It is a widely distributed species that inhabits both marshes and dryland forests. The long-nosed mongoose, *Herpestes naso*, is a nonforest species, found in market surveys (OELO); its presence was confirmed by Bahaa-El-Din (2013). The black-footed mongoose, *Bdeogale nigripes,* has been found in bushmeat markets (OELO).

Mustelidae

The honey badger, *Mellivora capensis,* occurs throughout Gabon. Its presence in the Lower Ogooué complex has only been reported by hunters, but it is very likely. In fact, its absence would be rather unexpected. The spotted-necked otter, *Hydrictis maculicollis,* was observed in Lake Evaro around 2004 (Vande weghe, pers. obs.), and according to fishermen, it occasionally gets caught in fishing nets.

The state of the populations

For most of the known species, population densities are low to very low, even though the living environments in which they live are nearly intact (Le-Duc Yeno *et al.*, 2015; Latour, 2005). This situation can only be explained by omnipresent and long-standing hunting activities (page 135). The only populations whose densities approximately approach "normal" numbers are those of buffalo and sitatunga, which inhabit the vast, almost inaccessible grassy marshes at the center of the deltaic complex. This observation indirectly confirms the scale at which mammals are being hunted: where hunters fail, game survives. Unfortunately, hunting comes with development, and each new project involves a new risk.

10.14. The servaline genet, Genetta servalina *(Photo: Laila Bahaa-El-Din).*

10.15. The marsh mongoose, Atilax palidinosus *(Photo: Laila Bahaa-El-Din).*

Importance of the Ogooué Delta for the conservation of mammals

According to the IUCN Red List (IUCN, 2017), of the 57 mammal species inhabiting the Lower Ogooué complex, 35 (61%) are not threatened, 5 (8.9%) are Near Threatened, 12 (21 %) are Vulnerable, 1 (1.7%) is Endangered, and 1 (1.7%) is Critically Endangered. Overall, about 30% of this mammalian fauna thus comprises species under various levels of threat. Compared to other vertebrates, this number is huge and should cause great concern. However, since most of these species are present in low density, the Lower Ogooué region is not included among Gabon's high-priority sites for mammal conservation. The region still shelters a large number of species, and for some of them, the populations present are large at the national level. The dwarf buffalo is a good example: it is not yet endangered but has already disappeared from a large part of its range and its numbers are inexorably declining. The populations of three species, listed below, are significant on an international scale.

The red-capped mangabey

The population in the Lower Ogooué could be one of the largest in Gabon, given the size of the complex. This species, endemic to Lower Guinea, is, however, classified as Vulnerable (IUCN, 2017). Its range includes coastal forests from southeastern Nigeria to southern Gabon, but it is increasingly fragmented. The species has also disappeared from certain regions where the hunting pressure was not particularly strong (Oates *et al.*, 2008). Whatever the cause, it is currently rare in Nigeria and Cameroon. The Gabonese population is therefore quite probably the largest for this species (Cooke & Moussopo, 2013).

The common hippopotamus

Until around 1700, this species still existed in the Nile Delta and in Upper Egypt, where it disappeared in the early 1800s (IUCN, 2017). Around 1960, it still occupied most of its sub-Saharan range, although it had already disappeared from South Africa, with the exception of Kruger National Park. Today, this animal is still widely distributed, although it is becoming very fragmented and has almost disappeared from sites where it was still very abundant 20 or 30 years ago. The classic example is that of Virunga National Park in the Democratic Republic of Congo, where the population declined from 26,530 in 1959 to 887 in 2005 (Languy, 2006). However, a study has shown that the overall genetic diversity of the species remains very high, although the genetic diversity of each individual population is low (Okello *et al.*, 2005). In other words, there is low gene flow between populations. It is therefore necessary, for the conservation of the species, to try to preserve as many populations as possible. In this context, the populations of Lower Guinea are very important. However, they are currently reduced to remnants, and some are probably no longer viable in the long term. The hippopotamus population in Lower Ogooué, which is probably the largest in Gabon, is therefore very important for the conservation of this species.

The African manatee

This species remains widely distributed, although according to the IUCN, fewer than 10,000 individuals currently survive in Africa. There are no currently recognized subspecies, but studies seem to indicate significant genetic diversity (Keith Diagne, 2014). The Lower Ogooué population is said to have the greatest genetic diversity of all, which suggests that in the recent past it may also have been the largest (Keith Diagne, 2014). Given the size of the Ogooué Delta, this would not be surprising.

This species remains poorly known. The threats to which it is exposed, however, seem to be increasing throughout its range and it is threatened in several regions (IUCN, 2017). This situation is attributed to the usual causes: poverty,* lack of protein,† and a lack of compliance with existing laws. The local destruction of coastal habitats associated with development activities further adds to this, along with the destruction of mangroves, the construction of dams, and water pollution caused by sediments. In Gabon, most of these threats are absent or minimal. Only hunting and fishing represent a real danger. The Lower Ogooué complex could, therefore, very easily lend itself to conservation efforts for this species, if hunting could at least be regulated, if not prohibited, and if mass-killings could be halted definitively.

Recommendations

Research to be undertaken

(1) The mammalian fauna of the Lower Ogooué region is still quite poorly known, and it is therefore important to continue conducting inventories, especially of micromammals.

(2) It is important to obtain population estimates of the three internationally important species: the red-capped mangabey, the common hippopotamus, and the African manatee.

* Poverty is often cited as a cause. However, it should be noted that in Gabon, manatee meat is mainly consumed by wealthy people. It is often a choice meat offered on special occasions.

† What prevents the rural populations of Gabon from practicing small-scale farming?

(3) The fauna of the Inguéssi Plain should be studied with means other than overflights. Many animal tracks are visible on the edges of small lakes, and it would be interesting to identify the species with certainty.

(4) It is necessary to obtain a more precise idea of the size of buffalo, sitatunga, and *Potamochoerus* populations living in the large marshes of the upper deltaic plain.

The most important areas

In terms of conservation measures, protection of the entire site should be improved, and the most critical areas should be placed under close surveillance (Figure 10.16):

(1) The deltas of the Ezanga River and the Rivière les Pélicans, which are probably home to the largest population of hippopotamus, not only in the Lower Ogooué complex but in Gabon and possibly even in Lower Guinea.

(2) The lakes that are home to the largest manatee populations.

(3) The large grassy marshes in the center of the complex, from the area of Lake Nyayé to the Nkengué marsh; they are difficult to access by humans and form a corridor between the regions located to the north (the Wonga-Wongué Reserve) and to the south of the Lower Ogooué marshes.

(4) The Inguéssi Plain, which constitutes an islet of very diverse environments, far from any village or permanent settlements and separated from Ozouri Island by a band of marshes and mangroves, which probably still shelters a good diversity of species and should be relatively easy to protect.

Acknowledgments

The authors thank Laila Bahaa-El-Din, who graciously authorized the use of her photos, as well as Cyrille Mvele, Marie-Claire Paiz, and Martial Angwe, who shared their manatee observations.

Bibliography

BAHAA-EL-DIN, L., P. HENSCHEL, R. ABA'A, K. ABERNETHY, T. BOHM, N. BOUT, L. COAD, J. HEAD, E. INOUE, S. LAHM, et al., 2013. Notes on the distribution and status of small Carnivores in Gabon. *Small Carnivore Conservation*, 48: 19-29.

Cooke, C.A. & R. Moussopo, 2013. GIS Analysis of the Ranging Behaviors of Red-capped Mangabeys (*Cercocebus torquatus*) from Sette-Cama, Gabon. *American Journal of Physical Anthropology*, 150: 101.

DU CHAILLU, P.B., (1863) 1996. *Voyages et aventures en Afrique Equatoriale*. Centre Culturel Français de Libreville - Sépia, Libreville.

DU CHAILLU, P.B., (1868) 2002. *L'Afrique Sauvage*. Editions du LUTO, Libreville.

KEITH, L., 2007. Report on West African Manatee (*Trichechus senegalensis*) Survey Activities in Gabon. WCS, Libreville.

Keith Diagne, L.W., 2014. Phylogentics and Feeding Ecology od the African Manatee (*Trichechus senegalensis*). Doctorat de l'université de Floride.

KEITH DIAGNE, L. 2015. *Trichechus senegalensis*. The IUCN Red List of Threatened Species 2015: e.T22104A97168578. Téléchargé le 6 mai 2017.

KINGDON, J., 2015. *The Kingdon Field Guide to African Mammals*. Second Edition. Bloombury, London.

LANGUY, M., 2006. Evolution des populations des grands mammifères. Pp. 141-151 in: M. Languy & E. de Mérode. *Virunga*. Lannoo, Tielt.

LATOUR, S., 2005. *Inventaire de faune autour des lacs du delta de l'Ogooué*. WCS, Libreville.

Le-Duc Yeno, S., S. Louembet, J.P. Vande weghe, P. Christy & S. Ratiarison, 2015. *Caractérisation de la faune et des menaces d'origine anthropiques sur la faune dans le Massif Forestier d'Evaro*. Rapport final. WWF, Libreville.

LOUEMBET, S., 2005. *Approche d'étude pour une connaissance de l'exploitation d'un mammifère aquatique: le lamantin (*Trichechus senegalensis*) dans la région des lacs du sud de l'Ogooué*. Mémoire ENEF.

LOUEMBET, S., 2012. *Quelques informations phares issues des études antérieures dans la zone d'Ebel Abanga et Lambaréné*. Rapport WWF, Libreville.

MBINA, C., 2001. *Evaluation and status of Manatee (*Trichechus senegalensis*) of Ogooué Basin in Gabon*. WCS, Libreville.

OATES, J.F., GIPPOLITI, S. & GROVES, C.P. 2008. *Cercocebus torquatus*. The IUCN Red List of Threatened Species 2008: e.T4201A10621068. http://dx.doi.org/10.2305/IUCN.UK.2008.RLTS.T4201A10621068.en. Téléchargé le 6 mai 2017.

OELO, 2013a, 2013b, 2014a, 2014b, 2015a, 2015b, 2016a, 2016b. *Rapport de vente illégale de la viande de brousse, Marché Isaac, Lambaréné*. Libreville, Gabon.

OKELLO, J.B.A., S. NYAKAANA, C. MASEMBE, H.R. SIEGISMUND & P. ARCTANDER, 2005. Mitochondrial DNA variation of the common hippopotamus: evidence for a recent population expansion. *Heredity*, 95: 206-215.

RAPONDA-WALKER, A., 1960. *Notes d'Histoire du Gabon. Brazzaville*.

STARKEY, M., LOUNDOU & BOUPOYA MAPIKOU, 2013. *Rapport de Synthèse. Complexe Evaro-Wonga-Wongué*. WCS, Libreville.

VANDE WEGHE, J.P., 2007. *Les Parcs Nationaux du Gabon: Loango, Mayumba et le bas Ogooué*. WCS, Libreville.

VANDE WEGHE, J.P., 2012. *Wonga-Wongué*. ANPN, Libreville.

VANDE WEGHE, J.P., 2017. Les populations humaines. Pp. 127-141 in : T. Stévart & J.P. Vande weghe (eds.) *Le delta de l'Ogooué*. ANPN, Libreville.

VANDE WEGHE, J.P., P. CHRISTY, M. DUCROCQ, M. LEE, G.R. VANDE WEGHE & O.S.G. PAUWELS, 2016. *Biodiversité des parcs nationaux et réserves du Gabon, 2: Espèces écosystèmes et populations*. ANPN, Libreville.

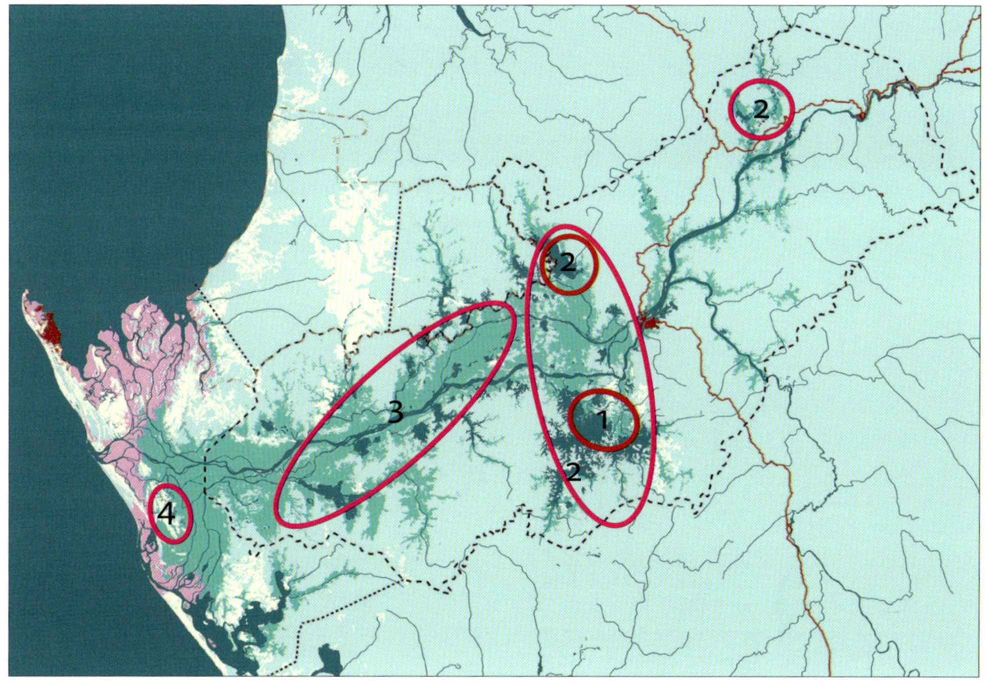

10.16. Important areas for mammal conservation:

1. *The lake deltas of the Rivière des Pélicans and the Ezanga River*
2. *The lakes that are probably home to the largest remaining manatee populations; could Lake Nyayé be one of them?*
3. *The central great plains of the upper deltaic plain*
4. *The Inguéssi Plain*

Appendix: List of mammals of the Ogooué Delta

Ecology: F = Forest; N = Nonforest; FN = Forest and nonforest; A = Aquatic; M = Marine[1]
Biogeography: B = Lower Guinean; BC = Lower Guinean and Congolean: GC = Guineo-Congolean; LD = Wide distribution
IUCN (level of threats): CR = Critically Endangered; EN = Endangered; VU = Vulnerable: NT = Near Threatened
Gabon: IP = Fully protected
Distribution: Zone A = Lower Ogooué valley; Zone B = Upper deltaic plain; Zone C = Lower deltaic plain; Zone D = Coastal sands; Zone E = Lagoons

	Ecology	Biogeography	IUCN	Gabon	Zone A	Zone B	Zone C	Zone D	Zone E
AFROSORICIDA									
TENRECIDAE									
Potamogalinae									
Giant Otter Shrew, *Potamogale velox*	AF	BC	NT		■	■			
HYRACOIDEA									
PROCAVIIDAE									
Western Tree Hyrax, *Dendrohyrax dorsalis*	F	GC		IP	■	■			
PROBOSCIDEA									
ELEPHANTIDAE									
Forest Elephant, *Loxodonta cyclotis*	FN	GC	EN	IP	■	■	■	■	
SIRENIA									
TRICHECHEDIDAE									
African Manatee, *Trichechus senegalensis*	A		VU	IP	■	■	■		
PRIMATES									
LORISIDAE									
Bosman's Potto, *Perodicticus potto*	F	GC		IP	■	■			
GALAGIDAE									
Demidoff's Galago, *Galagoides demidoff*	F	GC		IP	■	■			
[Thomas's Galago, *Galagoides thomasi*] [2]	F	GC			?	?			
[Gabon Galago, *Sciurocheirus gabonensis*] [3]	F	B		IP	?	?			
[Elegant Needle-clawed G., *Euoticus elegantulus*]	F	B		IP	?	?			
CERCOPITHECIDAE									
Cercopithecinae									
Mandrill, *Mandrillus sphinx*	F	B	VU	IP	■	■			
Red-capped Mangabey, *Cercocebus torquatus*	F	B	VU		■	■	■	■	■
Grey-cheeked Mangabey, *Lophocebus albigena*	F	B			■	■			
Northern Talapoin, *Miopithecus ogouensis*	F	B			■	■	■	■	
Crowned Monkey, *Cercopithecus pogonias*	F	B			■	■			
Putty-nosed Monkey, *Cercopithecus nictitans*	F	(B)			■	■		■	
Moustached Monkey, *Cercopithecus cephus*	F	B			■	■		■	
Colobinae									
Black Colobus, *Colobus satanas*	F	B	VU		■	■			
HOMINIDAE									
Homininae									
Western Gorilla, *Gorilla gorilla*	F	B	CR	IP	■	■			
Common Chimpanze, *Pan troglodytes*	F	GC	EN	IP	■	■			
RODENTIA									
NESOMYIIDAE									
Emin's Giant Rat, *Cricetomys emini*	F	GC			■	■			
MURIDAE									
Murinae									
Black Rat, *Rattus* sp. [4]	N	I			■			■	■
House Mouse, *Mus* sp. [4]	N	I			■			■	■
ANOMALURIDAE									
Lord Derby's Anomalure, *Anomalurus derbianus*	F	GC			■	■			
Beecroft's Anomalure, *Anomalurus beecrofti*	F				■	■			
SCIURIDAE									
Perinae									
[Biafran Bight Palm Squirrel, *Epixerus wilsoni*] [5]	FN	B			?				
[Thomas's Rope Squirrel, *Funisciurus anerythrus*] [5]	F	B			?				
[Lady Burton's Rope Squirrel, *Funisciurus isabella*] [5]	F				?				
[Ribboned Rope Squirrel, *Funisciurus lemniscatus*] [5]	F	B			?				
[Fire-footed Rope Squirrel, *Funisciurus pyrropus*] [5]	F				?				
Green Squirrel, *Paraxerus poensis*	F				■	■			
African Giant Squirrel, *Protoxerus stangeri*	F	GC			■	■			
GLIRIDAE									
Graphiurinae									
[Woodland Dormouse, *Graphiurus murinus*] [6]	F	LD			?	?			
HYSTRICIDAE									
Brush-tailed Porcupine, *Atherurus africanus*	F	GC			■	■			■
THRYONOMYIDAE									
Marsh Cane-rat, *Thryonomys swinderianus*	FN	LD			■	■			
CETARTIODACTYLA									
SUIDAE									
Suinae									
Red River Hog, *Potamochoerus porcus*	FN	GC			■	■		■	■

1. "Marine" refers to species that live on the open sea and are completely independent of the coasts, which they often avoid.
2. These lesser galagos are widely distributed in Guineo-Congolean forests but were described only quite recently. They are very similar Demidoff's galago; distinguishing it is difficult in the field, and few specimens are available in collections (Wilson & Reeder, 2017). Its presence in the Lower Ogooué region is very likely, and it was mentioned by the OELO team. To be accepted, these data should be better detailed.
3. These lesser galagos are found in Pongara and Loango national parks; it is very likely that they are present in the forests of the Lower Ogooué complex, especially in its eastern forests.
4. These two introduced species are found in Port-Gentil and probably in Lambaréné, but have not been identified with certainty.
5. These species of squirrels are found both north and south of the Lower Ogooué complex, and some species have been observed in the Crystal Mountains. Their presence in the Lower Ogooué is very likely, especially in the eastern forests; they have been reported by hunters, but this requires confirmation.
6. This woodland dormouse is found in Pongara and Loango national parks. Its presence in the Lower Ogooué complex is very likely but remains to be confirmed.

	Ecology	Biogeography	IUCN	Gabon	Zone A	Zone B	Zone C	Zone D	Zone E
DELPHINIDAE									
Bottlenose Dolphin, *Tursiops truncatus*	M								
Atlantic Humpback Dolphin, *Souza theuszii*	M		VU					■	■
HIPPOPOTAMIDAE									
Common Hippopotamus, *Hippopotamus amphibius*	A	LD	VU		■	■	■		
TRAGULIDAE									
Water Chevrotain, *Hyemoschus aquaticus*	F	GC			■	■			
BOVIDAE									
Bovinae									
Dwarf Buffalo, *Syncerus nanus*	FN	GC		IP	■	■	■	■	
[Bushbuck, *Tragelaphus scriptus*]	N	LD							
Sitatunga, *Tragelaphus spekii*	FN	LD			■	■	■	■	■
Cephalophinae									
Peters's Duiker, *Cephalophus callipygus*	F	B			■	■			
Bay Duiker, *Cephalophus dorsalis*	F	GC	NT		■	■			
Black-fronted Duiker, *Cephalophus nigrifrons*	F	BC			■	■			
Ogilby's Duiker, *Cephalophus ogilbyi*	F	B		IP	■	■			
Yellow-backed Duiker, *Cephalophus silvicultor*	F	GC	NT		■	■			
Blue Duiker, *Philantomba monticola*	F	BC			■	■			
CHIROPTERA									
PTEROPODIDAE									
Straw-coloured Fruit Bat, *Eidolon helvum*	FN	LD	NT		■	■			
Egyptian Fruit Bat, *Rousettus aegyptiacus*	FN	LD			■	■			
Hammer Bat, *Hypsignathus monstrosus*	FN	GC			■	■			
Wahlberg's Epauletted F. B., *Epomophorus wahlbergi*	FN	BC			■				
PHOLIDOTA									
MANIDAE									
Long-tailed Pangolin, *Phataginus tetradactyla*	F	GC	VU		■				
Tree Pangolin, *Phataginus tricuspis*	F	GC	VU		■	■			
Giant Pangolin, *Smutsia gigantea*	FN	GC	VU	IP	■	■			
CARNIVORA									
FELIDAE									
Felinae									
Golden Cat, *Caracal aurata*	F	GC	VU		■				
Pantherinae									
Leopard, *Panthera pardus*	FN	LD	VU	IP	■	■			
VIVERRIDAE									
Viverrinae									
African Civet, *Civettictis civetta*	FN	LD			■	■	■	■	■
Blotched Genet, *Genetta maculata*	FN	LD			■	■	■	■	■
Servaline Genette, *Genetta servalina* [7]	F	BE			■	■			
Central African Oyan, *Poiana richardsonii* [7]	F	BC				■			
NANDINIDAE									
African Palm Civet, *Nandinia binotata*	F				■	■			
HERPESTIDAE									
Marsh Mongoose, *Atilax paludinosus*	FN	LD			■	■	■	■	■
Long-snouted Mongoose, *Herpestes naso*	FN	BC			■	■			
[Slender Mongoose, *Galerella sanguinea*] [8]	N	LD							
Black-legged Mongoose, *Bdeogale nigripes*	FN	BC			■	■			
MUSTELIDAE									
Mustelinae									
Honey Badger, *Mellivora capensis* [9]	FN	LD			■				
Lutrinae									
Spot-necked Otter, *Hydrictis macullicollis*	A	LD	NT			■			

7. The presence of these two species was reported by Bahaa-El-Din *et al.* (2013).
8. The slender mongoose is found in Pongara and Loango national parks. Its presence in the Lower Ogooué complex is very likely but must be confirmed by dated observations from precise localities.
9. The honey badger is present throughout Gabon. It is not uncommon but nocturnal, and it is very typical. There is therefore no reason to doubt its presence in the Lower Ogooué region, as reported by hunters.

Notes

1. Hunters from the Great Lakes region also report the presence of the Cape porcupine, *Hystrix africaeaustralis*. This species is not found in Gabon, and in the Atlantic region of the continent, its distribution does not extend north of Cabinda. On the other hand, Kingdon (2015) mentions *Hystrix cristata* in northern Gabon, which is likewise not found in the country. Furthermore, if a species of the genus *Hystrix* were to exist in Gabon, one would inevitably find its very decorative "quills" in traditional headdresses and ornaments. Until proven otherwise, the presence of this genus in Gabon cannot be confirmed.
2. The white-bellied duiker, *Cephalophus leucogaster*, was mentioned to occur in the region on the basis of an animal sold at the Lambaréné market (OELO). On the one hand, the photograph of this butchered specimen is absolutely inconclusive; furthermore, there is no evidence that this individual was actually killed in the region. As this duiker is only known from eastern Gabon, its presence in the Lower Ogooué region cannot be validated without formal proof.

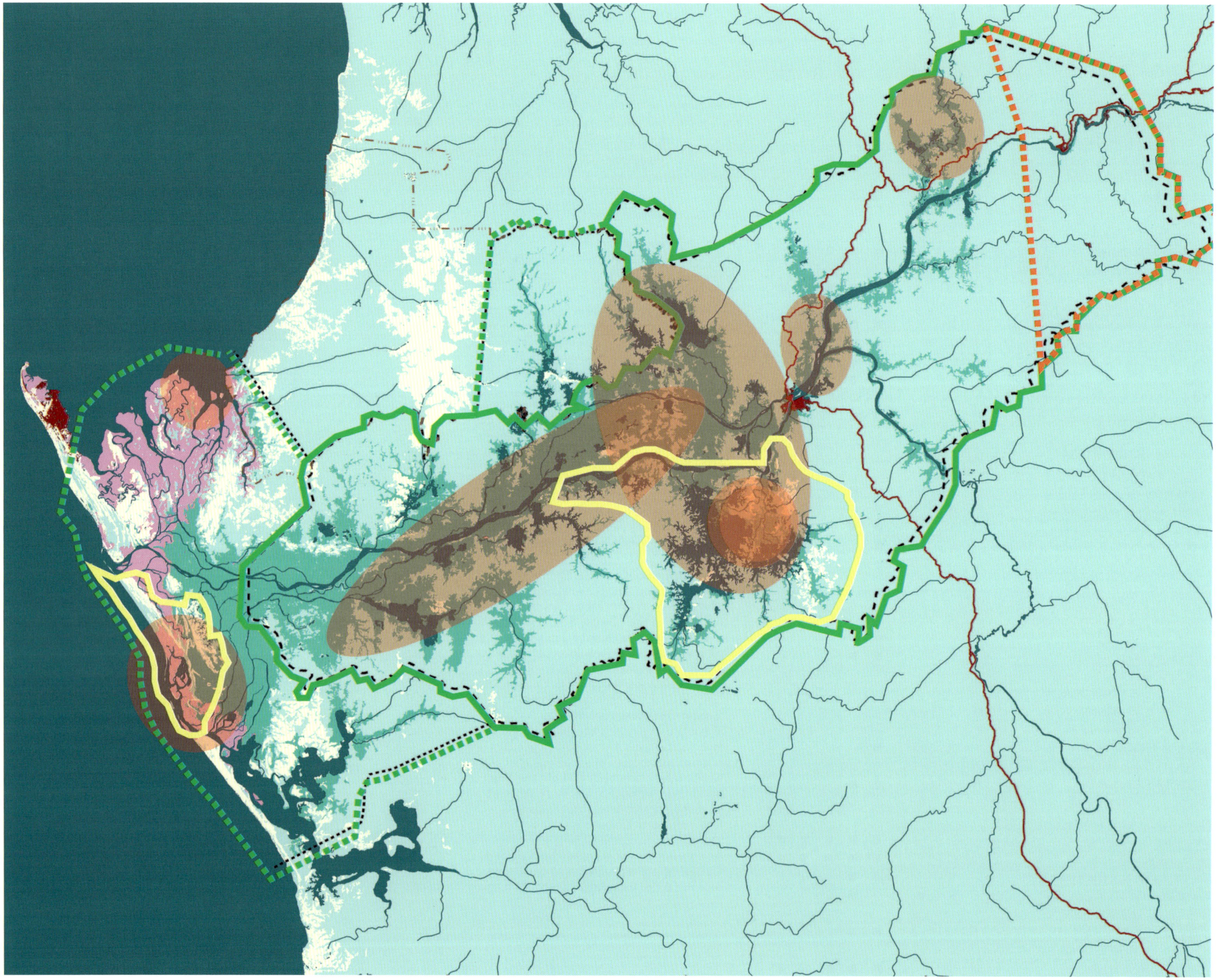

11. Conservation

Tariq Stévart, Olivier S. G. Pauwels & Jean P. Vande weghe

Historical context

The Ogooué Delta has long remained overlooked (page 11). Nevertheless, it has been the subject of several conservation initiatives since the middle of the 20th century.

First of all, the Wonga-Wongué Presidential Reserve was created between 1956 and 1976. This long process began in 1956 with the creation of the Wonga-Wongué hunting reserve, which already included the lakes in the northern part of the upper deltaic plain. In 1962, it was combined with the Petit Bam-Bam Wildlife Reserve and the Grand Bam-Bam Hunting Reserve to form a "rational wildlife exploitation area," the limits of which were, however, changed in 1966. In 1967, the wildlife reserve became a national park, while two new hunting reserves were established. Finally, in July 1972, the complex became the Wonga-Wongué Presidential Reserve. It was extended north in 1976 and now covers 428,188 ha. As a result, it includes the lakes in the north of the delta and their adjoining flood zones, from the western shore of Lake Azingo in the east, to Lake Gomé in the west, as well as the mangroves on the right bank of the Gongwé River, which belongs to of the northern part of the lower deltaic plain (Figure 11.1). In 1986, this presidential reserve was added to the list of Ramsar sites as number 351.

As for the Lower Ogooué Ramsar site, the first conservation initiatives date back to 1990. Based on the limited knowledge available at the time, as well as a few field expeditions, two sectors of the Lower Ogooué that contain wetlands were designated as critical for conservation in the IUCN report on forest ecosystems in Central Africa. They were quite logically proposed for national park status as a result (Wilks, 1990).

The first sector, centered on Lake Onangué (Figure 11.1), covered approximately 200,000 ha. Terrestrial species with a wide distribution in Gabon were cited among the main elements of interest for this area. Only the common hippopotamus, *Hippopotamus amphibius*, the African manatee, *Trichechus senegalensis*, and the slender-snouted crocodile, *Mecistops cataphractus*, were cited among aquatic animals. This site included the lake deltas of the Rivière des Pélicans and Ezanga River, as well as a portion of the main course of the Ogooué River. Unfortunately, this portion shelters only a few large sandbanks that can accommodate terns, African skimmers, *Rynchops flavirostris*, and gray pratincoles, *Glareola cinerea*, during their nesting periods. Furthermore, 75% of this sector was composed of dryland forests, the benefit of which had not been demonstrated, and included large human populations grouped along the lakes and the Ngomo mission. In fact, out of its 200,000 ha, it included at most 10,000 ha of truly interesting wetlands (lake deltas), since most of Lakes Onangué and Ezanga are neither of priority interest nor areas that require protective measures.

The second sector was centered on the Ozouri Island (Figure 11.1). It covered an area of 44,000 ha and stretched from the mouth of the Ogooué to the mouth of the Nkomi Lagoon. Its western edge was defined by the Atlantic coast, and its eastern limit by the Ogolélé and Mpolounié rivers. According to the IUCN report, the main points of interest of the site were its vegetation (wet savannas, coastal thickets, forests with *Chrysobalanus icaco*, floodable forests with *Haplormosia monophylla* and mangroves), its fauna (e.g., collared mangabey, *Cercocebus torquatus*, Gabon talapoin, *Miopithecus ogooensis*, and colonies of Caspian terns, *Hydroprogne caspia*), and especially the nesting beaches of the leatherback sea turtle, *Dermochelys coriacea*. This site covered the Inguéssi Plain, the Olendé Lagoon, and Ozouri Island, all of which are indeed very interesting environments. Unfortunately, the sandbars of the Nkomi Lagoon ("Barre de l'Arabe" and "Barre des

11.1. The Lower Ogooué Ramsar site:
Current boundaries
Boundaries of proposed additionnal areas
Boundaries of the area possibly to be removed from the current site
Boundaries of the protected areas proposed in 1990 by IUCN (Wilks, 1990)

Portugais") was not included within the proposed limits. The largest colonies of terns and skimmers were therefore outside the proposed protected area.

These proposals did not come to fruition until the year 2000, when the ministry responsible for biodiversity conservation asked WWF and WCS to carry out an assessment of sites that are critical for conservation. This evaluation, done in 2002, led to the creation of Gabon's 13 national parks. On this occasion, one of the authors of this book (JPV) was responsible for assessing the wetlands of the Lower Ogooué region. He highlighted the value of this complex as a whole, which included small yet ubiquitous human populations and widely dispersed areas of conservation importance. Given the complexity of managing this site and the lack of knowledge about it, no protected areas were designated in the Lower Ogooué region. Furthermore, at the time the attention of the conservation community in Gabon was focused on forests, so they were not concerned with savannas or wetlands. Savannas were in fact considered by many to be degraded forests.

In February 2009, despite the lack of knowledge, part of the Ogooué Delta was nevertheless added to the list of Ramsar sites (Figure 11.1). It comprises site number 1851, which has a total area of 862,700 ha (Mengué Medou *et al.*, 2008). In the absence of available data, the document that justified its Ramsar listing remained very incomplete and contained many inaccuracies. The delimitation of the site is curious, to say the least: it does not include the lower deltaic plain, with its immense freshwater swamp forests, mangroves, mudflats, and sandbars, regardless of the fact that many marine species were mentioned in various parts of the document, including whales and sharks. On the other hand, very little information was included about aquatic birds or other aquatic species (mammals, reptiles, amphibians, fish), which should constitute the main targets of any Ramsar site that is supposed to be focused, by definition, on wetlands. In fact, the current site covers only about 427,000 ha of wetlands, which accounts for 30% of all of the delta's wetlands.

At the time, no one realized that the Lower Ogooué wetlands constituted a unique deltaic complex that included the flooded valleys upstream from Lambaréné. Each zone was considered separately while, in fact, they are all closely linked. This conception—or rather, lack of conception—guided subsequent initiatives, which centered on Lake Evaro. The proposal to create a protected area centered around this lake had been made on several occasions since 2006, and underwent certain changes over the years. Various scenarios were thus put forward for the creation of a protected area, but nothing concrete has been achieved.

In 2013, studies were undertaken—mostly socioeconomic in nature—as part of the National Parks and Biodiversity Management Capacity Support Project (PARC) funded by GEF and executed by WCS-Gabon. These studies concerned the "Evaro-Wonga-Wongué" complex. It is worth noting that the first recent flora and vegetation inventories were carried out as part of this study.

In 2014 and 2015, other multidisciplinary studies were launched as part of a project to support the sustainable management of forest resources in Gabon. They were coordinated by WWF Gabon and, this time, focused on the "Evaro forest block." This project remained resolutely focused on forests and, in its initial conception, did not include a study of aquatic birds. The efforts were instead dedicated to the study of the forests that surround wetlands, and the section on "mammals" focused on transects made in the dryland forests on the perimeter of the wetland. Nothing was planned regarding hippopotamus or manatees, and no one was curious to know what lived in the large marshes in the center of the upper deltaic plain. As a result, wetlands and their fauna were only briefly mentioned in the various reports that resulted from these studies.

Finally, as part of the "Project to Support the Sustainable Management of Critical Wetland Ecosystems" (PAZH), conducted by the ANPN and the DGEPN, a decision was made to gather available knowledge regarding the Lower Ogooué wetland complex in order to report it in the form of the present book so that it would not be lost, and, for the first time, to offer an overview of this complex, which, all included, represents the third-largest delta in Africa and the only one that retains nearly intact landscapes.

Difficulties

State of knowledge

As described in the chapters of this book, knowledge about the Ogooué Delta remains incomplete, despite significant efforts. While the avifauna is well known and the flora fairly well known—many plant species still remain to be discovered—the biodiversity of other groups remains poorly understood or nearly unexplored. Almost nothing is known about the region's insects, and the small amount of available knowledge about butterflies seems to indicate that the delta is probably home to unexpected species (G. Vande weghe, pers. comm.). However, it is essential to have a good knowledge of the region's species and environments in order to identify conservation priorities and to develop an effective and realistic management plan. For example, a significant effort had previously been made to estimate mammal populations, but they were all limited to the dryland forests at the periphery of the wetlands, while observations made in extremis have revealed that the great alluvial plain at the center of the delta is home to large populations of buffalo, sitatunga, and probably also *Potamochoerus* and elephants. These limited observations considerably change the understanding of the delta's mammal population.

It should also be noted that most of the recent inventories were focused within the limits of the current Ramsar site and have neglected the lower part of the delta, which shelters a large part of the area's wetlands. It was not until the PAZH project was undertaken that the first inventories of flora and vegetation were conducted in this part of the delta, during which inventories on fish and birds were also conducted.

Wetland conservation

By definition, wetland conservation efforts face several constraints that are difficult to circumvent.

Water

A wetland complex, such as a delta for example, forms a unique ecosystem, regardless of its size. Water flows freely through this ecosystem, carrying not only nutrients but also pollutants. In addition, any change in water level in one sector inevitably affects all other sectors. In other words, all water development projects, even if very localized, affect the entire area.

Variability in water level

Periodic variation in water levels, whether related to tides or seasons, represents what could be referred to as the "engine" or the heart of the ecosystem. This variation maintains the water currents and therefore the distribution of nutrients, but it is also essential to the existence of certain environments, such as sandbanks and mudflats, as well as floodable forests and swamp forests and most open marshes. This variation in water level varies in amplitude and is not completely predictable. Wildlife is adapted to this phenomenon, but not humans, who dream of a world without change.

The dynamics of the environment

Over time, most of the environments that constitute a deltaic ecosystem are very dynamic: mudflats and sandbanks move and evolve, while mangroves, papyrus groves, and various reed beds are pioneer environments that inexorably transform into other types of environments, ultimately becoming terrestrial environments—all of this is only a matter of time. These transformations can be rapid, even visible on the scale of a human lifespan. Two satellite images taken a few years apart will very often reveal these changes. Therefore, the various types of environments that make up the Lower Ogooué complex are, for the most part, only transient stages of one or more types of ecological succession, certain stages of which can be very ephemeral, and which in the long term will inevitably change.

With regard to conservation, it can therefore be difficult, if not impossible, to include certain types of wetlands in a defined, limited space. However, many aquatic species—both animals and plants—are more or less restricted to these transient stages. Consequently, only very large spaces that include all of the habitats involved in the temporal dynamics of an area can sustain these types of environment along with their flora and fauna.

Human populations

Human populations make heavy use of most aquatic environments. Not only are open waters important for fishing, but marshes—the heart of production in aquatic ecosystems—are very often coveted by being drained and transformed for agricultural or agroindustrial development. Regardless of the needs of the local population, wetlands are also coveted by external interests for navigation and river transport or to produce hydroelectric power.

As a result, wetlands have become the most endangered environments on the planet—far more so than tropical forests or coral reefs—but this is almost invisible because most of them were already destroyed during the 19th and 20th centuries (page 13). It is for this reason that the Ramsar Convention was established in the early 1970s, at a time when conservation was just beginning to concern itself with forests and oceans.

Conservation of waterbirds

The conservation of waterbirds was largely responsible for the establishment of the Ramsar Convention and was its primary goal. These birds very often migrate across long distances and make use of a wide variety of environments through time and space, especially environments that may be very distant from one another. Many species only briefly reside at their nesting sites and range across a continental or "global" network of sites throughout the year. A good example of this can be seen in shorebirds, which nest in arctic tundra and stop in intertidal or continental wetlands in temperate and tropical regions for nine months a year. The ruff, *Philomachus pugnax*, which nests in Kamchatka and winters in the far south of the African continent, is an extreme example.

Apart from these remarkable cases, many African species also carry out trans-equatorial migrations or move over large distances within a country or a region, especially depending on water levels. This explains why birds that nest on the sandbanks of the Ogooué in the dry season move away in the rainy season—when water levels are high—toward the coast, in areas such as the Akanda, Pongara, Loango, and the mouth of the Nyanga (page 278). For these birds, the coastal sedimentary basin thus constitutes a single entity in which they inhabit different sectors at different times of year. If one of these sectors is lost, these birds will disappear from the entire complex.

Many species also have different resting places (or "dormitories") and feeding places, which means they prefer various environments throughout the day-night cycle. Skimmers are once again a good example: these birds spend the day on sandbanks or sandy mudflats and fish at dusk or at night (when there is moonlight) on large expanses of shallow water, sometimes many kilometers from their resting site.

Most aquatic birds thus suffer from the transformation of wetlands due to

Table 11.1. Ramsar criteria for identifying Wetlands of International Importance

Criterion #	General Ramsar criteria	Application in the case of the Ogooué Delta			
1	Contain a representative, rare, or unique example of a natural or near-natural wetland type found within the biogeographical region concerned.	The third-largest delta on the African continent and the only one to be well preserved.			
2	Support vulnerable, endangered, or critically endangered species or threatened ecological communities.		**CR**	**EN**	**VU**
		Mammals	1	2	11
		Birds	(1)	-	3
		Reptiles	2	1	3
		Amphibians	-	1	-
		Fish	1	7	7
		Plants	22	81	42
3	Support populations of plant and/or animal species important for maintaining the biological diversity of a particular biogeographic region.	Lower Guinea biogeographical region.			
4	Support plant and/or animal species at a critical stage in their life cycles, or provides refuge during adverse conditions.	Giant African threadfin, *Polydactylus quadrifilis* (growth area for immature individuals).			
5	Regularly support 20,000 or more waterbirds.	Without question.			
6	Regularly support 1% of the individuals in a population of one species or subspecies of waterbird.	The African skimmer, *Rynchops flavirostris* (10-15%), the gray pratincole, *Glareola cinerea* (10%), the *mechowi* subspecies of white-fronted plover, *Ochthodromus marginatus* (10%), the "Ogooué" tern, *Sterna* sp. (95%), and the royal tern, *Thalasseus maximus*, when wintering (...%).			
7	Support a significant proportion of indigenous fish subspecies, species, or families, life-history stages, species interactions, and/or populations that are representative of wetland benefits and/or values and thereby contributes to global biological diversity.	19 fish species endemic to Gabon or the Ogooué; 6 species endemic to the Lower Ogooué.			
8	Be an important source of food for fish, spawning ground, nursery, and/or migration path on which fish stocks, either within the wetland or elsewhere.	The giant African threadfin.			
9	Regularly support 1% of the individuals in a population of one species or subspecies of wetland-dependent, non-avian animal species.	The manatee (probably 1–3%, possibly more). 2 endemic amphibians (?).			

human activities, at least as much if not more than forest species, and they are particularly threatened by the omnipresent and very diverse forms of pollution that affect wetlands more than terrestrial environments. Moreover, shorebirds in the Arctic tundra are seriously threatened by the disappearance of their nesting environments as a result of global warming, which is very pronounced in this region.

Finally, many aquatic birds, often of medium or large size, suffer from very significant hunting pressure in certain parts of their annual habitat. This is not just a direct threat to their survival; in many places, it means that they can no longer frequent the remaining wetlands that have escaped being drained. These long-distance migrants are very sensitive to disturbance because they have to conserve their energy reserves for flying over great distances. Globally, many species of aquatic birds have experienced a noticeable population decline in the last few decades and some are truly threatened. Of the 16 globally threatened species of the Gabonese fauna, eight (50%) are aquatic (IUCN, 2014).

The long-term conservation of aquatic birds thus does not only require protecting certain sites but also creating a network of sites each of which protects certain populations at certain times of the day or year. This is the primary goal of the Ramsar Convention, which was extended to cover other aquatic groups (mammals, fish, amphibians, reptiles, invertebrates, flora, and vegetation).

Conservation of fish species

The fish fauna also deserves much more consideration than it generally receives. It is indeed rare for this group to be taken into consideration during the creation of protected areas, which generally focus on terrestrial environments and species. Very often, wetlands do not form the core of protected areas but are rather relegated to the limits, and fishing remains authorized. Most humans understand the protection of mammals and birds, but estimate that fish are just a food resource.

Furthermore, humans continually introduce fish to areas where they were not previously present, most often with the intention of increasing the yield of fisheries. However, these introductions generally have the opposite effect and constitute a real threat to fish faunas. While this has been known for a long time, it has not stopped these introductions, which are harmful to the environment.

The human populations

As just mentioned, the wetlands of the Ogooué, as elsewhere, are widely used by the human population. In the specific case of the Lower Ogooué complex, humans are very dispersed and act as they please. In fact, families most often live alone or in small, very independent and individualistic familial communities. They have not learned to recognize centralized and hierarchical power, and they are resistant to any attempt at outside regulation.

Furthermore, the local population remains more or less convinced that natural resources (fish, game) are inexhaustible. Although many women in Lambaréné complain that there is a shortage of fish, they find it difficult to admit that overfishing is the cause; rather, they tend to accuse fishermen of not properly doing their job. In certain places, some groups organize in an attempt to save their resources by limiting access, but very often this is more to deny other communities access to these resources than to protect them.

Finally, the limits of hunting and fishing grounds are fairly fluid, and they cover the entire deltaic complex. Even the most inaccessible areas will be claimed by one community or another as soon as someone shows interest. Understanding how these communities work is therefore very difficult and cannot be done through short field expeditions, especially since many residents know that there are laws they do not respect.

Values of the Ogooué Delta

The Ogooué Delta meets the nine criteria that define wetlands of international importance (Table 11.1). Below is a summary of the essential points.

Global values

First of all, it is important to realize that the Lower Ogooué wetlands represent a single, indivisible entity that extends from the point where the river enters the coastal basin to the Gulf of Guinea. The Nkomi Lagoon is also part of these wetlands, and this vast ensemble constitutes a very complex delta with all the classical structures (flooded valley, upper deltaic plain, lower deltaic plain, coastal sand structures, and associated lagoons). It may seem superfluous and redundant to reiterate this point, but in the minds of many people, the Ogooué Delta is limited to the mangroves of the lower deltaic plain, whereas it in fact begins in Lambaréné, and even the lower part of the valley further upstream was flooded during the formation of the delta, and is therefore part of the system.

Next, it is important to note that this delta is still very well conserved and that the human population is small, unlike in the Nile and Niger deltas (page 20). These two points make the Ogooué Delta a unique site in Africa, and indeed in the world, and that clearly deserves to become a World Heritage site.

Likewise, the Ogooué Delta is home to an ever-changing set of habitats, and it would be futile to try to preserve some of the specific habitats without protecting the overall dynamics of the environment. Consequently, we must aim to preserve the set of habitats that form its gradient, such as the transition from pure mangrove to closed swamp forest.

Specific values

Birds

The Ogooué Delta shelters more than 1% of the global populations of four species that nest on sandbanks: the "Ogooué" tern (95%), the African skimmer (8-13%), the gray pratincole (3-9%), and the *mechowi* subspecies of white-fronted plover (7-10%). In its marine portion, the delta has 10% of the royal tern population outside its nesting period.

Furthermore, the delta is home to the largest nesting populations in Gabon, and perhaps in all of Lower Guinea, of the reed cormorant, *Microcarbo africanus*, the African darter, *Anhinga rufa*, the great egret, *Casmerodius albus*, the purple heron, *Ardea purpurea*, and the gray-rumped swallow, *Pseudhirundo griseopyga*.

For Palearctic species, the complex serves as an important wintering place for shorebirds (page 277) and the western osprey, *Pandion haliaetus* (page 278).

Mammals

The area is home to large populations of manatees and hippopotamuses, which are the largest in Gabon and probably of international importance. For manatees this probably represents 1-3% of the global population, which is estimated at less than 10,000 individuals (page 307). As for the hippopotamus population, it is genetically significant since it is probably the only population still viable in Lower Guinea (page 309).

On a national level, the population of dwarf buffalo that lives in the central part of the upper deltaic plain is unique in Gabon because it lives in a vast marshy plain. All the other large populations are centered on savannas and dryland forests in Lopé and Moukalaba-Doudou national parks, as well as in the Wonga-Wongué Reserve.

Reptiles

The population of slender-snouted crocodiles, *Mecistops cataphractus*, which must include at least a thousand individuals, represents perhaps more than 1% of the population of this endangered Central African species, which still remains to be named (page 261).

Amphibians

The Lower Ogooué region may be home to an endemic species, *Phrynobatrachus ogooensis*.

Fish

The waters of the Lower Ogooué are home to a very diverse fish fauna, with 27 endemic species (page 230), not counting the many species that remain to be described or discovered.

The delta is an important region for the growth and maturation of the giant African threadfin, *Polydactylus quadrifilis*, a species of high commercial value. Clichids (Cichlidae) are also very important.

Flora and vegetation

1. The delta shelters a complex gradient from the coastal meadows and the pioneering mangroves to the swamp forests, flood forests, and ultimately dryland forests. Thus, it harbors an unaltered succession unique within Gabon and especially in the context of the Lower Ogooué. Note that this dynamic and the resulting gradient of habitats are situated in the lower part of the delta, an area that is not yet included in the Ramsar site.
2. Overall, although the Lower Ogooué only shelters a small number of threatened habitats, it probably contains 80% of Gabon's wetland habitats. Among the rarest and most endangered are: the *Ctenolophon* swamp forests, which are unique for their role in maintaining the forest vegetation of the coastal part of Gabon; the vegetation of the coastal dune ridges, which includes a set of dryland forests that have a limited distribution and are highly threatened; humid forests and dry savannas, including savannas with *Brachystelma letestui*; and flooded savannas. These last habitats are not represented within the current limits of the Ramsar site.
3. Finally, the delta includes at least 167 potentially threatened plant species (pages 169–171). Although most of them have been documented to occur in dryland forests, around 30 are found in wetland habitats.

Conservation strategies

Taking into account the constraints inherent to wetland conservation and those of the Lower Ogooué wetlands in particular (pages 318–321), and also factoring in the size of the delta, its complexity, and the great diversity of its specific features, conservation of the area should inevitably rely on a set of complementary and integrated strategies and measures.

Preserving the entire site

Any conservation measure must be based on the protection of the whole of the Lower Ogooué complex. With this in mind, the 1990 IUCN proposal, which involved the creation of two reserves (one by the sea, the other centered around Lake Onangué), indicated good intentions but was unrealistic and ill-adapted to the context of the delta. It was akin to saying that only the Eiffel Tower's tip and one of its "legs" need protecting. It is therefore above all necessary to:

1. preserve water quality;
2. maintain the system of seasonal floods (even a dam far upstream along the Middle Ogooué could have dire consequences);
3. regulate hunting and fishing, and possibly adapt Gabon's legislation to include endangered species that are not currently protected by law;
4. create "sanctuaries" in which all hunting and fishing are banned in order to ensure that the most important groups of animals can reproduce;
5. prohibit economic activities that could impact the dynamics of the environment and thereby conserve the set of habitats that form the gradients of vegetation, such as the transition from pure mangrove to closed swamp forest.

Reviewing the limits of the Ramsar site

This book deals with an area larger than the current Ramsar site (#1851), which was created in 2009 and extends from the transition between the marine delta and the alluvial plain to far upstream from Lambaréné (Figure 11.1). Thus, it does not include the lower deltaic plain nor the coastal sand structures. On the other hand, it encompasses huge expanses of dryland forest that have been extensively impacted by humans (villages, missions, camps, food crops, fallows, secondary forests, forests undergoing secondarization, areas of industrial palm oil and rubber exploitation, logging, mining, and petroleum), most of which have no connection to the aquatic environments targeted by the Ramsar Convention and a large part of which is located outside of the coastal sedimentary basin. In fact, in its current configuration, the Ramsar site does not even cover half of the wetlands of the deltaic complex.

This delimitation is supposed to have been based on a desire to include the watersheds of rivers that flow into the Lower Ogooué. This approach seems to be logical, but it does not take into account the fact that the vast majority of the waters of Lower Ogooué come from throughout Gabon, far upstream from the delta proper. Even pollution generated by mining activities in Haut-Ogooué province ultimately has a (probably significant) impact on the Lower Ogooué complex.

In order to preserve the entire deltaic complex, it will therefore be necessary to modify the boundaries of the Ramsar site, as proposed in this book (Figure 11.1). Based on the knowledge of authors who have contributed to this book and the information it contains, the following additions are suggested:

1. To the west, include the lower deltaic plain and coastal sand formations (Mandji and Ozouri islands, the Animba Plains, Ompinga, Inguéssi, and Koroso, as well as shoals or sandbars). The western limit should follow a line in the sea corresponding to a depth of 6 m, as suggested by the Ramsar Convention.
2. To the north, the lakes that border the upper deltaic plain should be included along with their associated marshes, from Lake Azingo in the east to Lake Gomé in the west. A part of these lakes is located in the Wonga-Wongué Reserve and is therefore included in that Ramsar site. There is no need to redefine Wonga-Wongué, but the boundaries of the two Ramsar sites should be updated. The Wonga-Wongué Ramsar site should be limited to the lakes of the plateau and the small coastal rivers, the largest of which is the Awagné.
3. To the east, it may be necessary to extend the site's boundary up to the edge of the coastal basin, excluding areas that are part of the Crystal Mountains ecosystem.

Within this context, the initiative proposed by the ANPN to create a UNESCO Biosphere Reserve is noteworthy as it would cover, among other things, the Lower Ogooué Ramsar site. This initiative involves the preparation of a management plan for the entire reserve, including recognizing different levels of conservation. The data presented in this book will be used to justify its recognition as a new Biosphere Reserve and should be used to establish the management plan. Similarly, the recommendations presented in this chapter could be used to guide this process.

Specific measures

For the manatee

1. Enforce existing laws (the species is fully protected).
2. Create sanctuaries around lakes where the species is still "abundant," so as to prevent the last individuals from becoming collateral victims of fishing.

For the hippopotamus

1. Enforce existing laws (the species is fully protected).
2. Create sanctuaries around the lake deltas of the Rivière des Pélicans and Ezanga River.

For the slender-snouted crocodile

1. Enforce existing laws (the species is fully protected).
2. Create sanctuaries around the lakes and stretches of rivers where the species is still sufficiently "abundant," in order to prevent individuals from becoming collateral victims of fishing.

For birds from sandbanks

1. Guarantee the maintenance of the system of the rise and fall of floodwaters.
2. Protect the sandbanks during nesting periods (June–September) and prohibit access to the sandbars located at sea.
3. Design education programs for children as well as fishermen and women.

For bird colonies

1. Protect these breeding colonies.
2. Protect the feeding grounds of species that nest in these colonies.

For the fish fauna

1. Organize true fishery monitoring.
2. Stop any new species' introduction.
3. Regulate fishing both in terms of means being used and fishing areas.
4. Create fishing sanctuaries.

For the flora and vegetation

1. The vast majority of species potentially threatened, according to the Red List criteria, are from dryland forests. Currently, these forests, which are among the most exploited in Gabon, are poorly represented in the network of protected areas. The future site management plan will have to take this into account, for example by classifying certain areas present in forest concessions, even small ones, as parts of conservation sites.
2. In the coastal region, beta diversity (the variation of habitats in space) is higher than elsewhere in Gabon, probably a consequence of the variety of soils and water regimes. A conservation strategy aimed at protecting all of the habitats of the Lower Ogooué region should therefore aim to protect a large area in order to guarantee the preservation of the largest number of habitats. It will be difficult to implement such a strategy. However, the data that have been collected in preparing this book can be used to identify and target a few areas with high beta diversity.
3. The aim should be to conserve the most threatened habitats, which implies the creation of a protected area that includes the coastal dune ridges and the swamp forests south of Wonga-Wongué.

Acknowledgments

The publication of this book represents the culmination of a data collection effort that began more than 10 years ago at the time of the "Support Project for the Reinforcement of the Management Capacity of Parks and Biodiversity," financed by the Global Environment Facility (GEF) and managed within the World Bank, and more recently the "Support Project for the Sustainable Management of Forest Resources," financed by the Congo Basin Forest Fund (FFBC) within the African Development Bank (AfDB). The authors would like to thank the people and NGOs who implemented these projects and greatly facilitated our work, in particular Malcolm Starkey, at the time working for WCS Gabon, and Idriss Deffry, Bas Verhage, and Sandra Ratiarison of WWF Gabon.

The initiative that led to the completion of this book belongs, however, to the "Projet d'Appui à la gestion durable des écosystèmes des Zones Humides critiques" (PAZH), financed by the GEF within the World Bank, implemented by the Direction Générale de l'Environnement et de la Protection de la Nature (DGEPN) and the Agence Nationale des Parcs Nationaux (ANPN). We would like to express our deep gratitude to the entire PAZH project team, and in particular to Faustin Ondamba Ombanda, Solange Ngouessono, Régis Ulrich Nguéma, and Aimé Roger Nzigou, who coordinate the project, as well as to the World Bank team led by Salimata Diallo Follea, and in particular Romain Beville, who provided support throughout the project and accompanied us in the field. The field expeditions were facilitated by ANPN agents, who helped with logistics during trips to the Ogooué Delta, and also by Solange Ngouessono, who brilliantly organized the field work and was always readily available.

Field work to collect botanical material was made possible thanks to the support of the National Herbarium of Gabon (IPHAMETRA, CENAREST). The authors would like to thank all the botanists associated with this institution who helped, and more specifically Ludovic Ngok Banak and Nestor Engone Obiang, successive curators of the National Herbarium, as well as Henri Paul Bourobou Bourobou, Director of IPHAMETRA.

It would be impossible to compile an exhaustive list of the numerous inhabitants of the Great Lakes region, the Fernan Vaz Lagoon region, and of the maritime delta, who offered precious help during the field expeditions. May the authorities and the inhabitants of Omboué, Port-Gentil, and Lambaréné receive the authors' sincere thanks for their unfettered hospitality.

The authors also thank the photographers who allowed the use of their images: Christophe Aubin, Cyrille Mvele, Gaël R. Vande weghe, Laila Bahaa-El-Din, Olivier Buisson, Olivier Lucanus, and Quentin Meunier.

The authors also thank Porter P. Lowry II, Director of the Africa and Madagascar Program of the Missouri Botanical Garden, Steven Dessein, Director of Botanical Garden Meise, Pierre Meerts, Director of the Herbarium of the Free University of Brussels, all of whom assisted certain authors, as well as Valérie Cawoy for her precious help with proofreading the French version of this book.

Many organizations and individuals have contributed to the making of this book in one way or another. These contributions were financial, logistic, and intellectual. It is not only a duty but also a pleasure to thank them: the National Center for Scientific Research (CENAREST), the Maboumine Company (ERAMET), the French National Research Institute for Sustainable Development (IRD), the French National Center for Scientific Research (CNRS), the Free University of Brussels (ULB), the Wildlife Conservation Society (WCS Gabon), the U.S. National Science Foundation (NSF), and the World Wide Fund for Nature (WWF Gabon). The staff of Maurel & Prom are warmly thanked for providing access to their concession, as well as for logistical help.

Jean Philippe Biteau, Director of the company Jardi-Gab, provided logistical assistance during the many years required to collect data regarding the flora, and his judicious advice was very useful for the success of the inventories; the authors express their sincere gratitude.

The data regarding flora were collected with the financial assistance of the Prince Albert II of Monaco Foundation as part of the project to develop a High Values for Conservation (HVC) approach for the flora and vegetation of Gabon.

Finally, the authors address their special thanks to Lee White, formerly Executive Secretary of the National Agency for National Parks (ANPN) and currently Minister of Water & Forests, without whom this work would never have seen the light of day.